Biotransformations

Volume 5

Biotransformations

*A survey of the biotransformations
of drugs and chemicals in animals*

Volume 5

Edited by

D. R. Hawkins

Huntingdon Research Centre Ltd

ROYAL
SOCIETY OF
CHEMISTRY

A catalogue record for this book is available from the British Library

ISBN 0-85186-147-4

Published by The Royal Society of Chemistry,
Thomas Graham House, The Science Park, Cambridge CB4 4WF

Set by Unicus Graphics Ltd, Horsham, West Sussex
Printed in Great Britain at the
Bath Press, Bath

Preface

This series encompasses biotransformation of chemical entities whether they are pharmaceuticals, agrochemicals, food additives, or environmental or industrial chemicals in vertebrates within the animal kingdom, a group which includes mammals, birds, and fish. Each volume of the series in general includes material published during a calendar year, this fifth volume covering mainly 1991. An attempt has been made to include a comprehensive coverage of the scientific literature but due to the great diversity of journals where reports on biotransformations appear there will undoubtedly be some omissions. Any notable omissions communicated to the Editor could be included in a subsequent volume. The incorporation of cumulative indexes first introduced in Volume 3 is being continued in this and subsequent volumes, a feature which will facilitate access to key information on structurally related compounds.

Arrangement of material and access

An overview chapter has been prepared which contains highlights such as novel biotransformation, mechanisms of toxicity and notable species differences. The abstracts are arranged according to compound class, although there may be cases where allocation to one or another class is somewhat subjective. It has been considered valuable to be able to access information on the biotransformation of compounds with similar structural features. For this purpose the concept of key functional groups has been developed. Selection and naming of these groups has evolved during preparation of the material. For each compound functional groups have been selected where biotransformation has been shown to occur but in addition groups have also been included where biotransformation has not taken place. The same functional groups may not necessarily be included in all abstracts of the same compound since some papers may be confined to specific aspects of biotransformation and here only the relevant groups are included. A list of the functional groups follows which may be referred to before proceeding to the corresponding index. Two other indexes have been included containing compound names and types of biotransformation processes respectively.

In the precis for each compound certain key information has been included when available. Where radiolabelled compounds have been used the position(s) of labelling have been indicated on the structure. Comments on the source of metabolites and information on the quantitative importance of individual metabolites such as percentage material in the sample or percentage administered dose are given where possible. Also in order to provide a perspective on the criteria for identification the procedures used for separation and isolation of metabolites and structural assignments such as chromatographic and physico-chemical techniques and use of reference compounds have been discussed.

D. R. Hawkins

Contents

Contributors

I. Blagbrough *School of Pharmacy and Pharmacology, University of Bath, Claverton Down, Bath, BA2 7AY, UK*

M. Dickins *Wellcome Research Laboratories, Langley Court, Beckenham, Kent, BR3 3BS, UK*

D. R. Hawkins *Huntingdon Research Centre, Huntingdon, Cambs, PE18 6ES, UK*

D. Kirkpatrick *Huntingdon Research Centre, Huntingdon, Cambs, PE18 6ES, UK*

I. Midgley *Huntingdon Research Centre, Huntingdon, Cambs, PE18 6ES, UK*

D. Needham *Schering Agrochemicals Ltd, Chesterford Park Research Station, Saffron Walden, Essex, CB10 1XL, UK*

M. Prout *ICI Central Toxicology Laboratory, Alderley Park, Macclesfield, Cheshire, SK10 4TJ, UK*

Key Functional Groups

R may be any unspecified group including H. Where two or more R groups are indicated these may be the same or different groups. Where aromatic rings or other cyclic systems are shown they may also contain substituents when they are not specified as part of the key functional group.

Name	Structure	Name	Structure
Acetal		Alicyclic ketone	$(CH_2)_n$ $C=O$
Acetamide	$CH_3\overset{O}{\overset{\|}{C}}NHR$	Alkadiene	$C=CH-CH=C$
N-Acetoxy	$R_2N-O\overset{O}{\overset{\|}{C}}CH_3$	Alkane	$CH_3(CH_2)_nR$
N-Acetyl aryl amine	$CH_3\overset{O}{\overset{\|}{C}}NH-$	iso-Alkane	RCH_2 RCH_2 $CH-$
Acetylene	$RC\equiv CR$	Alkene	$C=C$
N-Acetylimine	$\overset{R}{\underset{R}{}}C=N-\overset{O}{\overset{\|}{C}}CH_3$	Alkene carboxylate	$RCH=CHCO_2R$
Acetylthio	$CH_3\overset{O}{\overset{\|}{C}}SR$	Alkene carboxylic acid	$RCH=CHCO_2H$
Adenine		Alkenyl aldehyde	$RCH=CHCHO$
Alanine	$H_2NCH(CH_3)CO_2H$	Alkenyl ketone	$RCH=CH\overset{O}{\overset{\|}{C}}R$
tert-Alcohol	$\overset{R}{\underset{R}{}}R-C-OH$	Alkoxyphenyl	RCH_2O-

Alkyl alcohol	RCH_2OH	sec-Alkyl aryl amine	$R-N$ with $CH_2CH\overset{R}{\underset{R}{\big<}}$ on phenyl
sec-Alkyl alcohol	$\overset{R}{\underset{R}{>}}CH-OH$	Alkyl aryl ether	RCH_2O- phenyl
Alkyl aldehyde	RCH_2CHO	Alkyl aryl sulfoxide	$RCH_2\overset{\overset{O}{\uparrow}}{S}-$ phenyl
Alkyl amide	$R\overset{\overset{O}{\|}}{C}NHCH_2R$	Alkyl aryl thioether	RCH_2S- phenyl
Alkyl amine	RCH_2NH_2	Alkyl carbamate	$RCH_2O\overset{\overset{O}{\|}}{C}NHR$
Alkyl tert-amine	$\overset{R}{\underset{R}{>}}N-R$	Alkyl carbonate	$\overset{RCH_2O}{\underset{RCH_2O}{>}}C=O$
sec-Alkyl amine	$\overset{R}{\underset{R}{>}}CH-NHR$	Alkyl carboxamide	$RCH_2\overset{\overset{O}{\|}}{C}NHR$
tert-Alkyl amine	$R-\overset{\overset{R}{\|}}{\underset{R}{C}}-NHR$	iso-Alkyl carboxamide	$\overset{R}{\underset{R}{>}}CH\overset{\overset{O}{\|}}{C}NHR$
Alkylamino	RCH_2NH-	Alkyl carboxylate	RCH_2CO_2R
iso-Alkylamino/iso-alkylamine	$\overset{R}{\underset{R}{>}}CHNH-$	tert-Alkyl carboxylate	$R-\overset{\overset{R}{\|}}{\underset{R}{C}}-CO_2R$
Alkyl aryl amide	$\overset{R}{\underset{}{>}}\overset{\overset{O}{\|\|}}{C}-N\overset{CH_2R}{<}$ phenyl	Alkyl carboxylic acid	$RCH_2\overset{\overset{O}{\|}}{C}OH$
Alkyl aryl amine	$R-N\overset{CH_2R}{<}$ phenyl	iso-Alkyl carboxylic acid	$\overset{R}{\underset{R}{>}}CH\overset{\overset{O}{\|}}{C}OH$
iso-Alkyl aryl amine	$R-N\overset{CH\overset{R}{\underset{R}{<}}}{<}$ phenyl	N-Alkyl cycloalkylamine	$(CH_2)_n-NHCH_2R$

Alkylcyclohexane	RCH_2—(cyclohexane)	Alkyl *N*-oxide	RCH_2, RCH_2—N($\rightarrow$O)—CH_2R
Alkyl cyclopentane	RCH_2—(cyclopentane)	Alkyl peroxide	RCH_2O-OCH_2R
Alkyl ester	$RCOOCH_2R$	Alkylphenyl	RCH_2—(phenyl)
iso-Alkyl ester	$RCOOCHR_2$	iso-Alkyl phenyl	R_2CH—(phenyl)
Alkyl ether	RCH_2OR	Alkyl phosphate	$RCH_2OP(=O)(OR)OR$
Alkyl hydrazine	RCH_2NHNHR	*N*-Alkylpiperazine	RN(piperazine)NCH_2R
N-Alkylimidazole	(imidazole)—CH_2R	*N*-Alkylpiperidine	RCH_2N(piperidine)
Alkyl imide	$RCH_2C(=O)NC(=O)CH_2R$ (N–R)	*N*-Alkylpurine	(purine, NHR)—R
N-Alkyl imide	R–C(=O)–N(CH_2R)–C(=O)–R	Alkylpyridine	(pyridine)—CH_2R
Alkyl ketone	$RCH_2C(=O)R$	Alkylpyrimidine	(pyrimidine)—CH_2R
N-Alkylmorpholine	RCH_2—N(morpholine)O	Alkyl quaternary ammonium	$\equiv N^+$–CH_2R
Alkyl nitrate	RCH_2ONO_2	Alkyl sulfamate	RCH_2OSO_2NHR
Alkyl nitrile	RCH_2CN	Alkyl sulfate	RCH_2OSO_3H

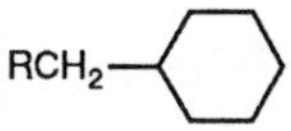

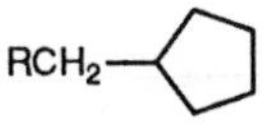

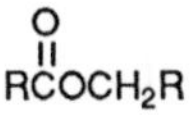

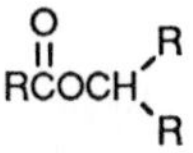

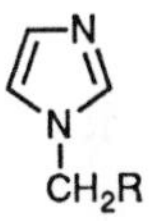

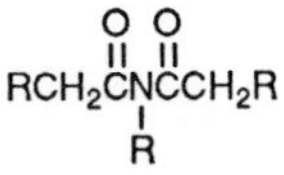

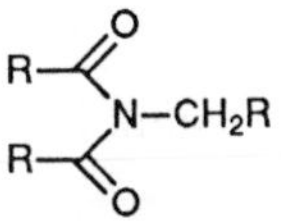

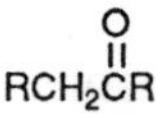

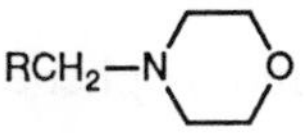

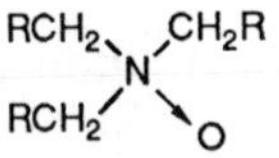

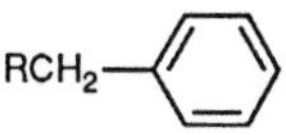

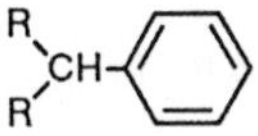

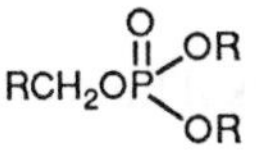

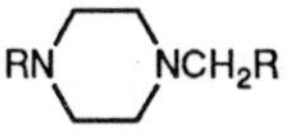

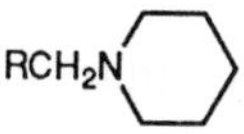

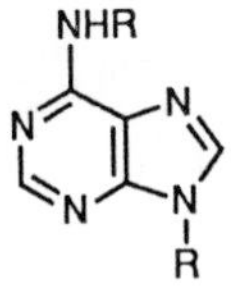

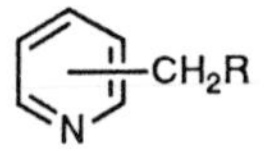

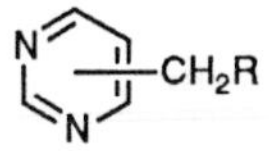

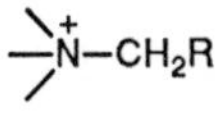

Alkyl sulfonate	RCH_2OSO_2R	Allylic methyl	$CH_3CH=CHR$
Alkyl sulfonic acid	RCH_2SO_3H	Amidine	$RC\begin{smallmatrix}NR_2\\NR\end{smallmatrix}$
Alkyl sulfoxide	$RCH_2S{\nearrow}^{O}_{\searrow R}$	Amidoxime	$RC\begin{smallmatrix}NH_2\\NOH\end{smallmatrix}$
S-Alkyl thiocarbamate	$RCH_2SCONHR$	Amino acid	$RCHNH_2$ CO_2H
Alkyl thiocarboxylate	$RCH_2C{\nearrow}^{O}_{\searrow SR}$	Aminoglycoside	—
Alkyl thioester	$R\overset{O}{\underset{}{C}}SCH_2R$	Aminoimidazole	
Alkyl thioether	RCH_2SR	Aminopurine	
Alkyl thiol	RCH_2SH	Aminopyridine	
Alkylurea	$RNH\overset{O}{\underset{}{C}}NHCH_2R$	Aminoquinoline	
Alkyne	$RC{\equiv}CR$	Aminothiazole	
Allyl	$RCH_2{-}CH=CH_2$	Aminothiophene	
Allyl amine	$RCH=CH{-}NH_2$	Aminotriazine	
Allylic alcohol	$RCH=CHCHOH$ with R	Androstadienone	

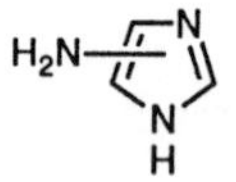

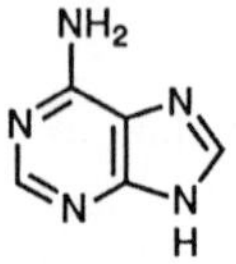

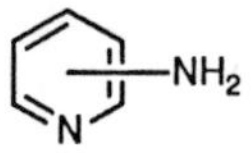

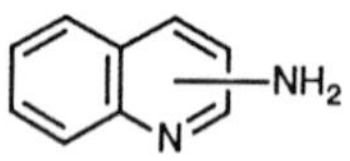

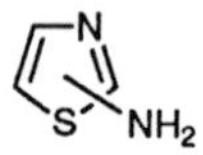

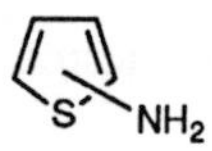

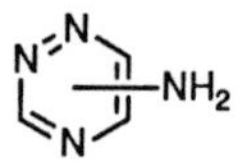

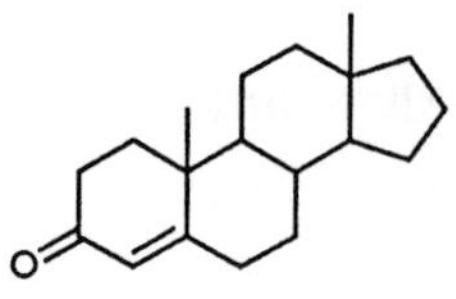

Androsten-3-one		Aryl carboxamide	
Anthracene		Aryl carboxylate	
Anthraquinone		Aryl carboxylic acid	
Aryl acetamide	—CH$_2$CONH$_2$	Aryl dihydrodiol	
Aryl acetic acid	—CH$_2$CO$_2$H	Aryl disulfide	—S—S—
Aryl aldehyde	—CHO	Aryl ester	—OCR
Arylalkene	—CH=CHR	Aryl ether	—OR
Arylalkyl	—CH$_2$R	Arylethylene	—CH=CH$_2$
Aryl tert-alkyl	—C(R)(R)R	Aryl hydrazine	—NHNH$_2$
Aryl amide	—NHCR	Arylhydroxylamine	—NHOH
Aryl amine	—NHR	Arylhydroxymethyl	—CH$_2$OH
Aryl amino acid	—CHNH$_2$ CO$_2$H	N-Arylimide	
Aryl carbamate	—OCNHR	N-Arylimine	—N=R

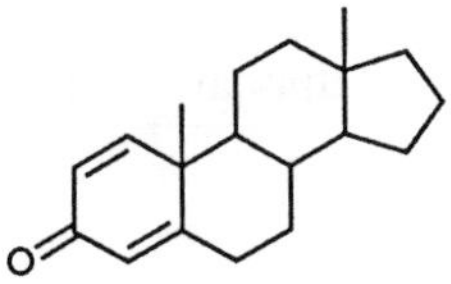

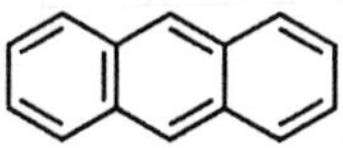

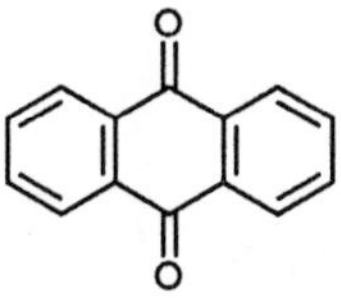

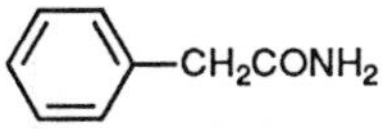

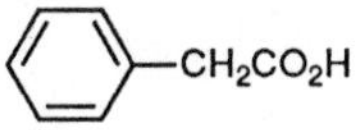

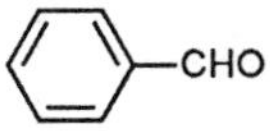

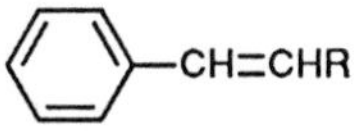

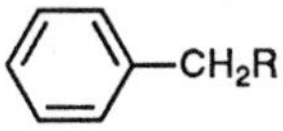

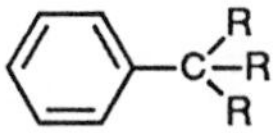

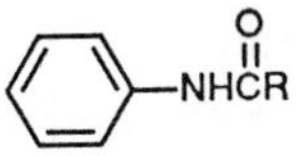

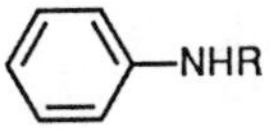

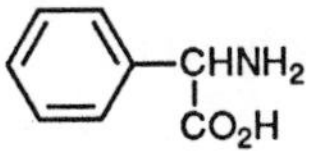

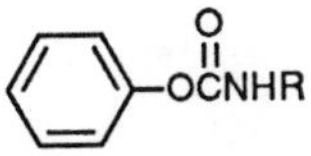

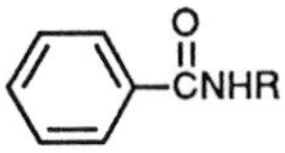

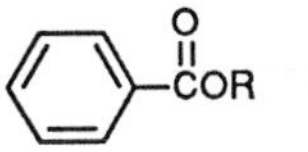

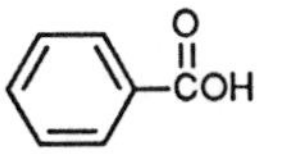

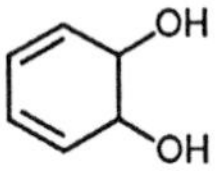

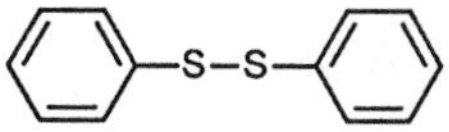

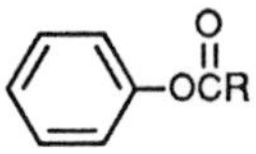

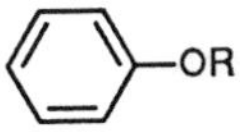

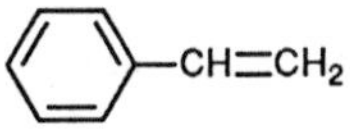

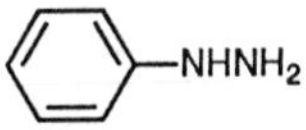

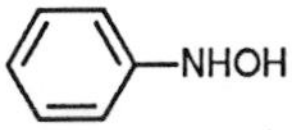

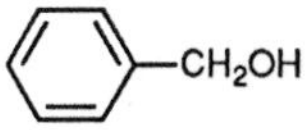

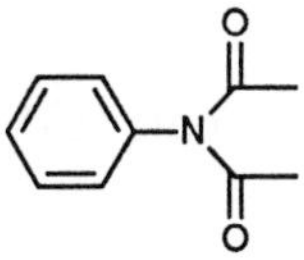

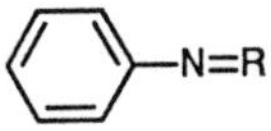

Aryl ketone		Aryl thiocarboxymide	
Arylmethyl		Aryl thiocarboxylate	
Aryl methyl ketone		Aryl thioether	
Aryl nitrile		Aryl thiol	
N-Arylnitrosamine		Aryl triazene	
Arylnitroso		Azide	$R-N=\overset{+}{N}=\overset{-}{N}$
Aryl N-oxide		Aziridine	
Aryloxacetic acid		Azobenzene	
Aryloxypropionic acid		Azoxy	
Aryl phosphate		Barbiturate	
Arylpiperidine		Benzamide	
Aryl propionic acid		Benzanthracene	
Aryl sulfonic acid		Benzazepine	

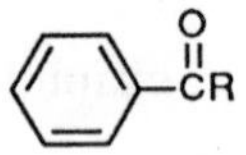

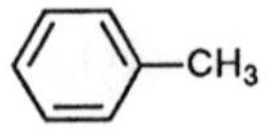

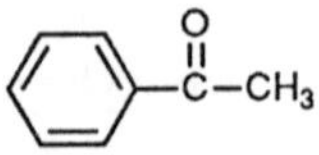

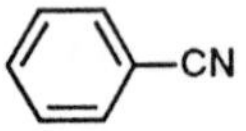

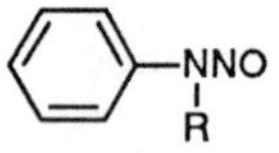

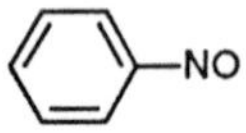

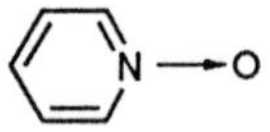

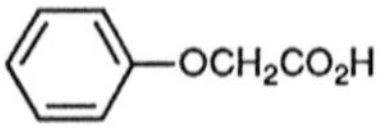

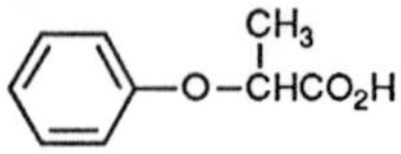

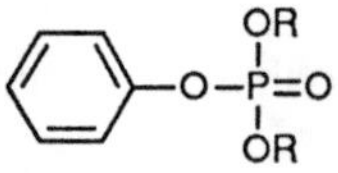

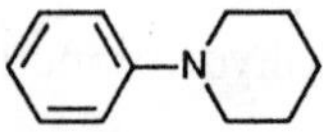

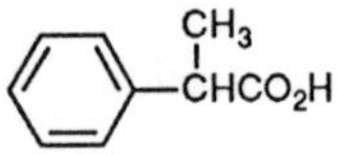

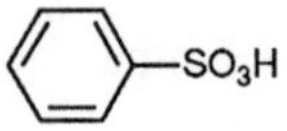

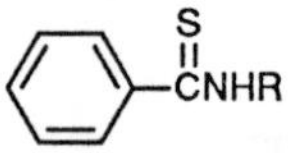

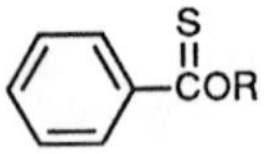

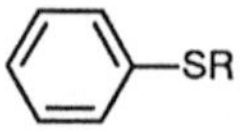

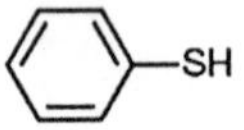

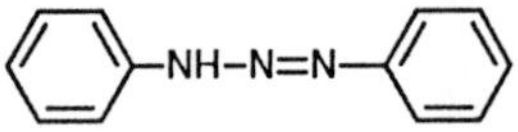

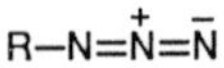

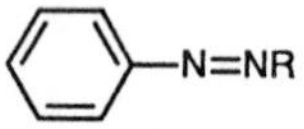

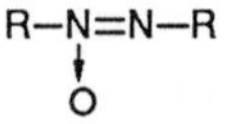

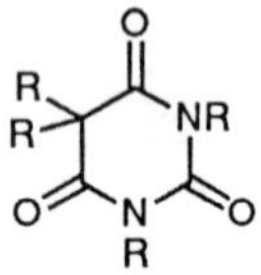

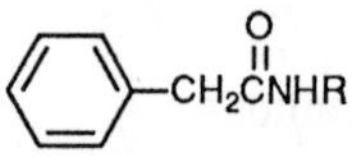

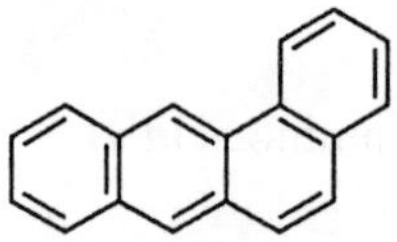

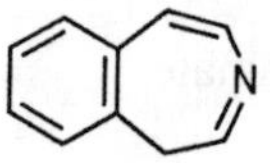

Benzhydrol		Benzo[c]phenanthrene	
Benzhydryl		Benzopyran	
Benzhydryl carboxylate	$RCO_2-\underset{C_6H_5}{\overset{C_6H_5}{C}}R$	Benzopyranone	
Benzidine	—NH–NH—	Benzo[a]pyrene	
Benzimidazole		Benzoquinone	
Benzodiazepine		Benzothiazine	
Benzodioxin		Benzothiazole	
Benzodioxolane		Benzotriazine	
Benzofuran		Benzoxazole	
Benzoisoselenazolone		Benzoxazoline	
Benzoisothiazole		Benzyl	—CH$_2$R
Benzoisoxazole		Benzyl alcohol	—CH$_2$OH
Benzonitrile		Benzyl amine	—CH$_2$N⟨^R_R

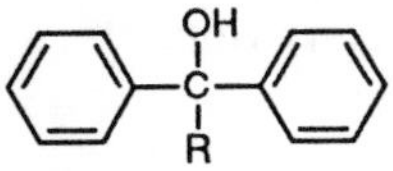

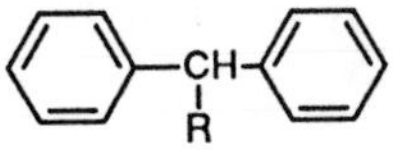

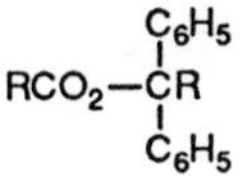

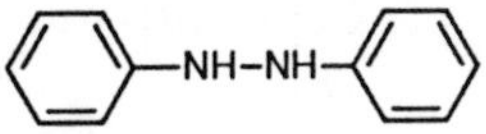

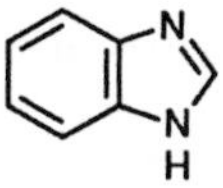

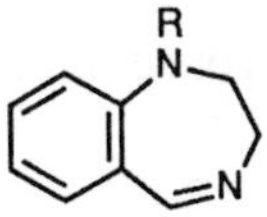

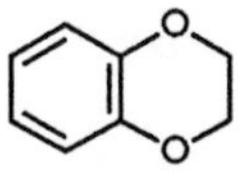

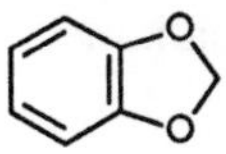

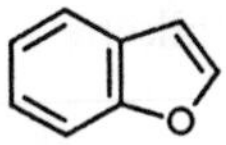

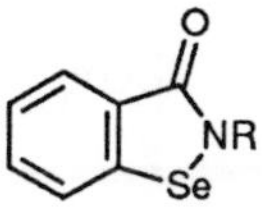

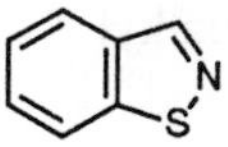

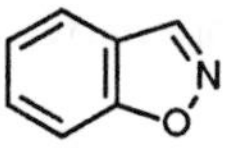

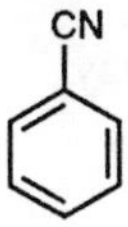

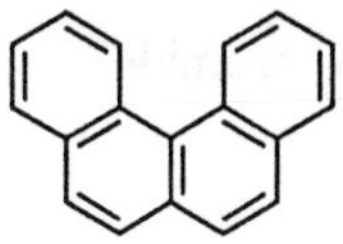

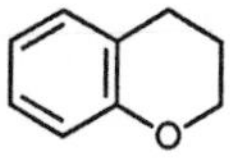

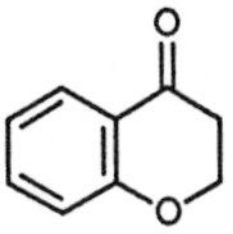

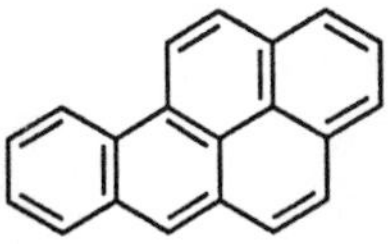

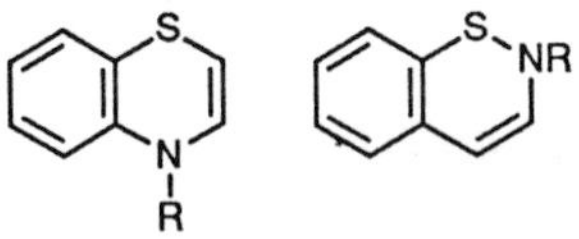

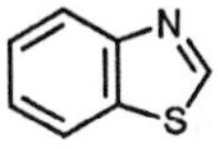

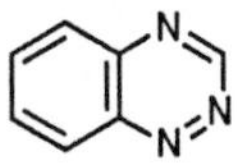

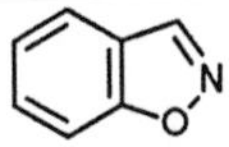

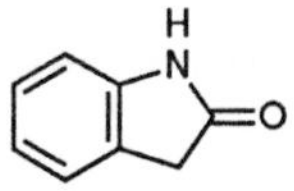

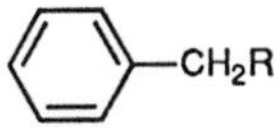

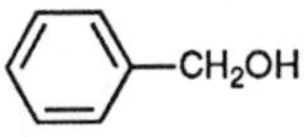

xv

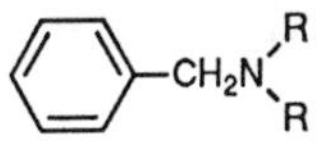

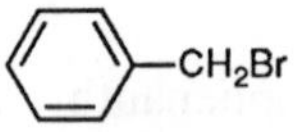

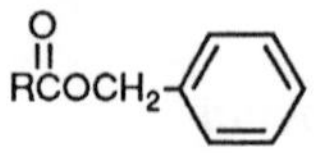

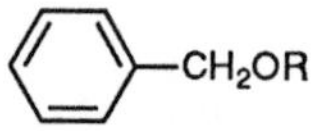

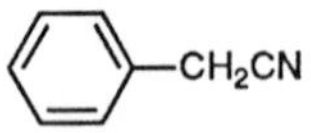

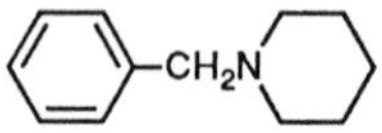

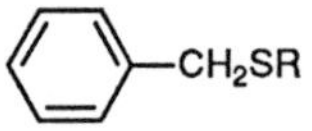

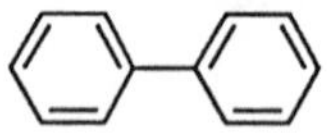

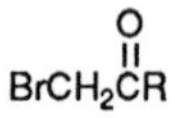

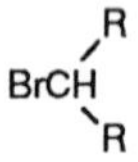

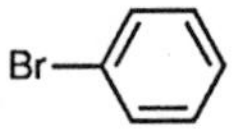

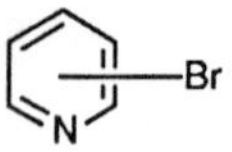

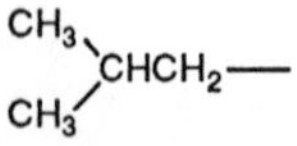

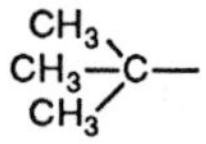

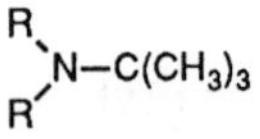

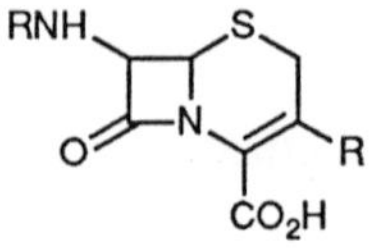

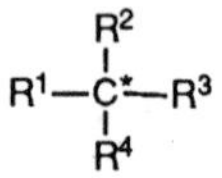

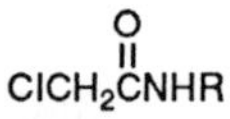

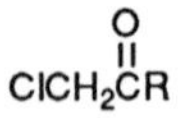

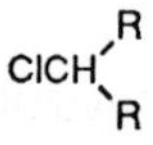

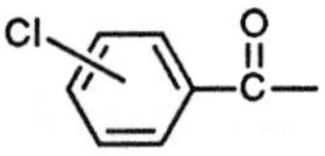

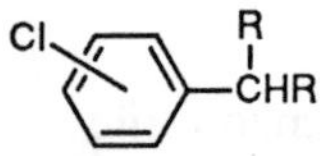

Benzyl bromide	—CH$_2$Br (benzene ring)	tert-Butyl	CH_3—C—CH_3 / CH_3
Benzyl ester	RCOCH$_2$— (benzene ring)	*N*-tert-Butyl	R—N—C(CH$_3$)$_3$ / R
Benzyl ether	—CH$_2$OR (benzene ring)	Carbamate	ROCNHR (C=O)
Benzyl nitrile	—CH$_2$CN (benzene ring)	Cephalosporin	(structure)
N-Benzylpiperidine	—CH$_2$N (benzene + piperidine ring)	Chiral carbon	R^1—C*—R^3 with R^2, R^4
Benzyl thioether	—CH$_2$SR (benzene ring)	Chloroacetamide	ClCH$_2$CNHR (C=O)
Biphenyl	(two benzene rings)	Chloroacetyl	ClCH$_2$CR (C=O)
Bromoacetyl	BrCH$_2$CR (C=O)	Chloroalkane	ClCH$_2$R
Bromoalkane	BrCH$_2$R	Chloroalkene	ClCH=CHR
Bromoalkyl	BrCH with R, R	Chloroalkyne	ClC≡CR
Bromophenyl	Br— (benzene ring)	Chloroalkyl	ClCH with R, R
Bromopyridine	(pyridine ring)—Br	Chlorobenzoyl	Cl—(benzene)—C— (C=O)
iso-Butyl	CH$_3$—CHCH$_2$— / CH$_3$	Chlorobenzyl	Cl—(benzene)—CHR with R

Chlorobiphenyl		Coumarin	
Chlorocyclopropane		N-Cyanoguanidine	RNHCNHR ‖ NCN
N-Chloroethyl	ClCH₂CH₂N<	Cycloalkane/cycloalkyl	$(CH_2)_n$
Chlorophenyl		Cycloalkene	$(CH_2)_n$ CH‖CH
Chloropyridazinone		Cycloalkylamine	$(CH_2)_n$ NR
Chloropyridine		Cycloalkyl amine	$(CH_2)_n$ CH—NH₂
Chlorotriazine		Cycloalkyl ester	$(CH_2)_n$ CHOCR
Chlorouracil		Cyclohexadienone	
		Cyclohexane/cyclohexyl	
Cholanic acid		Cyclohexanol	
		Cyclohexanone	
Cholestenone		Cyclohexenone	
Chrysene		Cyclohexyl amide	

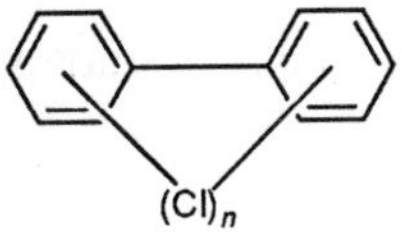

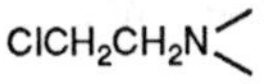

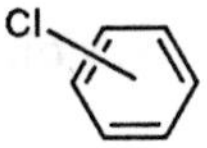

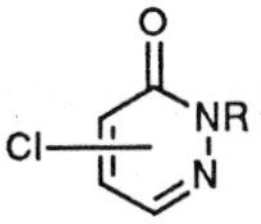

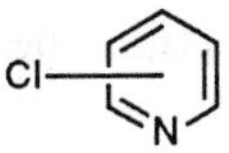

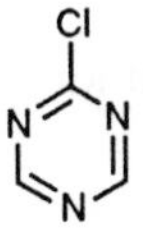

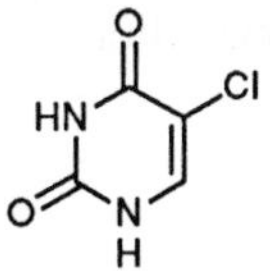

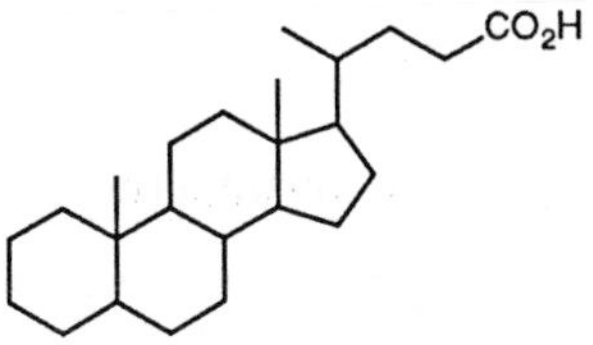

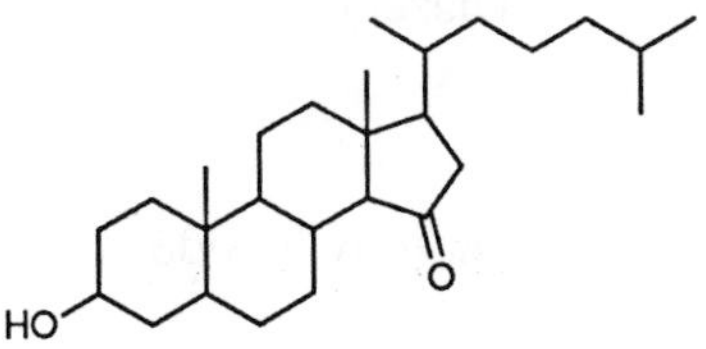

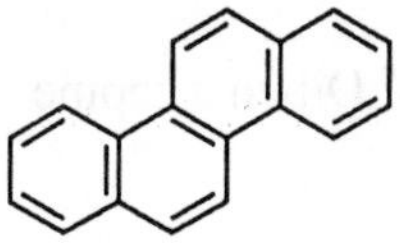

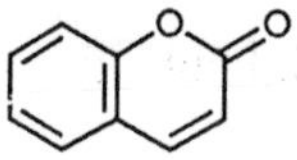

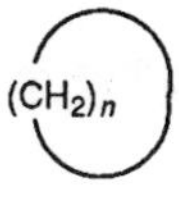

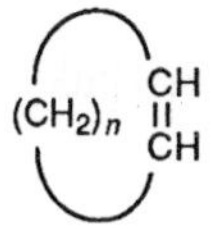

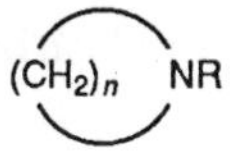

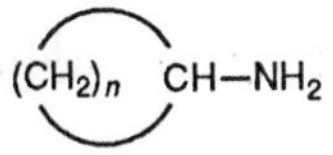

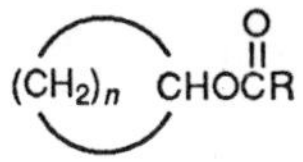

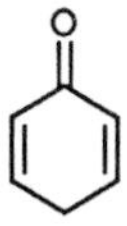

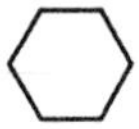

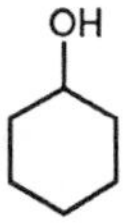

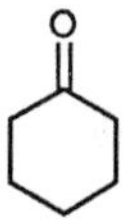

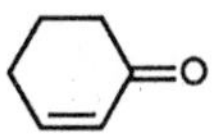

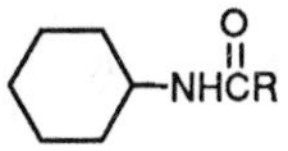

Cyclopentene		Dialkylaminoalkyl	RCH_2—N—CH_2R / RCH_2
Cyclopentenone		Dialkyl aryl amine	
Cyclopropane		Dialkyl ether	RCH_2OCH_2R
Cyclopropylalkyl		Dialkylisoxazole	
Cyclopropyl carboxylate		Dialkyl sulfoxide	RCH_2SCH_2R
N-Cyclopropylmethyl	R—NCH_2	Dialkyl thioether	RCH_2SCH_2R
Cysteine	HO_2CCHCH_2SH / NH_2	Diaryl amine	
Cytidine		Diaryl ether	
		Diaryl sulfoxide	
Cytosine		Diaryl thioether	
Dialkyl amide		Diazepine	
Dialkyl amine	RCH_2—NH / RCH_2	Diazohydroxide	$RN{=}N{-}OH$
Dialkylamino	RCH_2—N—R / RCH_2	Dibenzazepine	

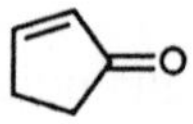

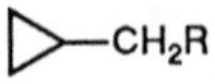

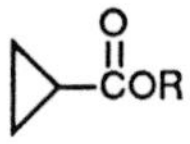

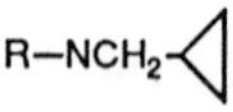

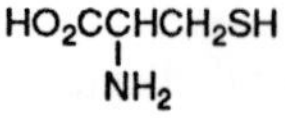

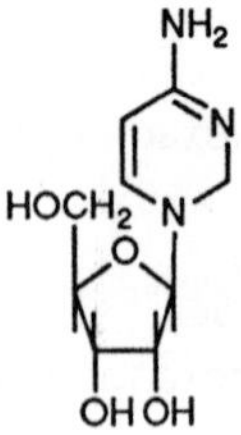

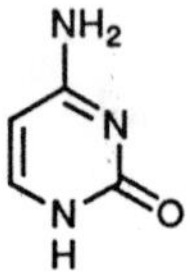

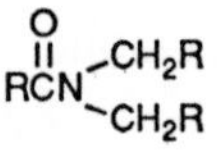

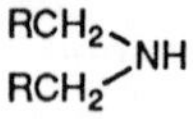

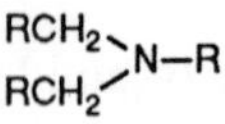

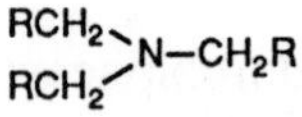

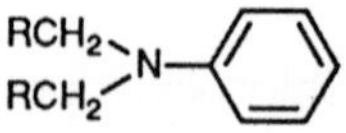

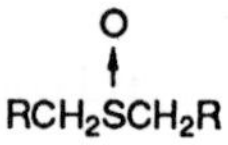

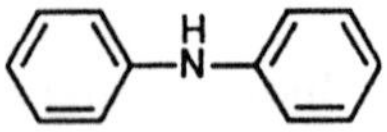

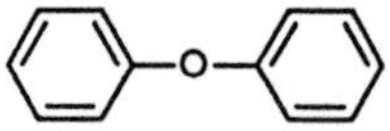

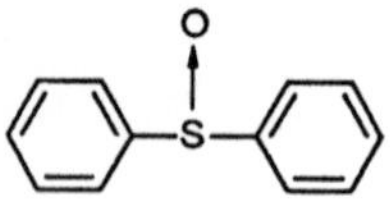

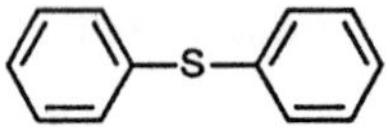

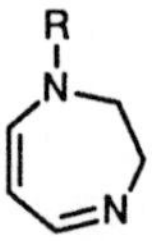

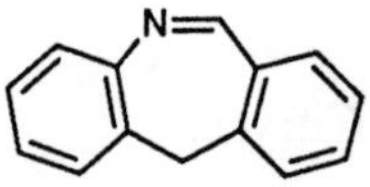

Diabenzocycloheptane		Dihydropyridine	
Dibenzofuran		Dihydroquinoline	
Dibenzothiazine		Dihydrothiazine	
Dibenzothiepine		Dimethylamide	
Dibenzoxepine		Dimethylamino	
Dichloroalkene		Dimethylaminoalkyl	
Difluoromethyl	F_2CH-R	Dimethylcarbamate	
Digitoxigenin	—	Dimethylcyclopropyl	
Dihydrobenzofuran		Dimethylsulfonamide	
Dihydrodiol		Dioxolane	
Dihydrofuran		Diphenyl ether	
Dihydropyran		Disulfide	$R-S-S-R$
Dihydropyridazinone		Dithiocarbamate	

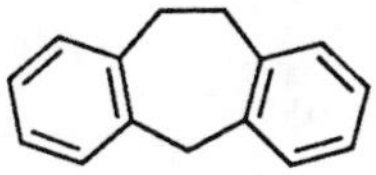

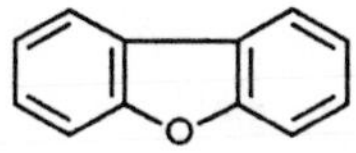

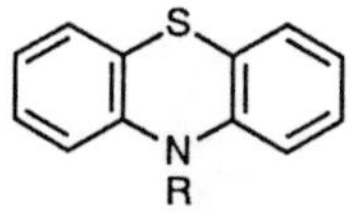

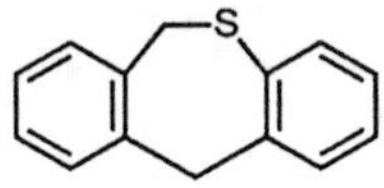

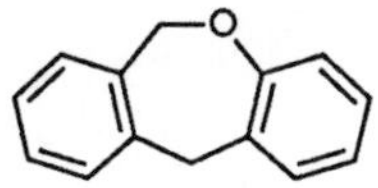

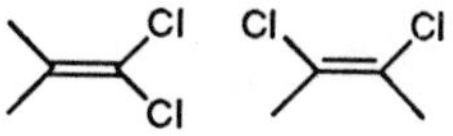

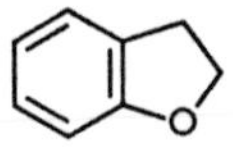

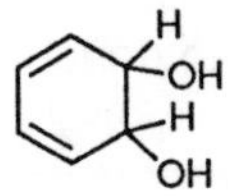

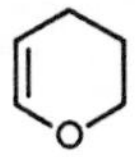

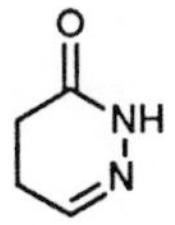

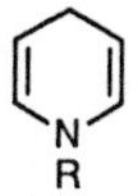

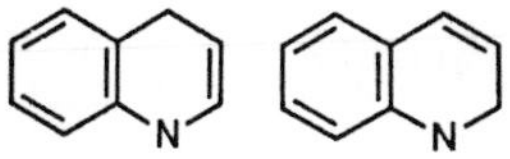

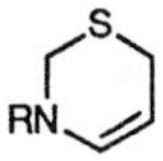

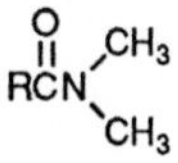

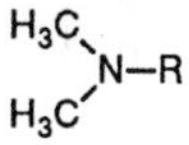

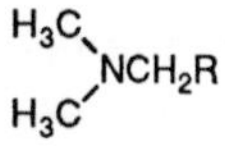

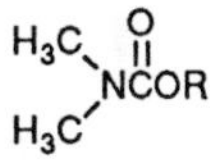

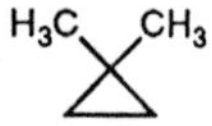

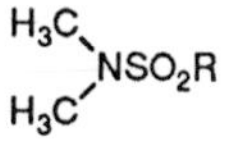

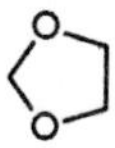

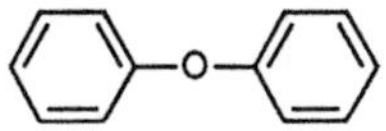

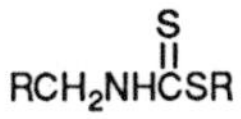

Dithiolane		Fluorophenyl	
Epoxide		Fluorouracil	
Ester glucuronide		Formamide	RNHCHO
Ethynyl	$R-C\equiv C-R$	*N*-Formyl	$(R)_2NCHO$
Fatty acid (saturated)	$R(CH_2)_nCO_2H$	Furan	
Fatty acid (unsaturated)	$R(CH=CH)_nCH_2CO_2H$	Furfural	
Fluoranthene		Furoyl carboxylate	
Fluorene		Germanorganic	
Fluoroacetyl		Glutamic acid	
Fluoroalkene		Glutathione	
Fluoroalkyl		Glycolamide	
Fluorocytosine		Glycoside	
Fluorohexose		Guanidine	

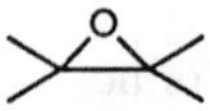

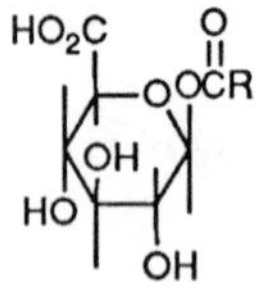

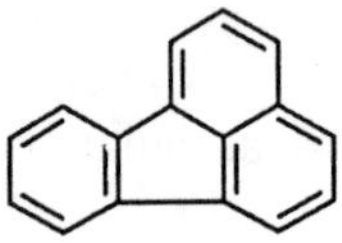

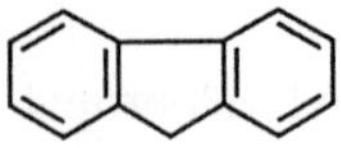

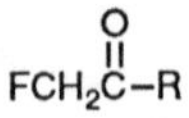

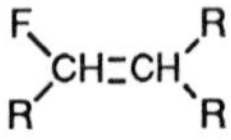

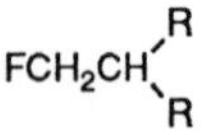

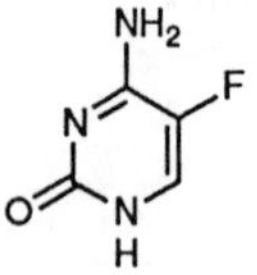

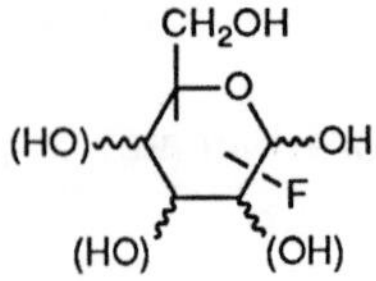

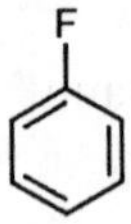

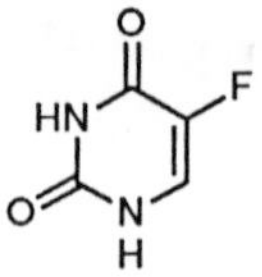

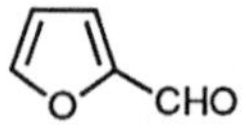

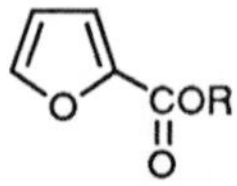

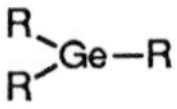

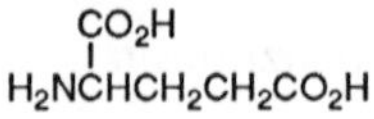

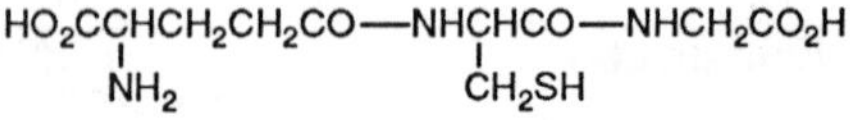

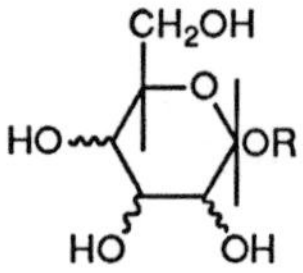

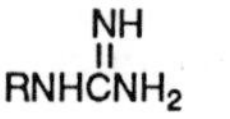

Haloalkyne   (Br)ClC≡CR

Hexahydroazepine

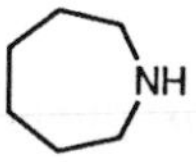

Hexahydronaphthalene

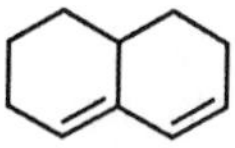

Hexose

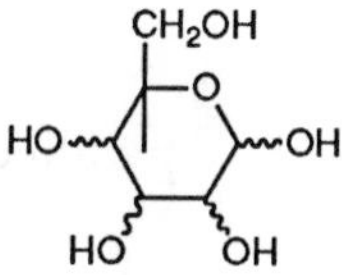

Hydantoin

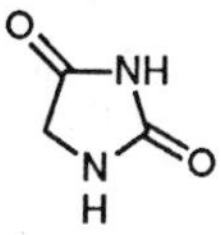

Hydrazide $\underset{\text{O}}{\text{RCNHNH}_2}$

Hydrazine   RNHNH₂

Hydrazone   RCH=NNHR

Hydroperoxide   ROOH

Hydroquinone

Hydroxamic acid $\underset{R}{\underset{\text{O}}{\text{RCNOH}}}$

N-Hydroxy $\underset{R}{\overset{R}{N}}$—OH

N-Hydroxy aryl amide

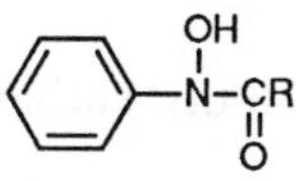

Hydroxycoumarin

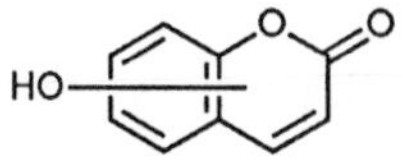

Hydroxyisoxazole

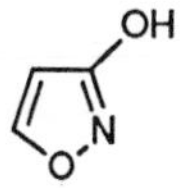

Hydroxylamine   RNHOH

Hydroxypiperidine

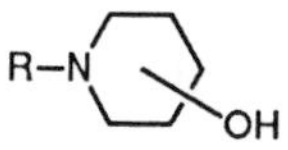

N-Hydroxy sulfate $\underset{R}{\overset{R}{N}}$—OSO₃H

Imidazole

Imidazolidinone

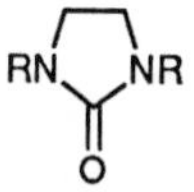

Imidazoline

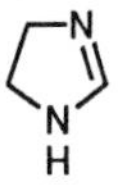

Imidazopyridine

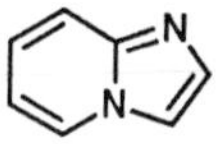

Iminecarboxylate   =N–C–OR

Indane

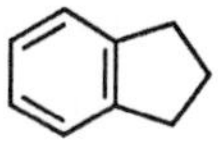

Indene

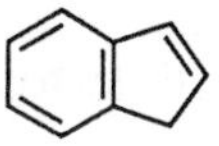

Indole

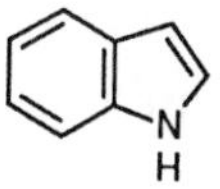

xxi

Indolone		Methanesulfonate	CH_3SO_2OR
Iodoalkyl	ICH_2R	o-Methoxyphenol	
Iodophenyl		Methoxyphenyl	
Isoprenoid	—	Methoxypyridine	
Isoquinoline		N-Methylalkylamine	
Isothiocyanate	$RN{=}C{=}S$	Methyl amide	
Ketal		N-Methylamidine	
Lactam		Methylamino	CH_3NHR
Lactone		N-Methyl aryl amine	
Leucine		N-Methylaziridine	
Lysine		N-Methylbenzylamine	
Macrocyclic lactone	—	Methyl carbamate	
Mercapturic acid		N-Methyl carbamate	

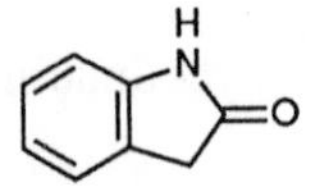

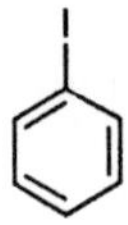

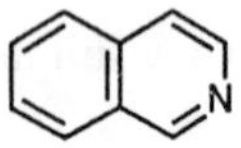

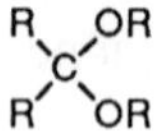

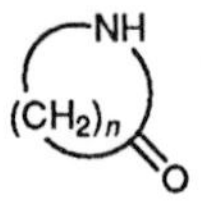

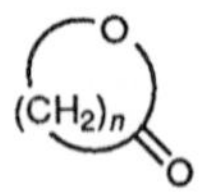

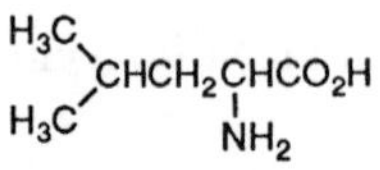

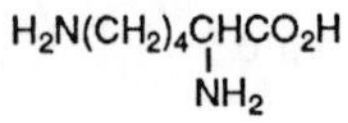

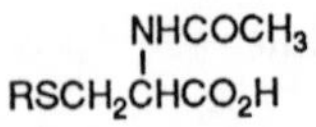

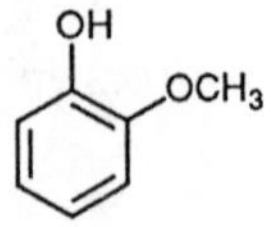

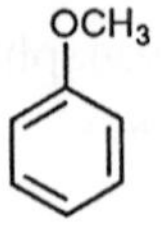

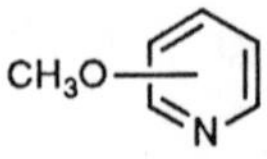

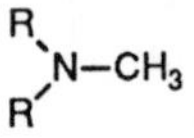

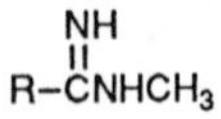

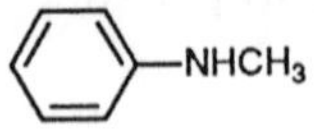

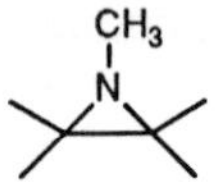

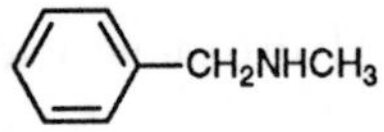

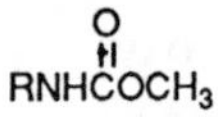

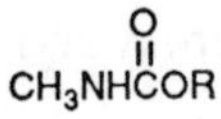

Methyl carboxylate	CH₃COR (=O)	Methyl ketone	CH₃CR (=O)
Methylcyclohexane	⬡—CH₃	N-Methylnitroamine	H₃C / R—N—NO₂
Methylcyclohexene	⬡—CH₃	N-Methylnitrosamine	CH₃NNO / R
Methylcyclohexenone	CH₃— (cyclohexenone)	Methyloxazole	(oxazole)—CH₃
N-Methyl diaryl amine	Ar—N(CH₃)—Ar	Methylphenyl	H₃C— (phenyl)
N-Methyldiazepine	(diazepine)NCH₃	Methylphosphinoyl	H₃C—P(OH)—OR
N-Methylperhydrodiazepine	RN (diazepine) NCH₃	O-Methylphosphorodithioate	CH₃O, CH₃O—P(=S)—SR
Methyl ester	RCOCH₃ (=O)	O-Methylphosphorothioate	CH₃O, CH₃O—P(=O(S))—SR (O)
Methyl ether	CH₃OR	N-Methylpiperazine	CH₃N (piperazine) NR
N-Methylhexahydroazepine	(azepine)NCH₃	N-Methylpiperidine	(piperidine)N—CH₃
N-Methylimidazole	(imidazole) N—CH₃	N-Methylpurine	RN (purine) R = H or CH₃
N-Methylimide	(imide)NCH₃	Methylpyran	(pyran)—CH₃
Methylindole	CH₃—(indole)—N—H	Methylpyrazolinone	CH₃, R—N—N(R)—(=O)

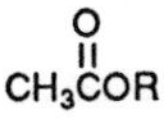

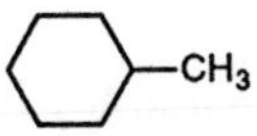

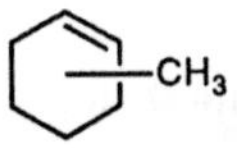

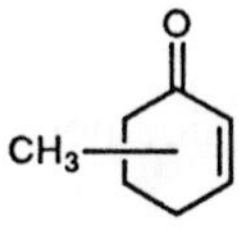

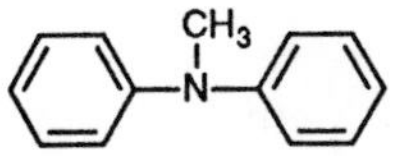

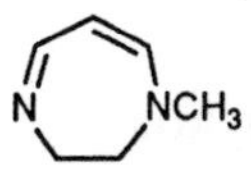

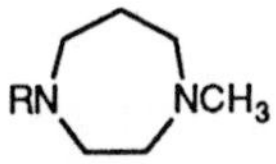

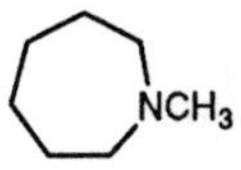

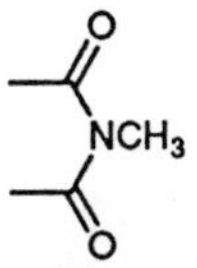

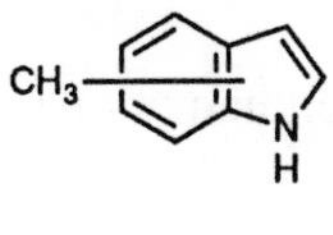

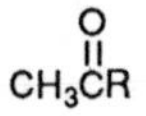

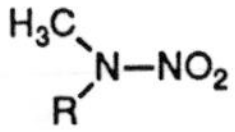

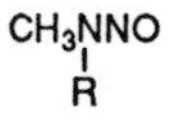

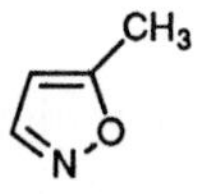

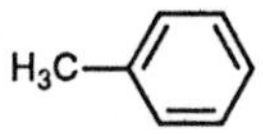

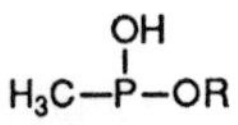

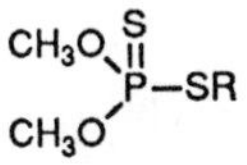

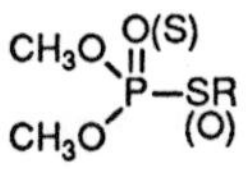

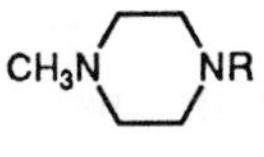

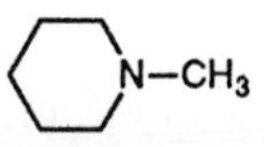

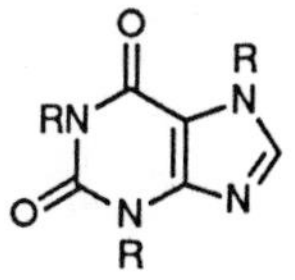

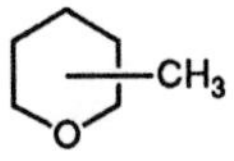

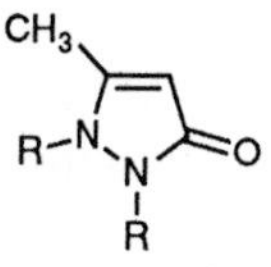

Methylpyridazine	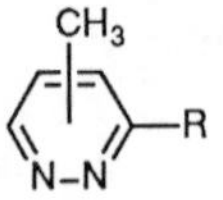	Methyl sulfonate	CH_3OSO_2R
Methylpyridine	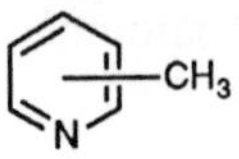	N-Methyltetrahydropyridine	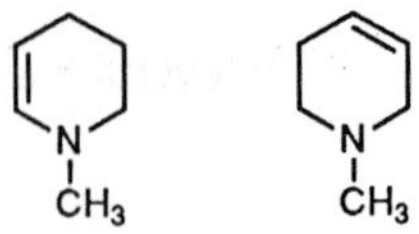
N-Methylpyridinium	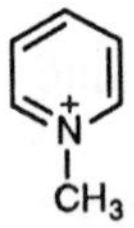	Methylthio ester	
N-Methylpyridone	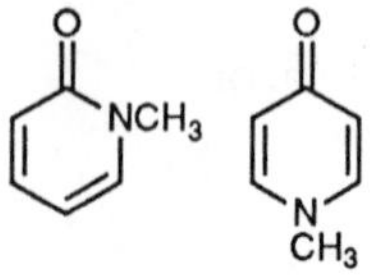	Methyl thioether	$RSCH_3$
Methylpyrimidine	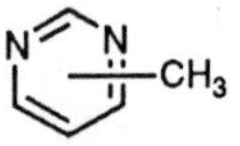	N-methyltriazene	
Methylpyrimidinone	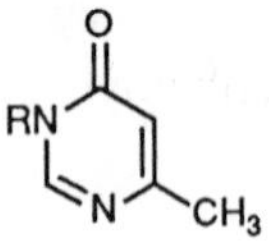	N-Methyltriazinone	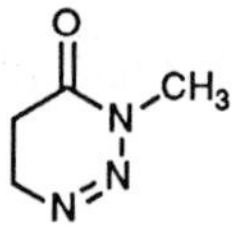
N-Methylpyrrole	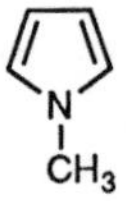	Methyltriazole	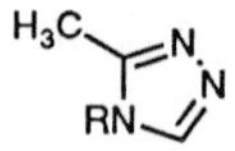
N-Methylpyrrolidine	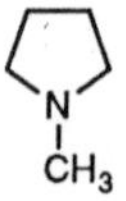	Methyluracil	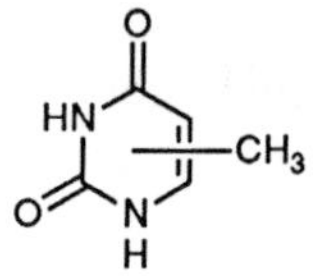
N-Methylpyrrolidone	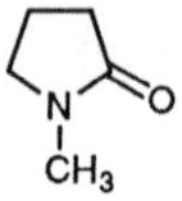	Monoclonal antibody	—
Methylquinoline	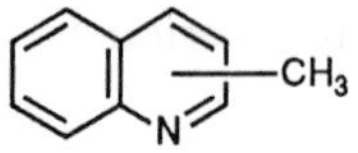	Morphinan	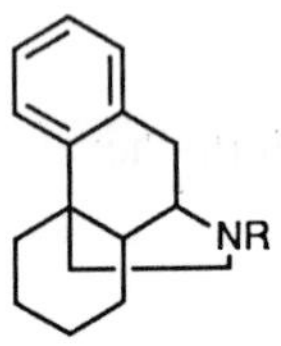
Methylquinoxaline	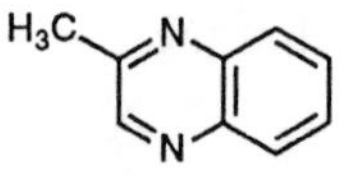	Morpholine	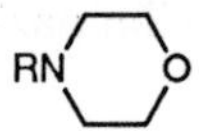
Methylsulfinyl		Naphthalene	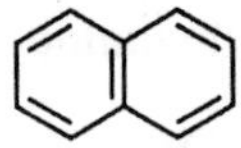
N-Methylsulfonamide		Naphthaquinone	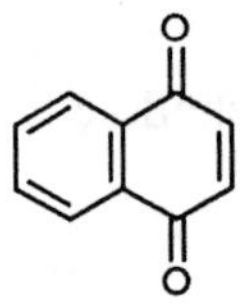

Nitramine	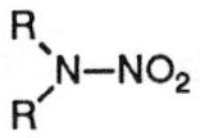 $\begin{smallmatrix} R \\ {} \\ R \end{smallmatrix}$N—NO$_2$
Nitrile	RCN
Nitroalkyl	RCH$_2$NO$_2$
Nitrofuran	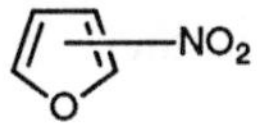 —NO$_2$
Nitroimidazole	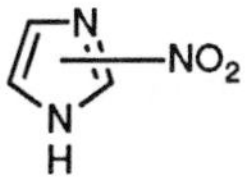 —NO$_2$
Nitrophenyl	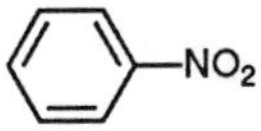 —NO$_2$
Nitroquinoline	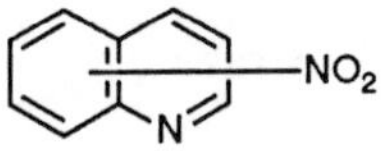 —NO$_2$
Nitrosamine	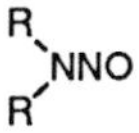 $\begin{smallmatrix} R \\ {} \\ R \end{smallmatrix}$NNO
Nitrosourea	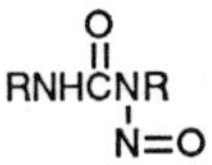 RNHĊNR N=O
Nucleoside	—
Octyl	CH$_3$(CH$_2$)$_7$—
Oestradiol	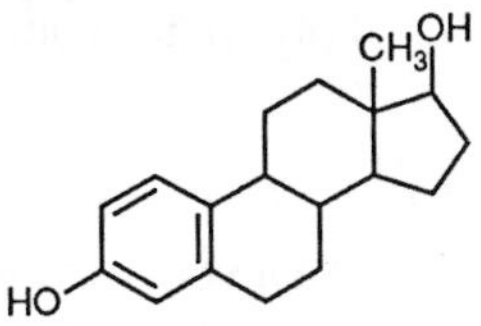
Oestren-3-one	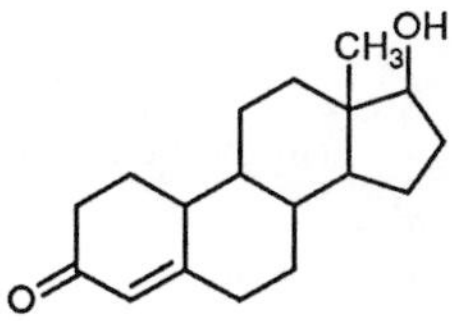

Oestrone	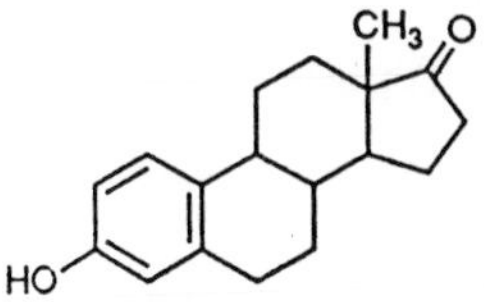
Organoarsenic	R$_3$As
Organometallic complex, platinum	—
Organoselenium	RSeR
Orthoester	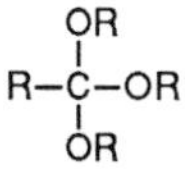 R—C—OR OR OR
Oxadiazolone	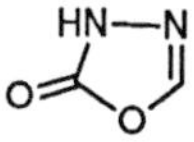
Oxathiolane	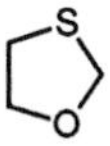
Oxazole	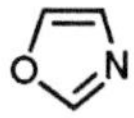
Oxazolidinone	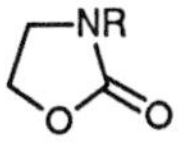
Oxime	RCH=NOH
Oximino	R=NOR
Ozonide	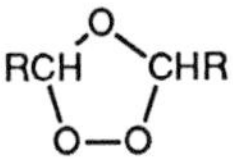 RCH—CHR O—O
Pentyl	CH$_3$(CH$_2$)$_4$—

Peptide	$RNH\overset{O}{C}CHNH\overset{O}{C}CHNH-$ with R groups	Phosphoramidothioate	$\underset{RNH}{\overset{RO}{\vphantom{.}}}\overset{S}{P}-OR$
Phenol	OH on benzene ring	Phosphorodithioate	$\underset{RO}{\overset{RO}{\vphantom{.}}}\overset{S}{P}-SR$
Phenol sulfate	OSO_3H on benzene ring	Phosphotothioate	$\underset{RO}{\overset{RO}{\vphantom{.}}}\overset{O}{P}-SR$
Phenothiazine	phenothiazine structure	Phthalazine	phthalazine structure
Phenoxyacetate	$-OCH_2\overset{O}{C}OR$ on benzene	Phthalimide	phthalimide structure NR
Phenoxybenzyl	$-O-$ $-CH_2R$	Phthalidyl	phthalidyl structure R
Phenoxypropionate	$-O\overset{R}{\underset{CH_3}{C}}CO_2R$	Piperazine	$R-N\quad N-R$
Phenyl	phenyl ring	Piperidine	piperidine NR
N-Phenyl	$\underset{R}{\overset{R}{N}}-$ phenyl	Piperidinedione	piperidinedione structure
Phenylethyl	$-CH_2CH_2R$ on benzene	Platinum	—
Phenylpyridine	$Ph-$ pyridine	Polycyclic aromatic	—
Phosphonate	$\underset{RO}{\overset{RO}{\vphantom{.}}}\overset{O}{P}-R$	Polycyclic aromatic amine	—
Phosphoramide	$\underset{RNH}{\overset{RO}{\vphantom{.}}}\overset{O}{P}-NHR$	Polycyclic aromatic hydrocarbon	—

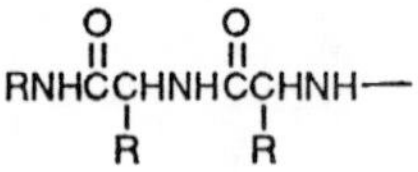

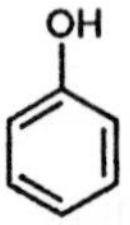

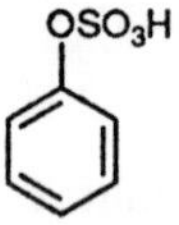

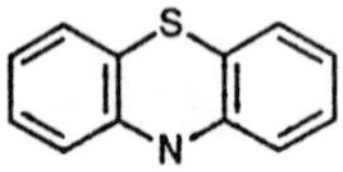

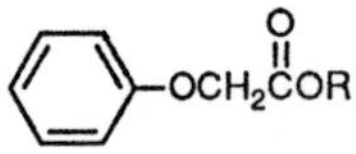

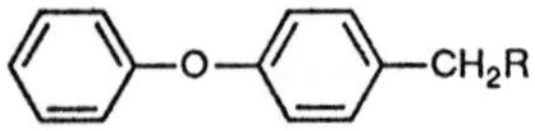

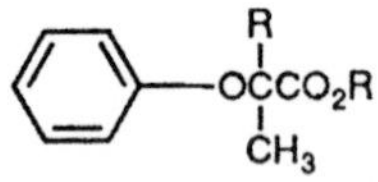

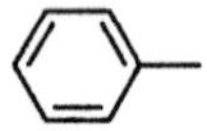

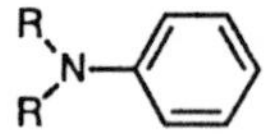

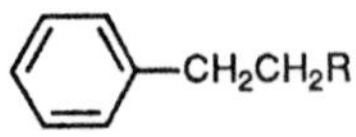

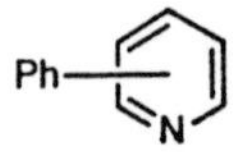

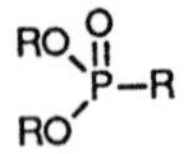

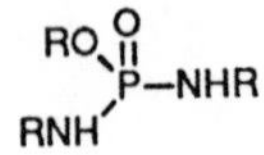

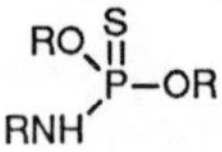

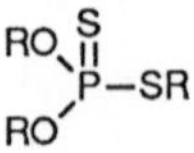

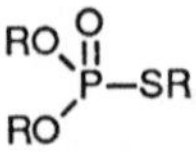

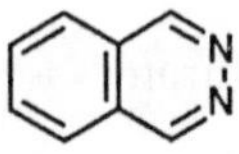

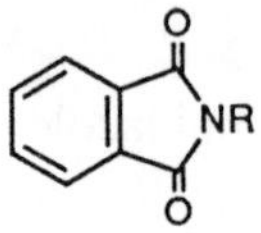

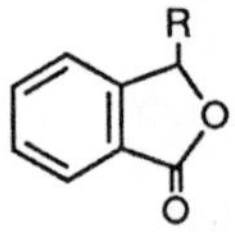

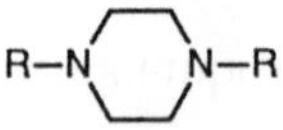

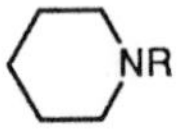

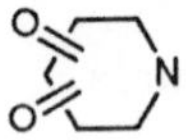

Prednisolone	(structure)	Pyrazolin-5-one	(structure)
Pregnadiene	(structure)	Pyrene	(structure)
Pregnene	(structure)	Pyridazine	(structure)
Prochiral carbon	(structure)	Pyridazinone	(structure)
N-Propyl	$R_2NCH_2CH_2CH_3$	Pyridine	(structure)
Propynyl	$CH_3C{\equiv}C-$	Pyridinium	(structure)
Prostanoid	—	Pyridopyrimidine	(structure)
Pteridine	(structure)	Pyrimidine	(structure)
Purine	(structure)	Pyrimidone	(structure)
Pyrazine	(structure)	Pyrrole	(structure)
Pyrazinone	(structure)	Pyrrolidine	(structure)
Pyrazole	(structure)	Pyrrolidine amide	(structure)
		Pyrrolidinone	(structure)

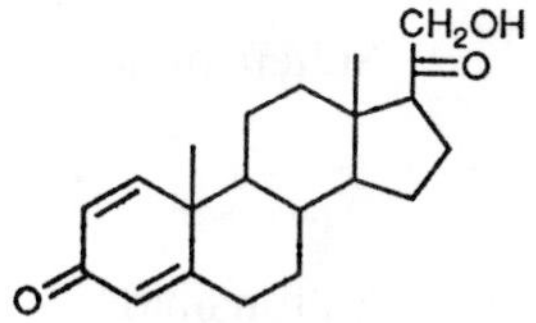

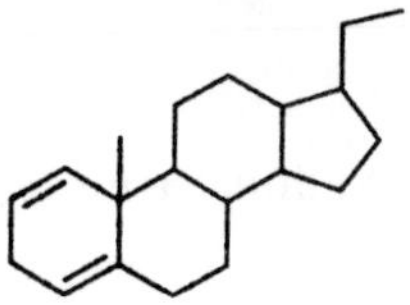

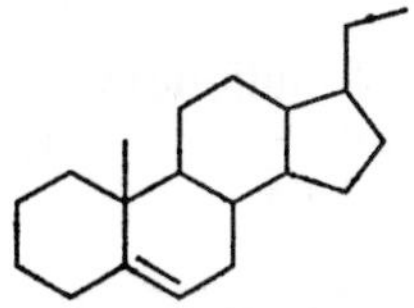

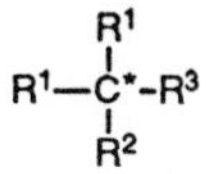

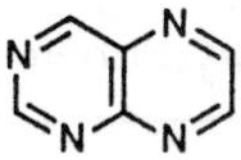

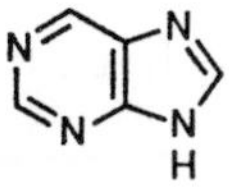

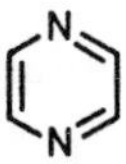

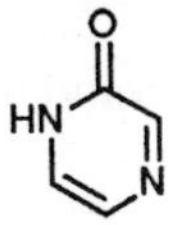

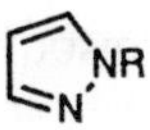

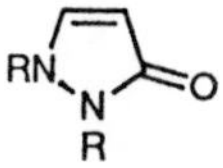

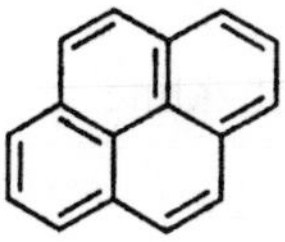

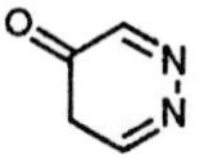

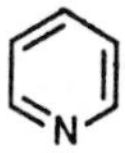

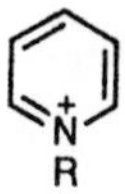

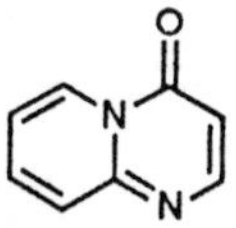

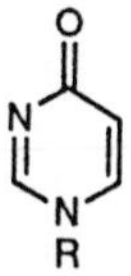

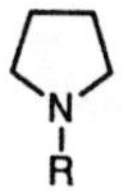

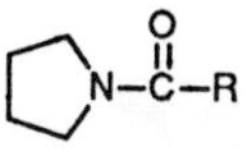

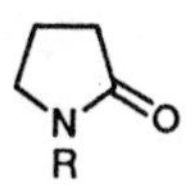

Pyrrolidone		Sulfenamide	$RS-NR_2$
Pyrrolizidine		Sulfonamide	RSO_2NHR
Quinazolinedione		Sulfonic acid	RSO_3H
Quinoline		*N*-Sulfonoxy	R_2N-OSO_3H
Quinolinone		Sulfonylurea	$RSO_2NHCNHR$ with O above C
Quinolizidine		Sydnone	
Quinolone		Terpene	—
Quinone		Tetrafluoroethyl ether	CHF_2CF_2OR
Quinonedi-imine		Tetrahydrobenzazepine	
		Tetrahydrocarboline	
Quinoneimine		Tetrahydrofuran	
Steroid	—	Tetrahydroindazole	
Sterol	—	Tetrahydroisoquinoline	

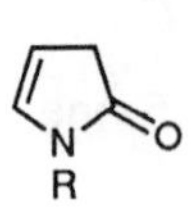

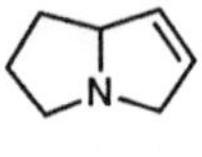

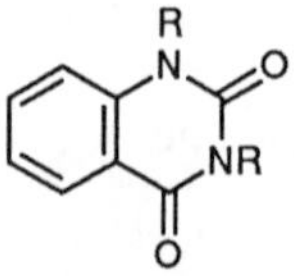

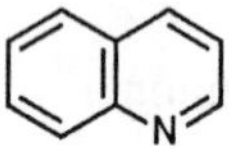

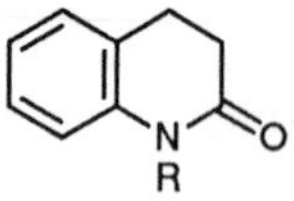

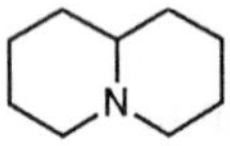

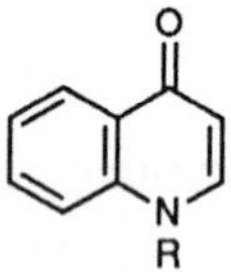

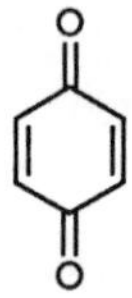

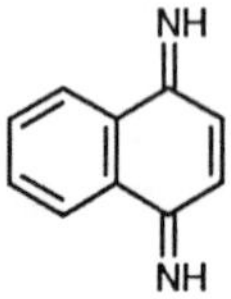

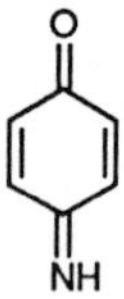

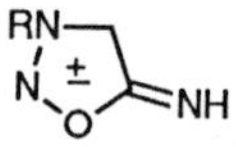

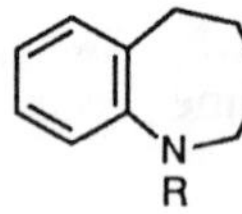

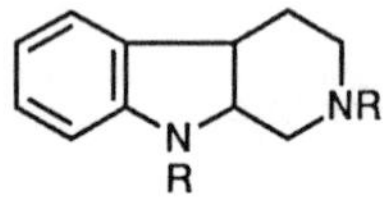

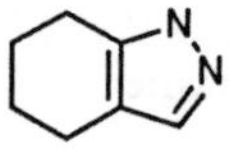

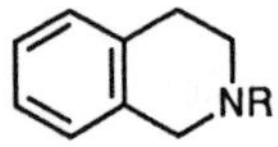

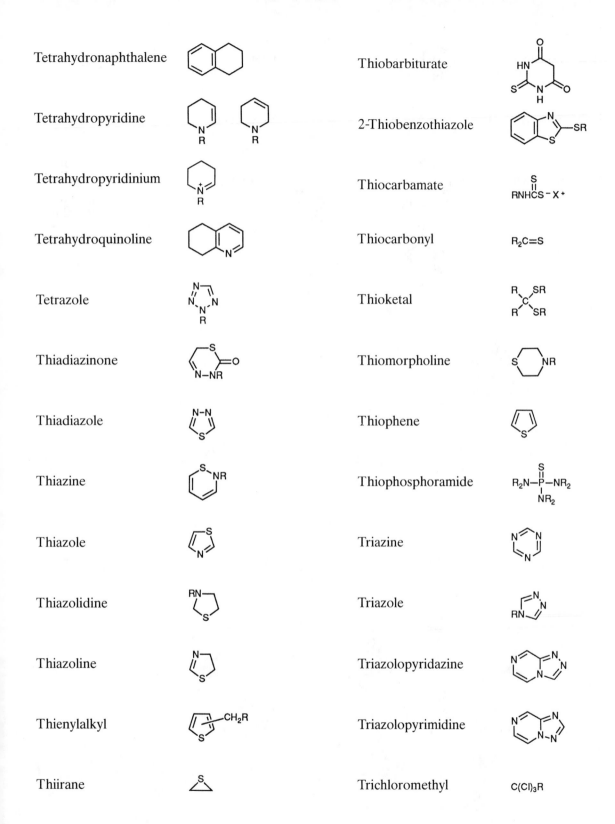

Tetrahydronaphthalene		Thiobarbiturate	
Tetrahydropyridine		2-Thiobenzothiazole	
Tetrahydropyridinium		Thiocarbamate	
Tetrahydroquinoline		Thiocarbonyl	$R_2C=S$
Tetrazole		Thioketal	
Thiadiazinone		Thiomorpholine	
Thiadiazole		Thiophene	
Thiazine		Thiophosphoramide	
Thiazole		Triazine	
Thiazolidine		Triazole	
Thiazoline		Triazolopyridazine	
Thienylalkyl		Triazolopyrimidine	
Thiirane		Trichloromethyl	$C(Cl)_3R$

Trichothecene

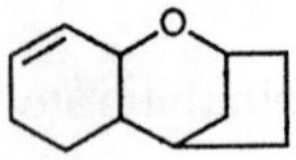

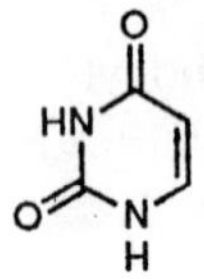

Ureide

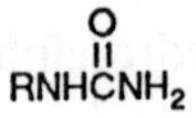

$RNH\overset{O}{\overset{\|}{C}}NH_2$

Trifluoromethylphenyl F_3C-

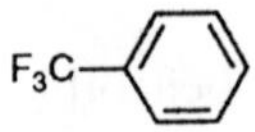

Vinca alkaloid —

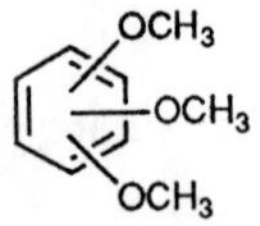

Trimethoxyphenyl

Xanthine

Uracil

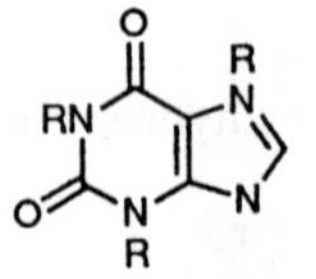

An Overview

The purpose of this chapter is to highlight some of the biotransformation studies included in this volume which report particularly interesting aspects. This includes novel biotransformations, stereoselective and stereospecific processes, and examples where mechanisms of toxicity have been attributed to specific biotransformation pathways. It is hoped that increasing the awareness of recent key information on biotransformations will ensure that it is utilized in making further advances in developing our knowledge of the subject.

1 NOVEL PATHWAYS

1.1 Oxidative

The metabolites of several fluorinated anilines have been compared *in vivo* and *in vitro* using rat liver microsomes (p. 158). Both pentafluoro- (1) and tetrafluoro-aniline (2) gave the sulfate conjugate (3) as a major urine metabolite. Evidence was obtained that whereas the phenol (5) was formed directly from (2) the quinone imine (4) was an intermediate in the biotransformation of (1). For *p*-fluoroaniline both *ortho* and *para* hydroxylation occurred and in the case of the latter there was no evidence for retention of fluorine via an NIH shift.

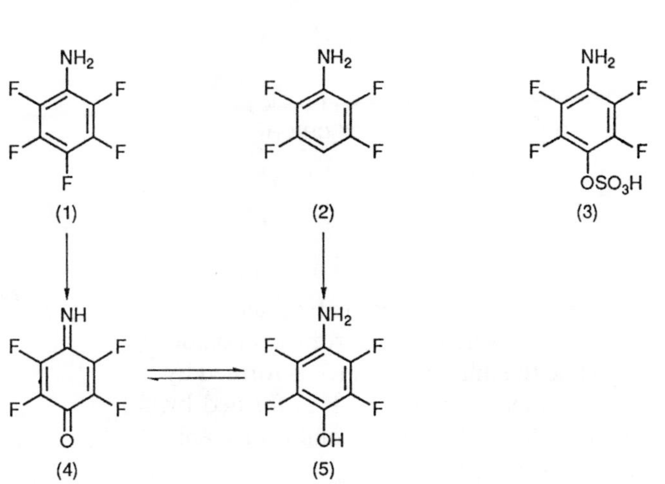

An analogue of 1 α-hydroxy-vitamin D_3 has been synthesized in which the three terminal side-chain carbons are incorporated into a cyclopropane ring (6). This compound has a potent calcemic effect *in vivo* and it was expected that metabolic activation by hydroxylation at C-25 would occur. However, *in vitro* experiments using human hepatocytes gave metabolites which were products of hydroxylation at C-24 although there was evidence that the C-25 hydroxylase enzyme was involved which has a broad substrate specificity (p. 208).

(6)

The metabolites of the halocarbon (7) which have been identified in rat urine are mainly predictable (p. 66). Hydroxylation at C-2 would give a chlorohydrin that would spontaneously yield the hydrated aldehyde (8) leading to the acid (9). Reduction to the alcohol (10) and formation of the glucuronide follows the same pathway as other trihalogenated compounds such as chloral hydrate. Interestingly two unknown conjugates of (8) and (10) were also detected.

$$ClCF_2CH_2Cl \quad \longrightarrow \quad \left[ClCF_2\overset{OH}{\underset{|}{C}}HCl \right] \quad \longrightarrow \quad ClCF_2CH\overset{OH}{\underset{OH}{\diagdown}}$$

(7)   (8)

$ClCF_2CO_2H$ $ClCF_2CH_2OH$

(9)   (10)

Two major metabolites of pravastatin (11) have been identified after incubation with rat hepatocytes (p. 209). These were derived from an epoxide intermediate (12). It is interesting that while the dihydrodiol (13) is formed by attack at the α,β-double bond the glutathione conjugate (14) is formed by direct nucleophilic attack of the epoxide at the bridge carbon and is probably the first example of this type of glutathione conjugate.

2

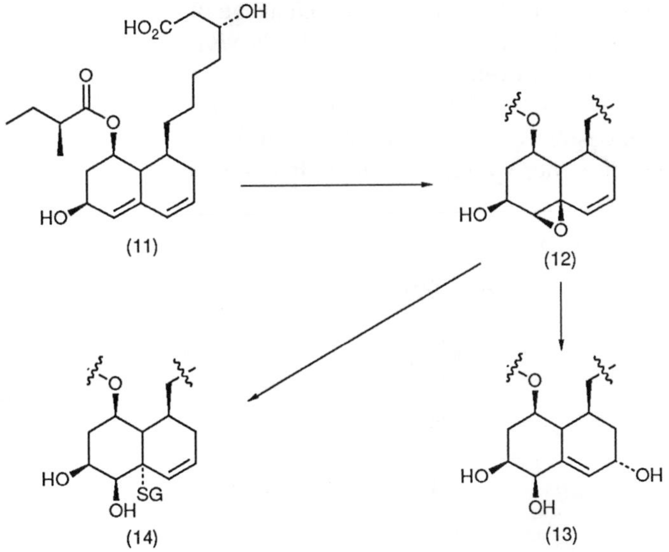

(11) → (12)

(14)   (13)

A single major metabolite of the fatty acid aniline (15), a plasma lipid regulator, has been identified in both rat and cynomolgus monkeys (p. 151). This metabolite was a major component in rat urine and bile and represented *ca.* 30% of an intravenous dose in monkey urine and was identified as the carboxylic acid (16) formed by oxidative loss of six carbons. There was no evidence for intermediate chain-shortened carboxylic acids or *O*-demethylated metabolites.

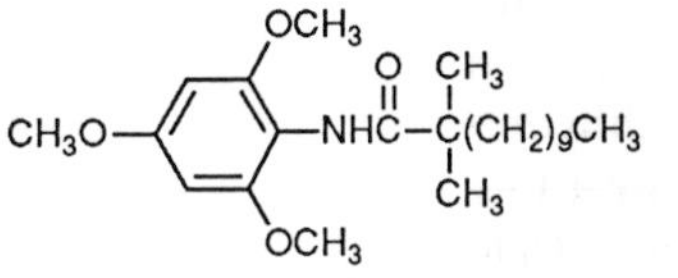

(15) 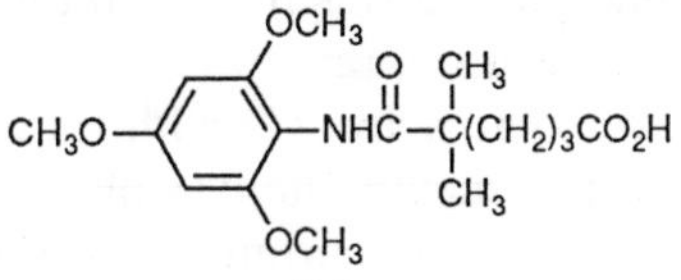   (16)

The major metabolites of the antiparasitic drug pentamidine (17) formed by rat liver microsomes were the result of hydroxylation of the alkyl bridge (p. 122). However, two minor metabolites were reported as the mono- and bis-amidoximes (18) formed by *N*-hydroxylation of the amidine group. There are few examples of the biotransformation of the amidine group and this indicates one reaction that should be considered.

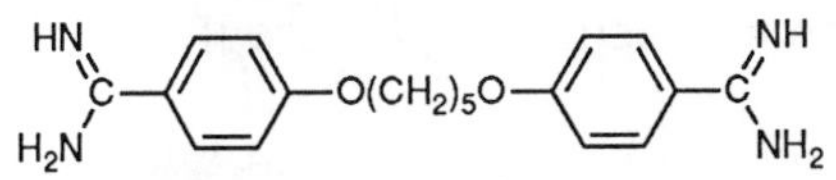

(17)

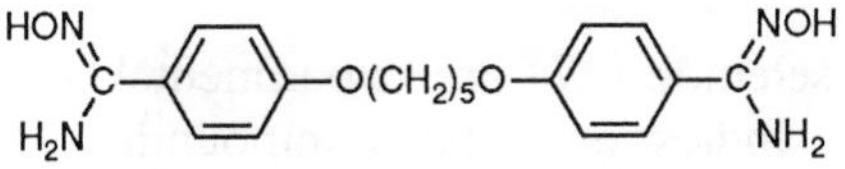

(18)

3

The main *in vitro* metabolites of dilthiopyr (19) are formed by cleavage of the thioester groups to give nicotinic acid derivatives such as (21). *In vitro* studies using cytosolic and microsomal rat liver preparations have shown that cleavage of the thioester functions occurs via oxidative processes and not direct hydrolysis by esterases. The thioester is activated by *S*-oxidation to give an unstable intermediate (20) which either spontaneously hydrolyses to the acid or reacts with glutathione (22).

(19)   (20)

(21)   (22)

1.2 Reductive

Studies on the metabolism of dihydropyridine antihypertensive drugs continue to provide interesting results due to the complex array of biotrans-formations which occur. Nitrendipine (23) is no exception and up to twenty metabolites have been identified in mice, rat, and dog (p. 273). A major rat plasma metabolite was the carboxylic acid (24) with a maintained dihydro-pyridine ring. A glucuronide of this metabolite was also detected in rat bile. Interestingly on incubation with β-glucuronidase/sulfatase from *Helix pomatia* none of the corresponding aglycone was detected, only the pyridine derivative (25). Metabolites formed by reduction of the nitro group were only important in mice. *In vitro* studies with liver microsomes showed that both conversion of the dihydropyridines to pyridines and ester cleavage were mediated by cytochrome P-450, indicating the formation of hydroxylated intermediates.

(23)   (24)   (25)

Benzyl selenocyanate (26) and dibenzyl selenide (27) are experimental antitumour agents and recent metabolism studies on these compounds provide useful background information on the biotransformation of selenium-containing functional groups (p. 81). Oral doses of both compounds to rats were mainly excreted in the urine and two major common metabolites were benzoic and hippuric acid. *In vitro* incubations of the selenocyanate with liver homogenates showed that the diselenide was formed, presumably via the selenide (28). The selenous acid (29) was also detected as an *in vitro* metabolite. There was a low recovery of oral doses to rats which may be attributed to formation of covalently bound adducts since a large proportion of the substrate was recovered in the protein fraction from *in vitro* incubations. The bound adducts could arise by formation of Se—S or Se—Se bonds involving the selenide (28). The biotransformation of these compounds is therefore similar to that expected for the corresponding sulfur analogues.

Epitiostanol (30) has an uncommon structural feature, namely a thiirane ring, the sulfur analogue of an epoxide. The compound is rapidly metabolized in rats undergoing extensive first-pass metabolism after oral dosing. Two metabolites have been identified in liver and plasma, one of these corresponding to the ketone formed by oxidation of the secondary alcohol. The other metabolite was the dehydro compound (31) which provides an insight into the major biotransformation pathway for the thiirane ring.

1.3 Ring Cleavage

The two major rat urine metabolites of the quaternary ammonium compound (32) were identified as the phenol (33) and quinone (34) (p. 316). Cleavage of the pyran ring to give the quinone is a novel type of process which is effectively *O*-dealkylation of a substituted phenol. Comparison of the tissue distribution of (32) and the corresponding tertiary amine (35) showed that there was a highly selective uptake of the quaternary compound by heart muscle with a heart:blood concentration ratio of > 20 compared with < 2 for the tertiary amine. The major component in the heart was the phenol (33). The selective tissue uptake is consistent with observations on quaternary anti-arrhythmic drugs whose long duration of action has been ascribed to the cardiac uptake.

(32) R = CH₃CO
(33) R = H

(34)

(35)

Several human metabolites of 6-benzoylbenzoxazolinone (36) resulting from cleavage of the heterocyclic ring have been identified (p. 299). The mechanism of ring-cleavage is not apparent since no intermediates between the parent and the aminophenol (37) were identified. It is unusual that a benzylic carbon has been replaced by a phenolic group. The *N*-methylated metabolite (38) is also novel, which also subsequently underwent ring-cleavage to give the metabolite (39).

(36)

(38)

(37)

(39)

1.4 Cyclization

Two major metabolites of the prostaglandin (40) have been identified in human plasma (p. 95). These metabolites were initially isolated and identified using a guinea-pig liver perfusion experiment and then co-chromatographed with human plasma extracts. One of these metabolites (41) was the PGA-analogue of the parent compound and the other was a novel cyclized compound (42). This metabolite was presumably formed following reduction of the Δ^{13} double bond by internal nucleophilic addition to the cyclopentenone ring.

(40)

(41)

(42)

6

A novel biotransformation of the anthraquinone mitoxantrone (43), an anticancer agent, has been reported (p. 214). The established pathways for this drug include oxidation of the alkyl alcohol side-chains to carboxylic acids but a less polar metabolite detected in the urine of rat, pig, and human cancer patients has been identified as the cyclized compound (46). This metabolite was formed on incubating (43) with hydrogen peroxide and horseradish peroxidase, and in the presence of glutathione two conjugates (44) and (45) were formed. A highly reactive intermediate quinone-diimine (47) was proposed as the precursor to these metabolites. Formation of the intermediate may be important in its mode of action.

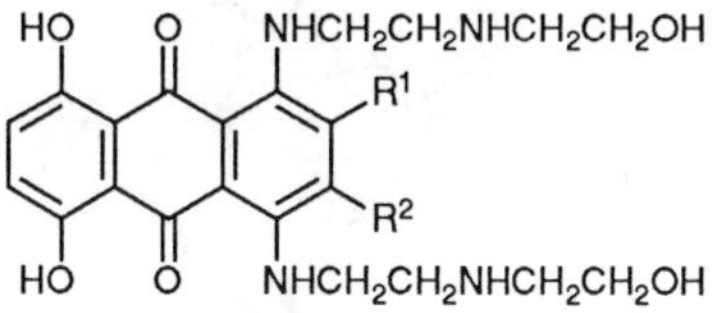

(43) $R^1 = R^2 = H$
(44) $R^1 = SG, R^2 = H$
(45) $R^1 = H, R^2 = SG$

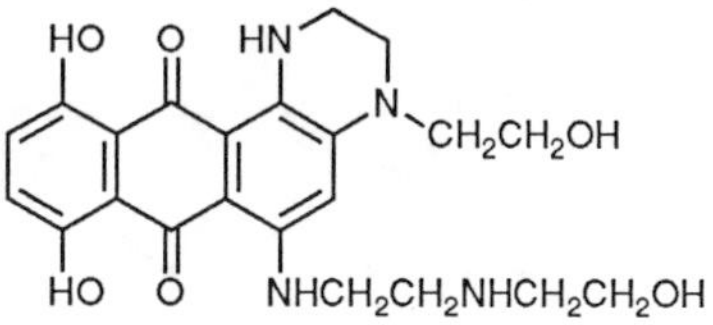

(46)

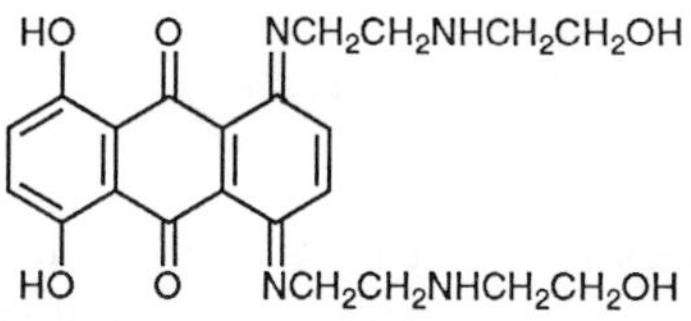

(47)

1.5 Rearrangement

Up to nine metabolites of methenolone (48) have been detected in human urine samples after single oral doses (p. 387). These metabolites were excreted mainly as glucuronides and three of them as sulfates. A series of metabolites were derived by initial formation of an exocyclic methylene group. It was proposed that these were formed by rearrangement via (49) to (50) followed by hydroxylation (51) or reduction of the keto group (52).

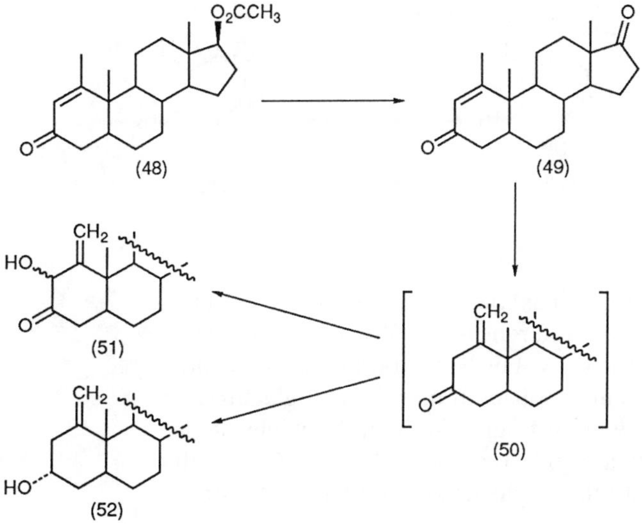

(48)

(49)

(51)

(50)

(52)

7

The mushroom toxin illudin S (53) possesses a novel structure consisting of a cyclopropane group substituted on a cyclohexanone ring. Two novel metabolites have been identified formed on incubation with rat liver preparations (p. 207). Both metabolites were formed by cleavage of the cyclopropane ring, and reduction of the α,β-unsaturated carbonyl system to give the reactive intermediate (54) was proposed. Nucleophilic attack on the cyclopropane could yield the identified metabolites (55) and (56). The insertion of chlorine in metabolite (56) is particularly novel.

Some *in vivo* rat metabolites of the antimalarial drug arteether (57), which has a novel peroxide bridge as part of a tricyclic system, have been identified by HPLC–MS (p. 223). Up to twelve metabolites were identified in rat plasma. Some of these metabolites resulted from loss of the ethyl group either by hydrolysis of the acetal function or oxidative *O*-de-ethylation and ring hydroxylation. Contraction of the peroxide bridge also occurred to give a metabolite such as (58) with an ether bridge. A further novel process involving ring contraction and formation of a tetrahydrofuran gave metabolites such as (59).

1.6 Conjugates

Formation of quaternary glucuronides has been established as a major biotransformation pathway for several different types of compound, occurring particularly in humans and monkeys. Further experiments have now been reported for amitriptyline (60) metabolism in man. The *N*-glucuronide (61) represented 3–21% of the dose in 24 hour urine samples while similar conjugates of 10-hydroxyamitriptyline (63) represented *ca.* 2–6% of the dose. From other experiments where the *N*-glucuronides were administered and rates calculated for excretion of metabolites it was proposed that *N*-glucuronidation of (62) and hydroxylation of (61) each contributed to the formation of the quaternary glucuronide (63).

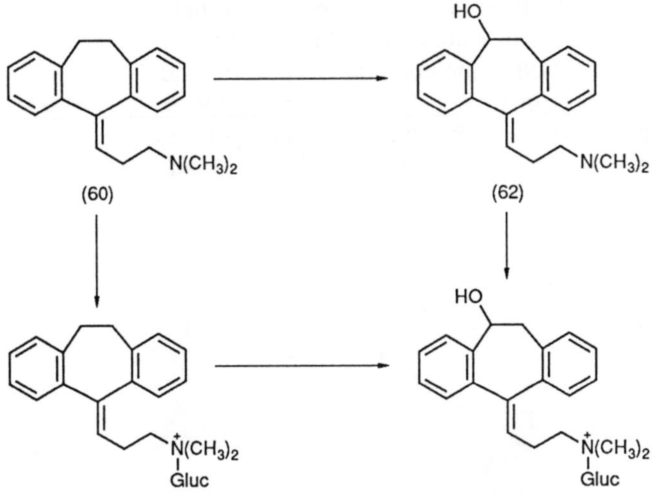
HO
N(CH3)2
N(CH3)2
HO
N(CH3)2
Gluc
N(CH3)2
Gluc

(60) (62) (61) (63)

The presence of these quaternary ammonium conjugates has now been investigated in human urine following administration of six different antihistamines which contain dialkylaminoalkyl groups (p. 175). The corresponding conjugate was identified for all six drugs, with diphenhydramine (64) producing the largest amount (4% of the dose).

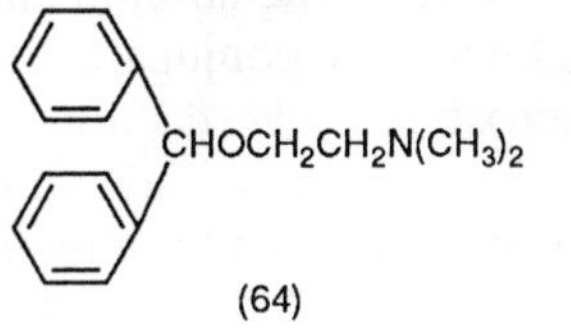
CHOCH2CH2N(CH3)2

(64)

Formation of glucuronide conjugates of the angiotensin receptor antagonist (65) has been investigated using monkey liver slices. The molecule contains two possible sites for direct glucuronidation, the hydroxymethyl group and the tetrazole ring. Interestingly the *N*-2-*β*-glucuronide (66) was identified as a major metabolite, which may represent the first example of conjugation with this type of heterocyclic ring.

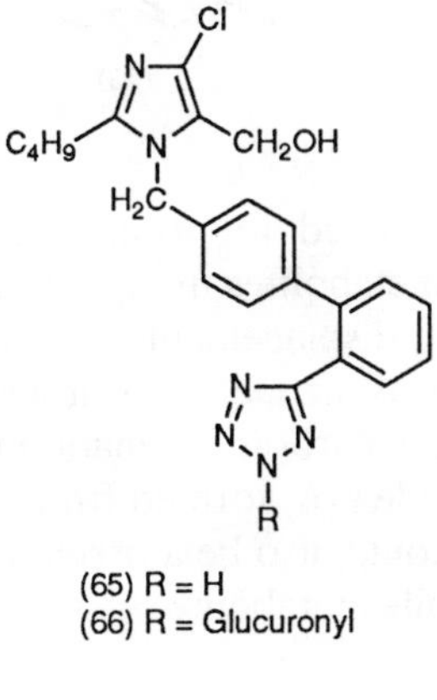
Cl
N
C4H9
N
CH2OH
H2C
N
N
N
N
R

(65) R = H
(66) R = Glucuronyl

Dichloroethyne (67) is a potent nephrotoxin in rats and causes nerve damage in humans. Metabolites of this compound have now been investigated after inhalation exposure of rats (p. 70). Direct glutathione conjugation is the major pathway and the mercapturic acid (68) represented more than 60% of the urinary metabolites. The only metabolite detected in bile was the glutathione conjugate. Other urinary metabolites included dichloroacetic acid (69) and dichloroethanol (70), a process involving oxidation and chlorine migration mediated by cytochrome P-450. As with other compounds of this type the nephrotoxicity is probably associated with thiols formed by degradation of the cysteine conjugate in the kidneys.

Metabolites derived by hydrolysis of the thioacetyl group in spironolactone (71) have previously been identified. These include the sulfinic and sulfonic acids formed by oxidation of the thiol. A glutathione conjugate (73) has also been identified, formed in rat liver microsomes by reaction of glutathione with an electrophilic species. Evidence was obtained indicating that the process may be mediated by a flavin mono-oxygenase and the intermediate (72) was suggested.

The trichloropropane (74) contains chlorine attached to primary and secondary carbon atoms and an investigation of its metabolites in the rat has shown that different conjugation reactions involving displacement of either chlorine can occur (p. 65). A major urine metabolite was the mercapturic acid (77) formed by glutathione conjugation at the primary carbon. Formation of this metabolite could conceivably occur via an epoxide (76) formed from an hydroxylated intermediate (75). The same epoxide could also be a precursor to the glutathione conjugate (78) which was a major bile metabolite.

10

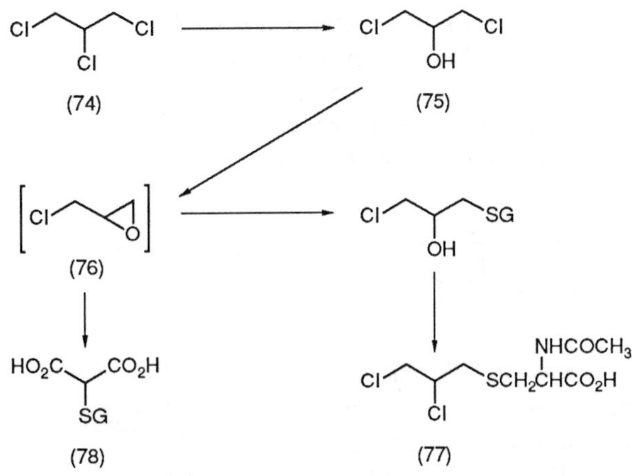

(74) → (75)

(76) → (77)

(78)

A novel glutathione conjugate (80) of the dihydrobenzofuran (79) has been identified after incubation with a rat liver preparation in the presence of glutathione (p. 302). The formation of this conjugate involves cleavage of the dihydrofuran ring but the nature of the intermediate which is susceptible to nucleophilic attack by glutathione is unknown. It was also found that the $R(+)$-enantiomer inhibited the conjugation of the $S(-)$-enantiomer.

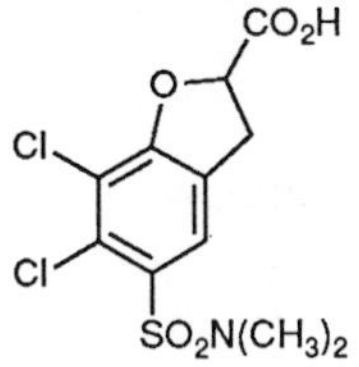

(79)

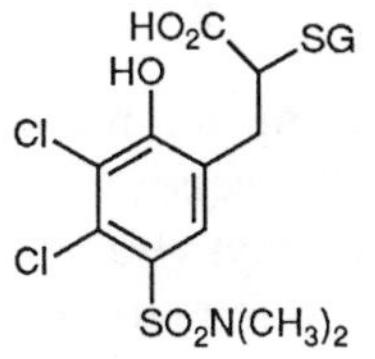

(80)

A major Phase I metabolite of the tricyclic *cis*-flupentixol (81) in the rat is the corresponding sulfoxide (p. 342). Rat bile was shown to contain mainly conjugates including an ether glucuronide of the parent drug and a sulfate of the sulfoxide. A further conjugate was identified as the unstable glutathione conjugate (82) formed by addition to the exocyclic double bond. This metabolite could also be synthesized non-enzymatically and represents the first example of its type.

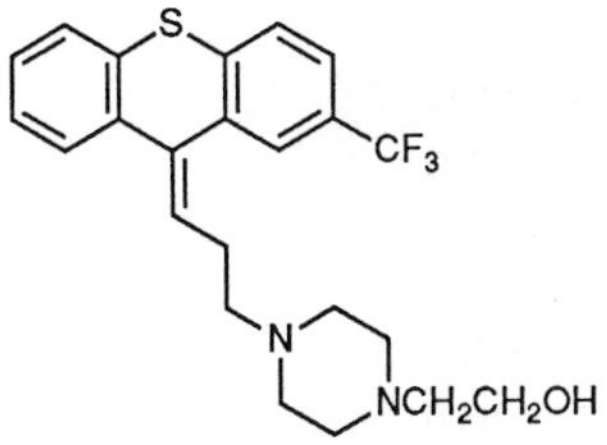

(81)

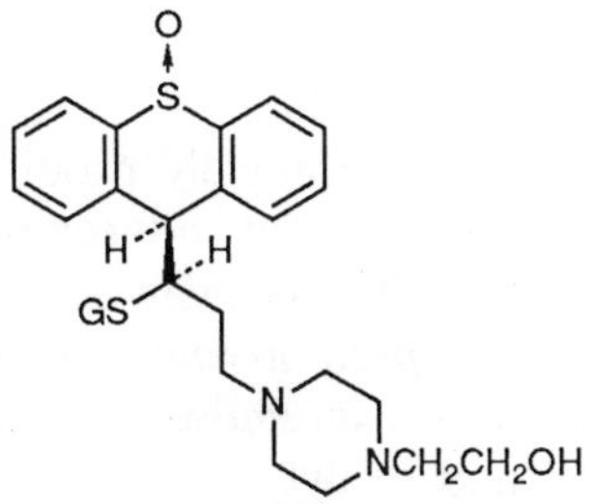

(82)

11

Formylation of aromatic amines has been reported for several compounds but it remains a relatively novel biotransformation. Another example has now been reported for 5-aminosalicylic acid (83) (p. 162). The corresponding *N*-formyl metabolite (84) has been identified in the urine of pigs after oral dosing and also in human plasma of subjects receiving intravenous doses.

(83)   (84)

The implication of pyridinum ion metabolites in the neurotoxicity of several compounds has prompted studies with the tetrahydropyridinium metabolite (85) of the psychotomimetic drug phencyclidine (p. 265). Incubation of the metabolite with rat, brain, or liver mitochondrial preparations produced the formylated compound (86). There is evidence that this was produced from a *trans*-formylation reaction.

(85)   (86)

Bioalkylation and dealkylation of chrysene and methylchrysene have been demonstrated in rat liver cytosol (p. 36). Chrysene (87) and 5-methylchrysene (88) formed 6-methyl (89) and 5,6-dimethyl (90) analogues respectively. 6-Methylchrysene underwent a dealkylation to reform chrysene. 5,6-Dimethylchrysene formed the cyclic derivative (91) in an enzyme-mediated process.

(87) $R^1 = R^2 = H$
(88) $R^1 = H, R^2 = CH_3$
(89) $R^1 = CH_3, R^2 = H$
(90) $R^1 = R^2 = CH_3$

(91)

1.7 Miscellaneous

The sulfonylurea sulofenur (91) can theoretically produce *p*-chloroaniline (92) as a metabolite which is a known potent inducer of methaemoglobin formation. In order to detect whether this pathway occurred urine samples from dosed animals were analysed for *p*-chloroaniline metabolites since less than 5% of a dose of this compound is excreted unchanged. The two major metabolites were the sulfate (93) and oxalyl conjugate (94), both of which were detected in urine of rats and monkeys dosed with sulofenur.

(91)

(92)

(92)

(94)

N-Acetylation of aryl amines is a well established and often important biotransformation but fewer examples have been reported for other amino functions. However, it has now been shown that acetylation is a major route of metabolism for cyanamide (95) with the urinary metabolite (96) representing up to 40% of an oral dose in rat and human (p. 351).

$$H_2N-C\equiv N$$

(95)

$$CH_3CONHC\equiv N$$

(96)

Comparative studies on the metabolism of the decapeptide (97) in rats and cynomolgus monkeys provide some interesting information on the fate of this type of compound for which there are as yet few examples (p. 379). After intravenous and subcutaneous doses urine and plasma from both species contained mainly parent drug although there was extensive biliary excretion and rat bile contained only traces of the drug and three major metabolites. These metabolites were identified as truncated peptides formed by cleavage of the linkages between L-amino acids Ser-Tyr and Leu-Arg. One of these metabolites was also detected in monkey plasma indicating that the same biotransformations probably also occurred in this species. The long half-life of the peptide was attributed to the restricted metabolism by endogenous peptidases due to the presence of the unnatural D-configuration of five amino acids.

(97)

2 STEREOSPECIFIC/STEREOSELECTIVE BIOTRANSFORMATION

Studies on the comparative metabolism of the two enantiomers of sparteine illustrate how extensive stereoselective biotransformation can occur (p. 345). For the (−)-enantiomer (98) two major urine metabolites were identified after oral doses to rats. These two metabolites were the didehydro compound (99) shown to be formed by stereospecific abstraction of the axial 2β

13

hydrogen atom and the ketone (100). The ketone was also shown to be formed as a metabolite of the dehydro compound. In contrast the major urinary metabolite of the (+)-enantiomer (101) was the 4S-hydroxy compound (102). Two dehydro metabolites (103) and (104) were identified as minor components in urine. Inhibition studies *in vitro* with rat liver microsomes showed that both sparteine enantiomers were metabolized by the same cytochrome P-450 isozyme (IID1) showing the marked substrate and product stereoselectivity of this enzyme.

(98)   (101)

(99)   (100)   (102)   (103)   (104)

Substituted dihydropyridines contain an asymmetric centre and are subject to stereoselective metabolism which can be species-dependent. Studies on the *in vitro* metabolism of felodipine to the pyridine derivative using liver microsomes have shown that for rat and dog the *R*-enantiomer (106) was metabolized slightly faster than the *S*-enantiomer (105), whereas with human microsomes the reverse occurred (p. 275). These results are consistent with those found *in vivo* comparing the bioavailability of each enantiomer, with $R > S$ in dog and $S > R$ in humans.

(105) X =   , Y = H

(106) X = H, Y =

Stereoselective metabolism of the dihydropyridine (107) occurs in rats but there is also a quantitative difference in the metabolism between males and females (p. 276). Plasma levels of the (+)-enantiomer were higher than those of the (−)-enantiomer in both sexes but levels of each enantiomer were higher in females than in males. Plasma levels of the carboxylic acid (108) were much higher after administration of the (−)-enantiomer. *In vitro* experiments with liver microsomes showed a marked sex difference in the formation of the metabolites (109) and (110).

14

(107) R =

(108)

(109)

(110)

The rates of hydrolysis of racemic and enantiomeric lorazepam 3-acetate (111) to lorazepam (112) have been determined in rat and human liver microsomes (p. 326). The *R*-enantiomer was hydrolysed by rat and human liver microsomes 2.7-and 6.8-fold faster respectively than the *S*-enantiomer when the racemate was the substrate. However, enantiomer interactions occurred and the presence of the *R*-enantiomer enhanced hydrolysis of the *S*-enantiomer while the presence of the latter inhibited hydrolysis of the *R*-enantiomer. When pure enantiomers were used as substrates the *R*-enantiomer was hydrolysed 8.4-fold and 166-fold faster in rat and human liver microsomes respectively compared with the *S*-enantiomer. Using rat brain S9 fraction the *S*-enantiomer was hydrolysed 2.3-fold faster.

† = chiral carbon
(111) R = COCH₃
(112) R = H

Stereoselective metabolism of three arylpropionic acids has been investigated in rats after intravenous doses (p. 126). The total clearances of the *R*-enantiomers of (113) and (114) were significantly greater than those of the corresponding *S*-antipodes whereas for (115) the clearance of each enantiomer was similar. The *R* to *S* inversion ratio was 0.54 for (114) but only 0.003 and 0.02 for (113) and (115) respectively. Biliary excretion of the parent drug and its glucuronide conjugate was extensive and stereoselective. Thus the excretion of the *R*-enantiomer of (113) was much greater than that of the *S*-enantiomer whereas the reverse situation occurred for (114) and (115).

(113)

(114)

(115)

15

Glucuronidation is a major pathway for the metabolism of the aryl-propionic acid naproxen (115), and the stereoselectivity of this reaction has been compared in gastrointestinal and liver cells of the guinea-pig (p. 135). The rate of conjugation was about thirty times greater in the liver cells compared with that in intestinal cells. In all gastrointestinal cells the rate of formation of the S-enantiomer conjugate was greater with gastric cells showing the highest ratio (S/R, 7.5). In contrast liver cells preferentially formed the R-glucuronide (S/R, 0.5).

The major metabolites of carvedilol (116) and (117) excreted in the bile of rats are the glucuronides of 1-hydroxy- (118) and 8-hydroxycarvedilol (119) (p. 194). The stereochemical composition of the conjugates was investigated by administration of radiolabelled pseudoracemates, namely [^{14}C]-R/[^{12}C]-S and [^{12}C]-R/[^{14}C]-S carvedilol respectively. The ratios of the two metabolites were similar after oral or intravenous administration but there were considerable differences in the S/R enantiomer ratios of each metabolite, namely 0.59 and 3.29 for the 1-hydroxy and 8-hydroxy metabolites respectively, after oral dosing. These results suggest that formation of the 1-hydroxy metabolite is stereoselective for the R-enantiomer (116) while 8-hydroxylation is stereoselective for the S-enantiomer (117).

(116) X = OH, Y = H, R^1 = R^2 = H
(117) X = H, Y = OH, R^1 = R^2 = H

(118) R^1 = OH, R^2 = H
(119) R^1 = H, R^2 = OH

Glucuronidation of the carboxyl group is a major biotransformation pathway for ofloxacin in rats but the conjugation appears to be stereoselective since the conjugate represents 31% and 7% of the dose in rat bile following administration of S(−)- (120) and R(+)-ofloxacin (121) respectively. This was confirmed from *in vitro* studies which also showed that the R(+)-enantiomer competitively inhibited glucuronidation of the S(−)-enantiomer. After intravenous administration of the racemate the areas under the serum concentration−time curves for the S(−)-enantiomer were 1.7-fold greater than those after administration of the S(−)-enantiomer alone.

16

(120)

(121)

The mechanism for the specific conversion of the *R*-enantiomer of aryl-propionic acids to the *S*-enantiomer but not the reverse has been an intriguing problem for many years. A recent study involving the use of deuterium-labelled ibuprofen and stereoselective GC–MS analysis has given further insight into a possible mechanism (p. 133). It was shown that in rats inversion occurred with loss of the C-2 deuterium label but this process was not subject to a measurable kinetic isotope effect and so was not the rate-limiting step. When the pentadeuterated compound (122) was administered together with unlabelled ibuprofen plasma was shown also to contain tetra-deuterated *R*-ibuprofen (123). This result indirectly confirms the existence of a non-asymmetric intermediate. Stereoselective formation of a coenzyme A thioester metabolite (124) of the *R*-enantiomer was proposed and subsequent reversible enzyme-catalysed transformation to the symmetrical enolate tautomer (125).

(122)

(123)

(124)

(125)

Flamprop-methyl (126) is a herbicide which is a different type of substi-tuted propionic acid. The main rat metabolite is the carboxylic acid and the enantiomer ratio of this urine metabolite has been measured after administra-tion of the *R*-enantiomer (p. 138). About 13% of the acid was excreted as the

17

S-enantiomer providing a further example of the generality of the inversion process and the broad substrate range.

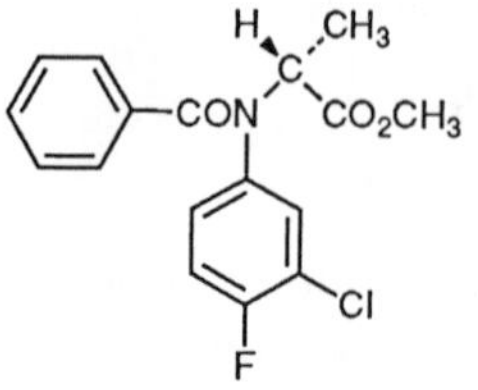

(126)

Formation of *N*-glucosides is a novel conjugation which has been shown to occur for several barbiturates. Barbiturates such as amobarbital (127) are prochiral compounds which on conjugation form chiral metabolites. The stereoselectivity of this reaction has been investigated by analysis of the glucosides excreted in human urine which showed a predominance of the *S*-enantiomer (128). A new study with cat liver microsomes has demonstrated that enzyme-catalysed *N*-glucosylation can also occur in this species and that product selectivity was 2:1 in favour of the *R*-glucoside (129).

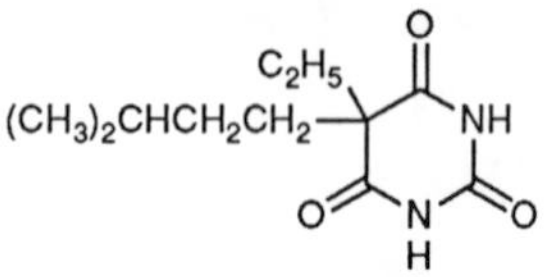

(127)

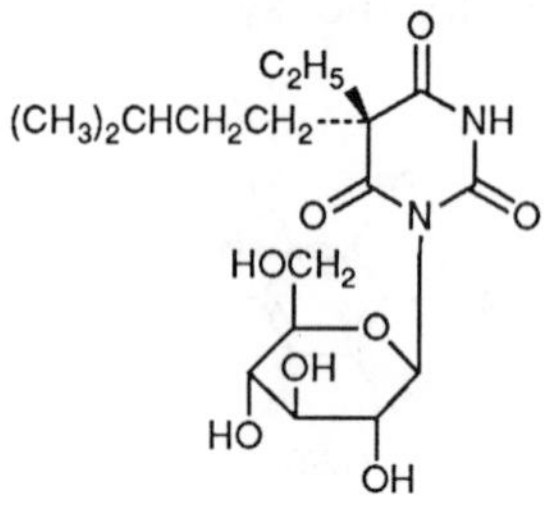

(128)

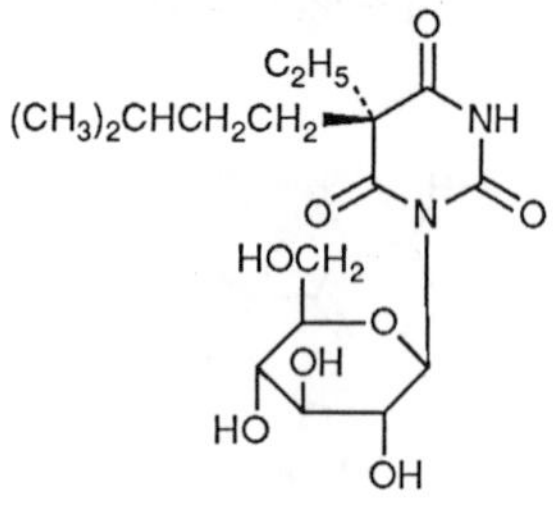

(129)

3 MECHANISMS OF TOXICITY

The toxicity of valproic acid (130) is believed to be due to formation of unsaturated metabolites such as (131) and (132). The metabolism of these two compounds has now been investigated in rats (p. 76). The glutathione conjugate (133) was detected in the bile of rats treated with either unsaturated compound and the corresponding mercapturic acid was also identified as a urinary metabolite representing 40% of the dose (132). The same metabolite was also detected in the urine of pediatric patients under-going valproic acid therapy providing evidence for the formation of an intermediate alkene.

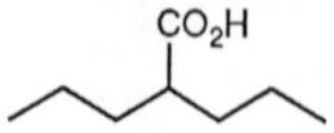

(130)

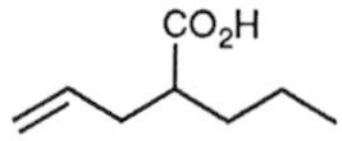

(131)

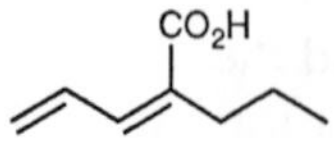

(132)

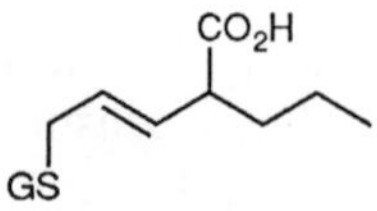

(133)

18

Further investigations have been reported on the nature of the ultimate metabolite responsible for the observed hepato- and pneumo-toxicity of pulegone (134) (p. 56). Menthofuran (136) was shown to be the major metabolite of pulegone on incubation with rat liver microsomes and a minor metabolite was identified as the allylic alcohol (135) which is the precursor to the furan. Similar incubation of isopulegone (137) showed formation of the allylic alcohol (138) which was more stable than (135) and was only slowly converted into menthofuran. Further incubation of menthofuran with liver microsomes resulted in formation of the aldehyde (139) by oxidative cleavage of the furan ring. The aldehyde has been proposed as the ultimate toxic metabolite.

4-Vinylcyclohexene (140) is an ovarian carcinogen whose toxicity has been attributed to the rate of formation of the 1,2-epoxide (141) although the 7,8-epoxide (142) is also a metabolite. A study using human liver microsomes has shown that formation of the 1,2-epoxide occurred but an order of magnitude slower than in the rat, a species in which ovarian tumours do not occur.

The hepatotoxicity of coumarin (143) and the associated species differences have been known for many years but the mechanism has remained unknown although bioactivation by P-450 dependent enzymes is known to be necessary. Recent studies in rats have revealed a new metabolic pathway and the mercapturic acid (145) has been identified as a urine metabolite. The presence of this metabolite suggests formation of the intermediate epoxide (144) which may be involved in the observed toxicity. The proportion of this metabolite was less at a low dose (10 mg kg^{-1}), representing ca. 3% dose, compared with a high dose (100 mg kg^{-1}) when it represented ca. 8% dose.

(143)　　　　　　　(144)　　　　　　　(145)

Propylthiouracil (146) is associated with idiosyncratic agranulocytosis and a study has been performed to investigate the formation of reactive metabolites in activated human neutrophils which might cause toxicity by covalently binding to protein. Three oxidized metabolites were identified as the disulfide (148) the sulfinic acid (149) and the sulfonic acid (150). Further experiments implicated the sulfenyl chloride (147) as an intermediate and both this compound and the sulfonic acid were shown to react with sulfhydryl-containing compounds such as N-acetylcysteine and thus had the potential to form protein adducts.

(146)　　　　　　　(147)

(149)　　　　　　　(148)

(150)

4 SPECIES DIFFERENCES

Nimodipine (151) is a representative of the dihydropyridine cardiovascular drugs which usually undergo extensive biotransformation resulting in a complex mixture of metabolites. Species differences in metabolism may play an important part in determining their pharmacological activity and recent studies with nimodipine have compared metabolism in rat, dog, and monkey (p. 271). Aromatization of the dihydropyridine ring is usually rapid and extensive. Two dihydropyridine metabolites (152) and (153) were detected in rat plasma at 1 hour after dosing and the only excreted dihydropyridine metabolite was a conjugate of (153) present in bile as a glucuronide. A major rat plasma metabolite was the pyridine carboxylic acid (154) and the main rat urine metabolites were (155) and (156). In the dog the main urine metabolites were (155), (157), and (158), the latter two formed by reduction of the nitro group. Monkey urine contained (156), (157), and (158) as major metabolites

20

and (159) which was only present in small amounts in rat and dog urine. Conversely the major rat and dog metabolite (155) was not detected in monkey urine. The total amounts of metabolites containing a reduced nitro group were in the order monkey > dog > rat.

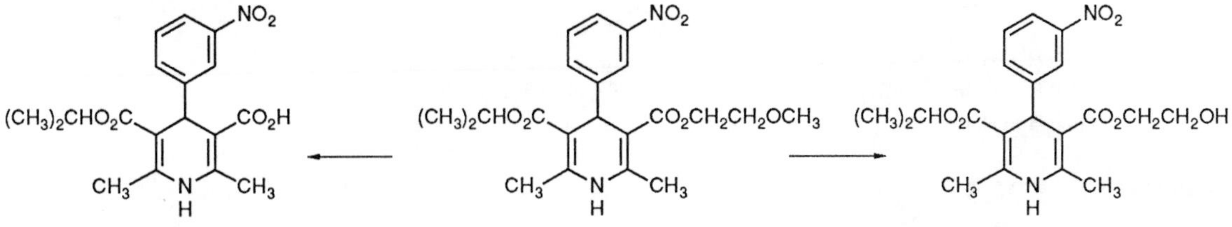

(152)   (151)   (153)

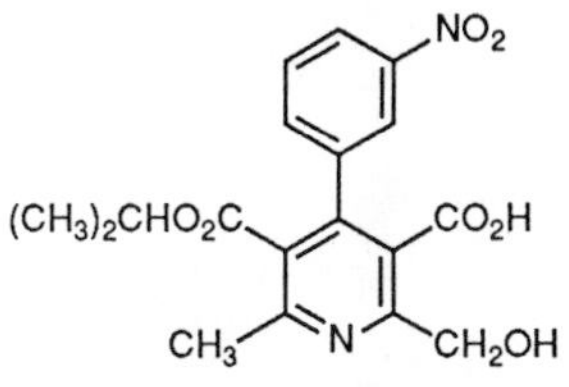

(154) 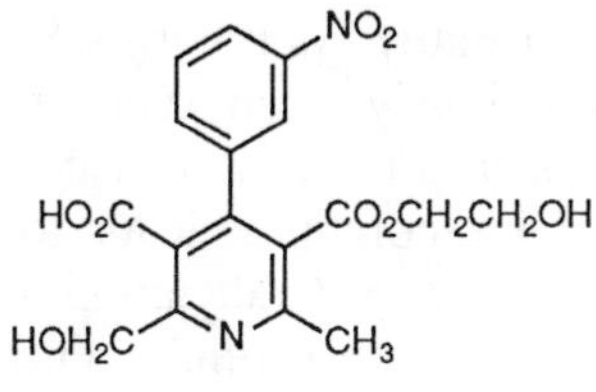   (155) 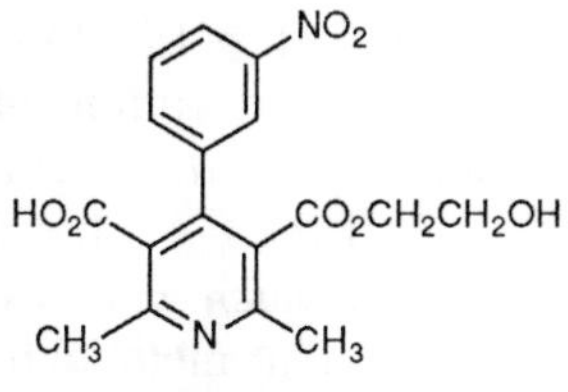   (156)

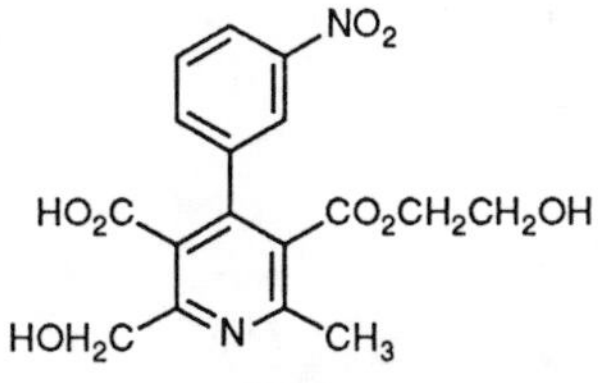

(157) 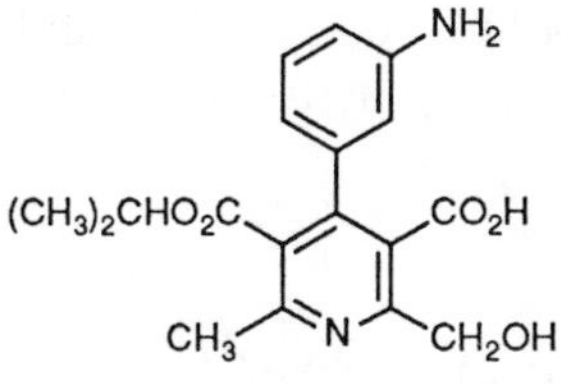   (158) 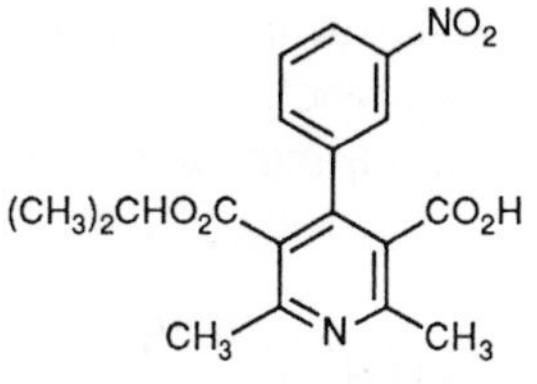   (159)

The main routes of metabolism of rimantidine (160) involve ring-hydroxylation. Two metabolites (16) and (162), identified in dog urine but not rat urine, were formed by replacement of the amino group by hydroxyl. These metabolites were presumably produced by oxidative deamination to give initially a ketone followed by reduction to the alcohol.

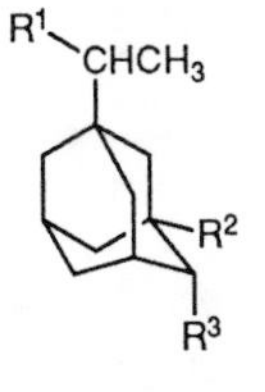

(160) $R^1 = NH_2$, $R^2 = H$, $R^3 = H$
(161) $R^1 = OH$, $R^2 = OH$, $R^3 = H$
(162) $R^1 = OH$, $R^2 = H$, $R^3 = OH$

Although codeine (163) is a long established drug its biotransformation in different species has not been rigorously compared. In a recent study the major rat urine metabolites were shown to be morphine (166) and its glucuronide (167), representing almost 30% of the dose. While the same metabolites were also detected in mouse urine norcodeine (165) was also a major metabolite that was not detected in other species. In guinea-pigs and rabbits the glucuronide of codeine (164) was the major metabolite, representing 40% and 25% of the dose respectively.

21

(163) R¹ = H, R² = CH₃
(164) R¹ = Gluc, R² = CH₃
(165) R¹ = H, R² = H

(166) R = H
(167) R = Gluc

5 MISCELLANEOUS

Although sutan (168) is a simple alkyl thiocarbamate at least 29 metabolites have been detected in rat urine after single oral doses and 18 of these have been identified (p. 90). One of the major pathways involves *S*-oxidation and glutathione conjugation which has previously been reported. The corresponding mercapturic acid (169) has been identified as a urine metabolite. Cleavage of the C—S bond in the cysteine conjugate leads to the carbamic acid (170) which can also be formed by *S*-dealkylation of the parent compound. This metabolite occurred as an *S*-glucuronide in urine. The other pathways involved hydrolysis of the carbamate to di-isobutylamine combined with hydroxylation in the *N*-butyl group. Hydroxylation of the butyl group also resulted in formation of cyclized metabolites such as (171) and (172) by intramolecular reaction of a primary or tertiary hydroxyl group with the carbonyl of the carbamate moiety.

(168)

(169)

(170)

(171)

(172)

Malonylcysteine conjugates have been identified as metabolites of xenobiotics in plants. Since these metabolites can enter the food chain consideration has to be given to their potential toxicity and a recent study has reported on the fate of a model compound, the *S*-benzyl analogue (173) in rats (p. 000). The major urine metabolites were identified as hippuric acid (177) and the sulfoxide (175) and sulfone (176). The results indicated that the *N*-

22

malonyl group is hydrolysed and the resulting cysteine conjugate (174) undergoes degradation by the action of C—S-lyase.

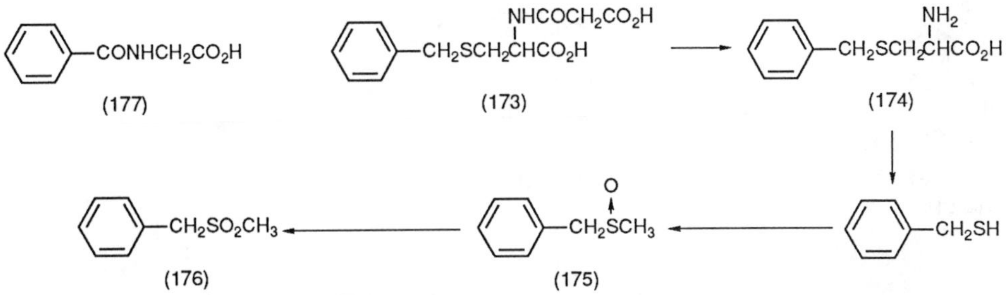

Although 4-aminobiphenyl (178) is a well known compound and there have been extensive investigations on its *in vitro* metabolism, until recently there had been no definitive study on its biotransformation *in vivo*. In this most recent study up to eighteen components were detected in urine of rats administered oral or intraperitoneal doses of [^{14}C]aminobiphenyl (p. 153). The main routes of metabolism involved N-acetylation and ring hydroxylation. N-Hydroxylation of the parent amine and the N-acetyl derivative (179) gave two metabolites (180) and (181) which were also interconvertible. Oxidation of the N-acetyl metabolite also occurred to give the alcohol (182) and acid (183). No quantitative data on the metabolites were available owing to the extensive chromatographic separations and purifications required.

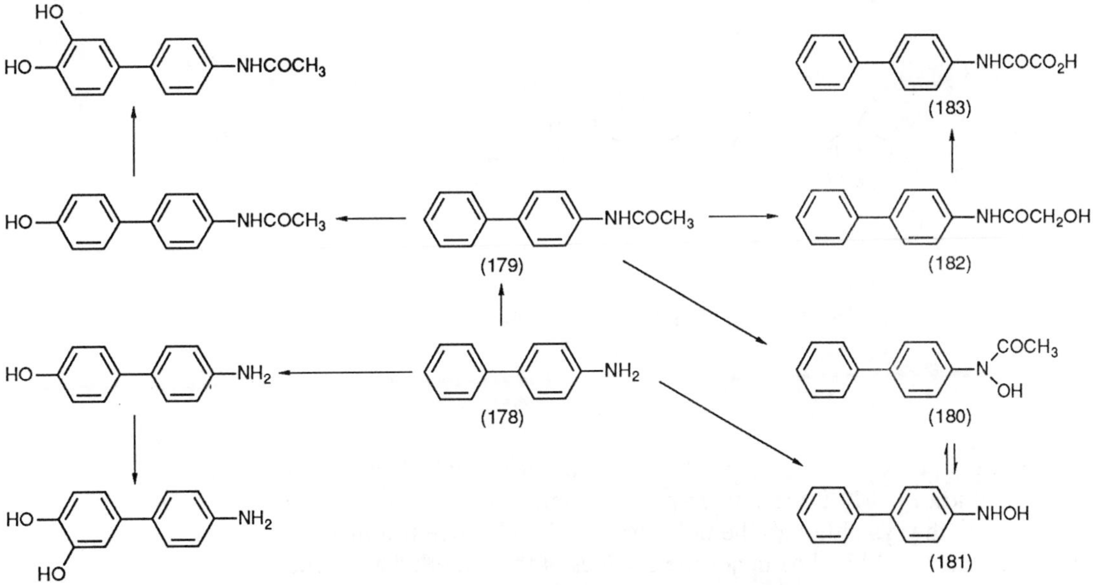

A major pathway for metabolism of doxepin is N-demethylation and it has been shown that some isomerization of the Z-isomer (184) occurs in humans. Although the isomers of doxepin do not undergo interconversion, E-doxepin (186) was metabolized to Z-N-desmethyldoxepin (185) in significant quantities. It was proposed that isomerization occurred during N-

23

demethylation and that a hydrated double bond intermediate could be involved.

(184) R = CH₃
(185) R = H

(186)

A study of the comparative metabolism of 13-*cis*-(187) and *trans*-retinoic acid (188) in cynomolgus monkeys has illustrated the effect of a small structural change on biotransformation (p.93). There were marked differences in pharmacokinetics and the *cis*-isomer was eliminated faster than the *trans*-isomer. A glucuronide of the parent and the ketone (189) were major metabolites of the *trans*-isomer whereas the glucuronide was the only major metabolite of the *cis*-isomer. The same *trans*-retinoic acid metabolites were also detected in serum samples from a human patient. Some isomerization of the *trans*-compounds also occurred.

(187)

(188)

(189)

Comparative studies on the metabolism of monomethyl- (190) and dimethyl-cannabidiol (191) using guinea-pig liver microsomes have shown different major pathways although the only structural difference is a methylated phenolic group (p. 227). The major metabolites of the dimethyl analogue were the 4-hydroxy (197) and 6-hydroxy (193) compounds, and the epoxide (196) was also detected. The monomethyl analogue (190) formed the 7-hydroxy (194) and the 6-hydroxy (192) compounds as major metabolites. A corresponding epoxide (195) was also presumably formed which underwent rearrangement by internal nucleophilic attack of the proximal free phenolic group to give the cyclized metabolite (198).

24

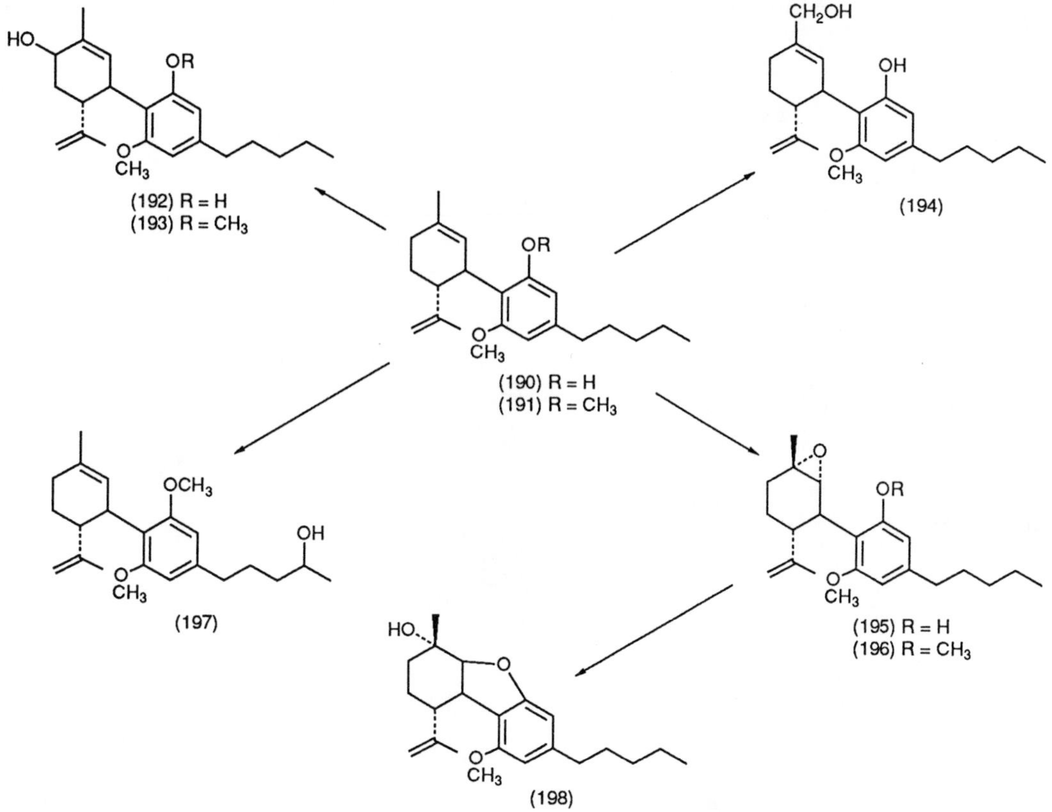

(192) R = H
(193) R = CH$_3$

(194)

(190) R = H
(191) R = CH$_3$

(197)

(195) R = H
(196) R = CH$_3$

(198)

The substitution of iodine in one of the aryl groups in tamoxifen has an appreciable effect on its metabolism as shown by a study using rat hepatocytes (p. 59). The iodo analogue (199) was metabolized four times more slowly than tamoxifen. The iodine atom not only blocks metabolism in its vicinity but also reduces the rate of side-chain N-demethylation and N-oxidation.

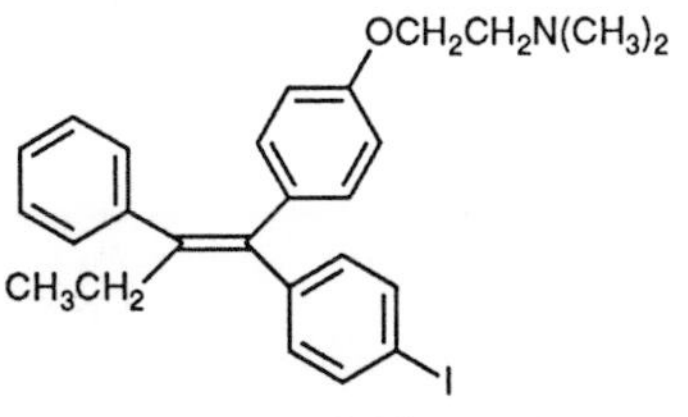

(199)

(200)

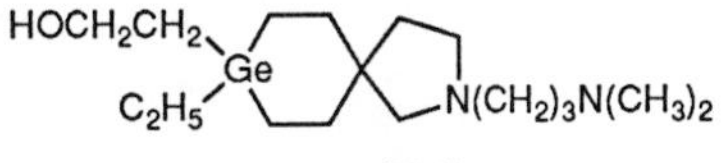

(201)

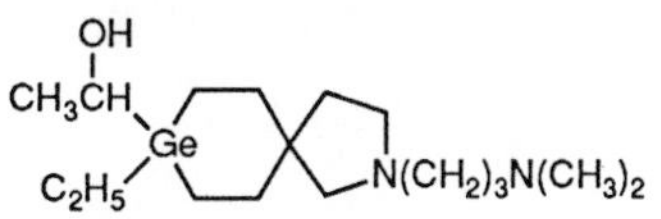

(202)

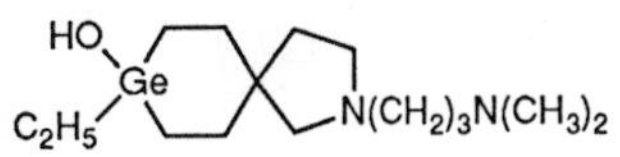

(203)

Studies with spirogermanium (200) represent one of the first investigations with an organogermanium compound (p. 401). Two hydroxylated metabolites (201) and (202) were identified in mouse urine after an oral dose and they were also formed by incubation with mouse liver microsomes. The same metabolites were also detected in a human urine sample. A further metabolite was tentatively identified as (203), formed by dealkylation.

Aromatic Hydrocarbons

Benzene

Use/occurrence:	Industrial chemical/solvent
Key functional groups:	Phenyl
Test system:	Mouse (oral, 10 and 200 mg kg^{-1})

Structure and biotransformation pathway

(1) * = ^{14}C   (2)   (3)   (4)

(2)   (6)   (7)

The objective of this study was to compare the disposition and metabolism of benzene in young and old male C57BL/6N strain mice, with a view to possibly relating any such age-related differences to the enhanced susceptibility of elderly humans to the toxic effects of environmental contaminants such as benzene. Following oral administration of [^{14}C]benzene (1) to 3 and 18 month old mice, urinary excretion of radioactivity accounted for at least 90% of the 10 mg kg^{-1} doses to animals of both ages, but this proportion was effectively halved when the dose level was increased to 200 mg kg^{-1}, and was accompanied by a corresponding increase in the proportion of the dose eliminated in the expired air as unchanged (1) from 1% at the low-dose level for mice of both ages to 56% and 46% at the high-dose level for 3 and 18 month old mice respectively. The proportions of the known benzene metabolites (2)–(7) in urine were determined by radio-HPLC. At the 10 mg kg^{-1} dose level, (2), (3), and (4) were the major metabolites in the urine of mice of both ages, accounting for *ca.* 40, 28, and 15% respectively of the total water-soluble metabolites, whereas (5)–(7) each accounted for only 0.5–5%. At the 200 mg kg^{-1} dose level, the proportion of (4) was significantly less than that found at the low dose level for mice of both ages; in addition, the proportion of (5) increased with dose in 3 month old mice and that of (3) increased with dose in 18 month old animals. These results demonstrate therefore that both the total metabolism of benzene and the formation of its principal metabolites are affected by ageing.

Reference

T. F. McMahon and L. S. Birnbaum, Age-related changes in disposition and metabolism of benzene in male C57BL/6N mice, *Drug Metab. Dispos.*, 1991, **19**, 1052.

Naphthalene, Naphthalene oxide

Use/occurrence:	Model compounds
Key functional groups:	Epoxide
Test system:	Isolated perfused mouse lung

Structure and biotransformation pathway

Gluc = glucuronyl
SG = glutathionyl

Naphthalene is a known pulmonary toxin and produces selective injury to the Clara cells *in vivo* and in the isolated perfused lung. The aim of the present study was to investigate the metabolism and role of naphthalene and naphthalene oxides in these toxic effects.

The predominant metabolites of [^{14}C]naphthalene (1) following perfusion of an isolated mouse lung preparation were the dihydrodiol (3) and glutathione conjugates (4)–(6) of (1) which composed 70% of the total metabolites. The ratio of the glutathione conjugates (6):(4)+(5), indicative of the ratio of (1R,2S)- vs. (1S,2R)-naphthalene oxide (2) was dependent on the concentration of (1), the ratio being greater at higher concentrations of (1).

The profile of metabolites isolated from lungs perfused with [^{3}H]naphthalene oxide (2) showed a different pattern from that observed with (1). The $t_{1/2}$ of (2) in the culture medium was shown to be increased from 4 to 11 minutes in the presence of serum albumin and bovine serum albumin was added in the experiments with (2). The major metabolites derived from (2) were 1,4-naphthoquinone (7) and naphthol glucuronide (8), which represented 31.9% and 22.6% of the total metabolites, respectively. Glutathione conjugates (4)–(6) and the dihydrodiol (3) accounted for 16% and the cysteine conju-

30

gates 3.7% of the total metabolites respectively. Two unidentified peaks accounted for the remaining 14% of the isolated metabolites, but these did not co-elute with 1,2-naphthoquinone, 1-naphthol, or mercapturic acid conjugate standards.

The authors suggest that (2) may be formed (but not detoxified) in the liver and circulated (with enhanced stability due to the presence of serum albumin in the blood) to the lung. Such circulating metabolites may be important in the selective toxicity of (1) to pulmonary Clara cells.

Reference

S. Kanekal, C. Plopper, D. Morin, and A. Buckpitt, Metabolism and cytotoxicity of naphthalene oxide in the isolated perfused mouse lung, *J. Pharmacol. Exptl. Therap.*, 1991, **256**, 391.

Benzo[a]pyrene

Use/occurrence:	Environmental carcinogen
Key functional groups:	Polycyclic aromatic hydrocarbon
Test system:	Rat lung microsomes

Structure and biotransformation pathway

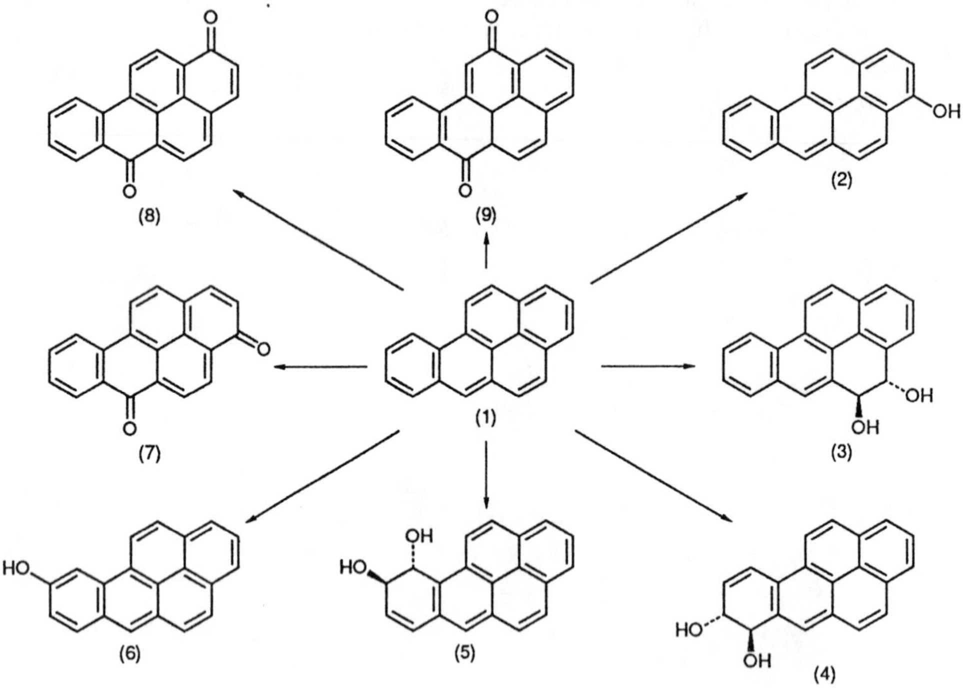

The presence of superoxide or hydrogen peroxide inhibits the activity of benzo[a]pyrene (1) hydroxylase. Analysis of the metabolic profile after incubation with rat lung microsomes using HPLC showed that the formation of the 3- and 9-hydroxybenzo[a]pyrene [(2) and (6)] was significantly inhibited by 20–70% by the presence of superoxide or hydrogen peroxide. There was no significant effect on the formation of the 4,5-diol (3), the 7,8-diol (4), the 9,10-diol (5), or the total quinones [(7)–(9)], which were measured together.

The reduction in overall metabolism of (1) by the reactive oxygen species may lead to a greater amount of the dihydrodiols or quinones being produced which would lead to an increase in the carcinogenic potential of (1).

Reference

N. L. Flowers and P. R. Miles, Alterations of pulmonary benzo[a]pyrene metabolism by reactive oxygen metabolites, *Toxicology*, 1991, **68**, 259.

Benzo[a]pyrene

Use/occurrence:	Environmental carcinogen
Key functional groups:	Polycyclic aromatic hydrocarbon
Test system:	Rat lung; rat lung microsomes

Structure and biotransformation pathway

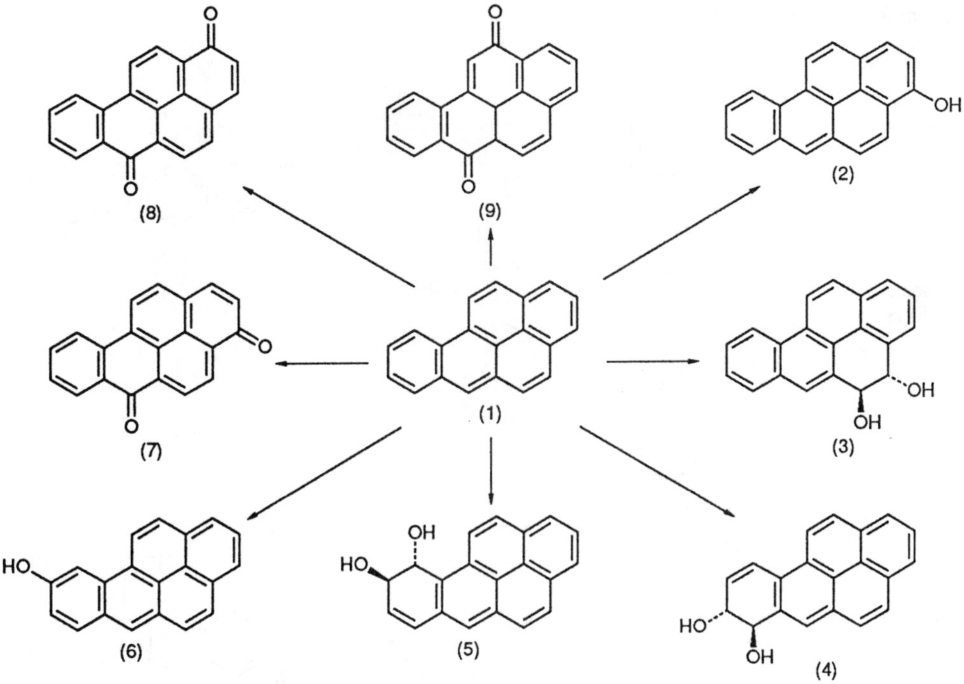

Lung microsomes were obtained from rats sacrificed 4 hours after pre-treatment with toluene (1 g kg^{-1}), and were incubated with benzo[a]pyrene (1). The metabolites were analysed by HPLC. The results showed that there was a 36% inhibition in the formation of the 3-OH compound (2), the major metabolite, whilst the formation of the 4,5-diol (3), 7,8-diol (4), 9,10-diol (5), 9-OH compound (6), 3,6-quinone (7), 6,12-quinone (8), and 1,6-quinone (9) remained unchanged. The formation of (2) is a major route of detoxification of (1) and inhibition by toluene may adversely affect the cancer risk of (1).

Reference

G. M. Furman, D. M. Silverman, and R. A. Schatz, The effect of toluene on rat lung benzo[a]pyrene metabolism and microsomal membrane lipids, *Toxicology*, 1991, **68**, 75.

Dibenz[*a,c*]anthracene, Dibenz[*a,h*]anthracene, Dibenz[*a,j*]anthracene

Use/occurrence:	Model compounds
Key functional groups:	Polycyclic aromatic hydrocarbon
Test system:	Rat liver microsomes (0.2 mM substrate)

Structure and biotransformation pathway

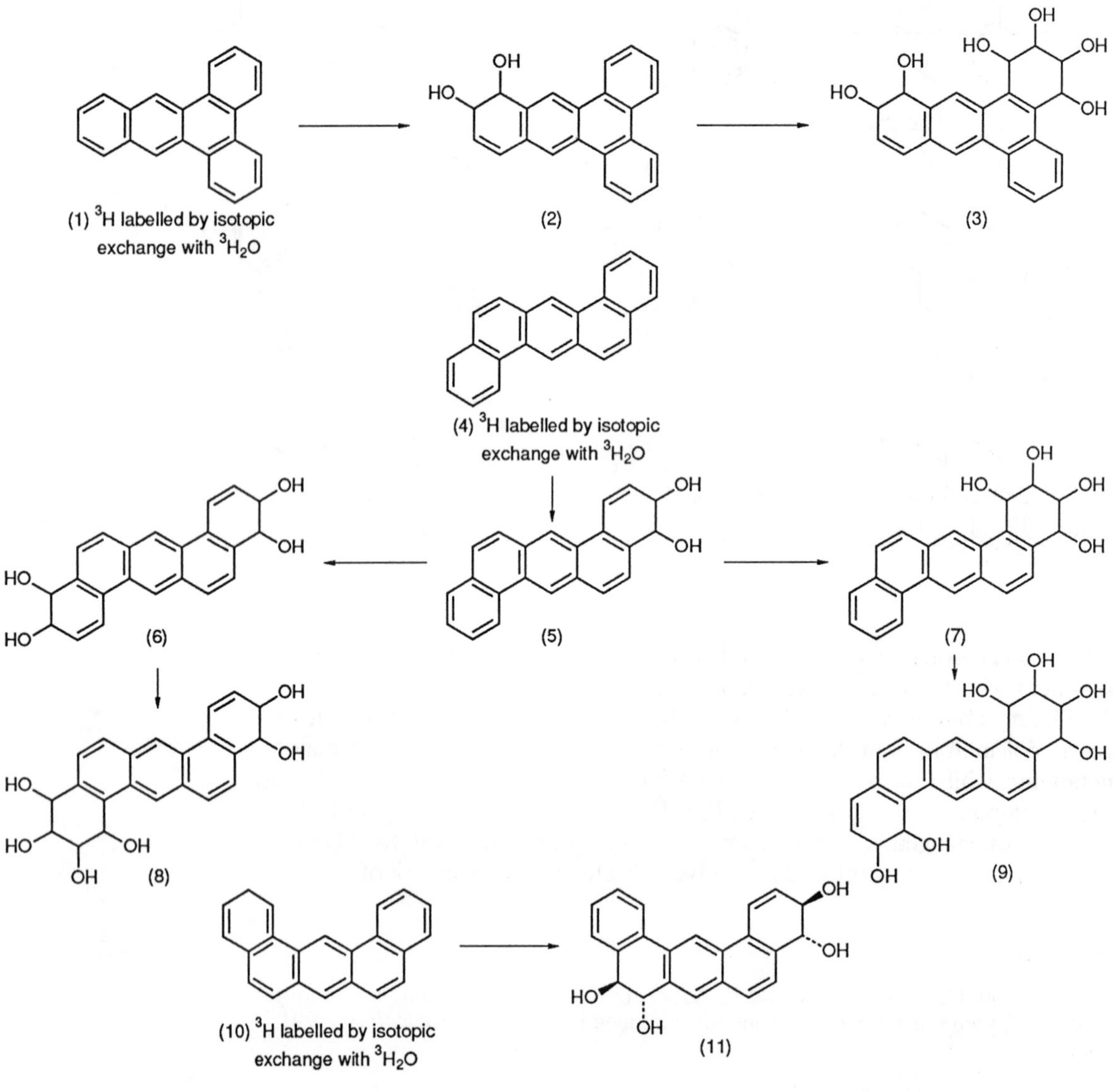

(1) ³H labelled by isotopic exchange with ³H₂O

(2)

(3)

(4) ³H labelled by isotopic exchange with ³H₂O

(6)

(5)

(7)

(8)

(9)

(10) ³H labelled by isotopic exchange with ³H₂O

(11)

34

Ethyl acetate soluble metabolites formed during the incubations of (1), (4), (5), and (10) with 3-methylcholanthrene-induced microsomes were separated by HPLC, and identified following fluorescence, UV, and NMR spectroscopy. The possible role of the bis-dihydrodiols in the metabolic activation of the three dibenzanthracenes was discussed.

Reference

S. Lecoq, O. Chalvet, H. Strapelias, P. R. Grover, D. H. Phillips, and M. Duquesne, Microsomal metabolism of dibenz[a,c]anthracene, dibenz[a,h]anthracene, and dibenz[a,j]anthracene to bis-dihydrodiol and polyhydroxylated products, *Chem. Biol. Interact.*, 1991, **80**, 261.

Chrysene, 5-Methylchrysene, 6-Methylchrysene, 5,6-Dimethylchrysene

Use/occurrence:	Model compounds
Key functional groups:	Arylmethyl, polycyclic aromatic hydrocarbon
Test system:	Rat liver cytosol (*in vitro*, 0.2 mM), rat subcutaneous tissue (*in vivo*, *ca.* 3 μmol kg^{-1})

Structure and biotransformation pathway

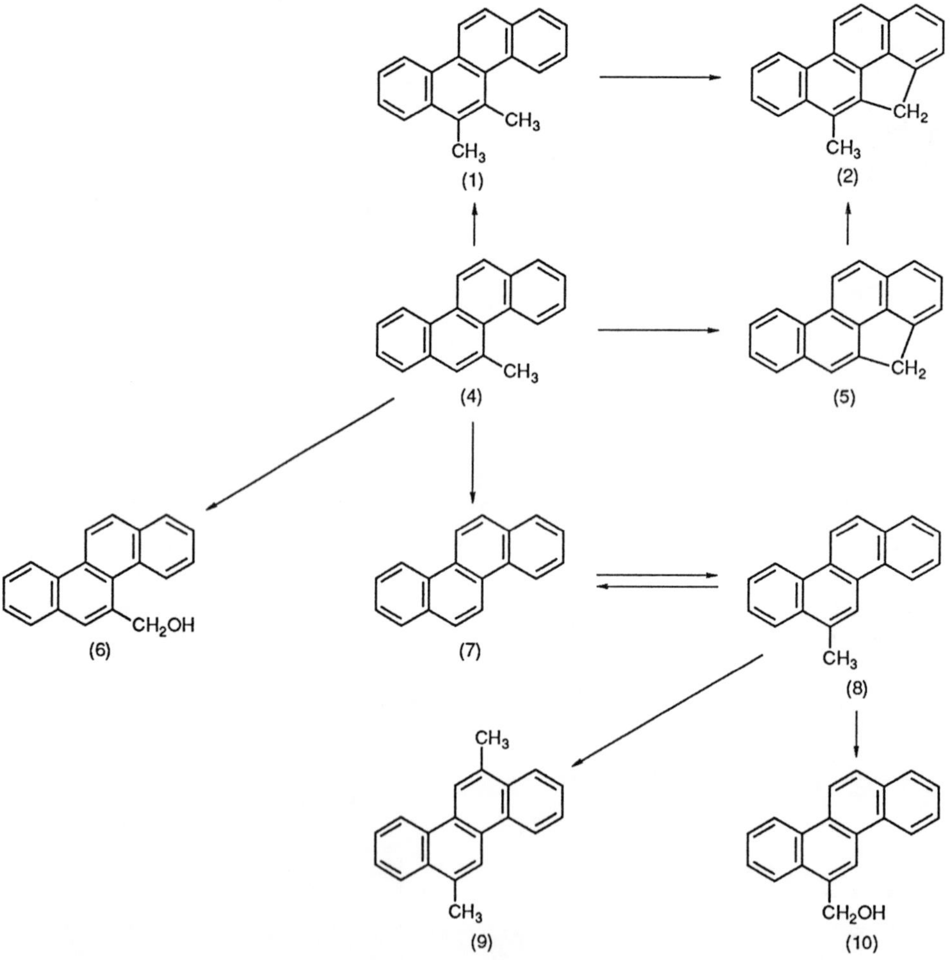

Chrysene (7) undergoes a bioalkylation substitution reaction *in vitro*, in rat liver cytosol preparations, and *in vivo*, in rat dorsal subcutaneous tissue, to yield 6-methylchrysene (8) as a metabolite. Both 5-methylchrysene (4) and

6-methylchrysene (8) were found to undergo a dealkylation reaction to yield chrysene (7). Both (4) and (8) were also metabolized to produce the corresponding hydroxyalkyl metabolites (6) and (10) and to corresponding dimethylchrysenes (1) and (9). Both 5-methylchrysene (4) and 5,6-dimethylchrysene (1) enzymically cyclized to give (5) and (2) respectively. Other transformations for 5,6-dimethylchrysene (1) included the dealkylation and alkyl hydroxylation reactions previously described. Metabolites were analysed by HPLC, and were identified by GC–MS.

Reference

S. R. Myers and J. W. Flesher, Metabolism of chrysene, 5-methylchrysene, 6-methylchrysene, and 5,6-dimethylchrysene in rat liver cytosol, *in vitro*, and in rat subcutaneous tissue, *in vivo*, *Chem. Biol. Interact.*, 1991, **77**, 203.

7,12-Dimethylbenz[*a*]anthracene

Use/occurrence:	Experimental carcinogen
Key functional groups:	Arylmethyl, polycyclic aromatic hydrocarbon
Test system:	Rat liver perfusion

Structure and biotransformation pathway

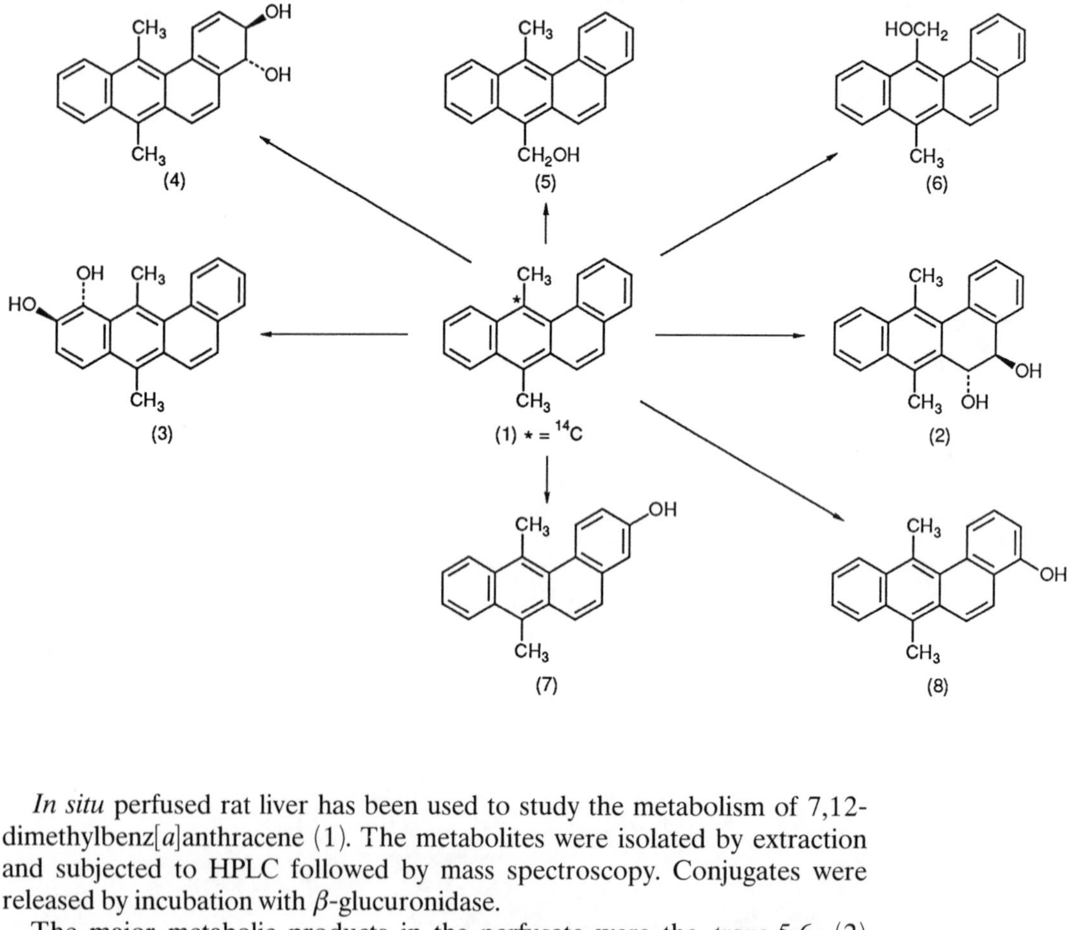

In situ perfused rat liver has been used to study the metabolism of 7,12-dimethylbenz[*a*]anthracene (1). The metabolites were isolated by extraction and subjected to HPLC followed by mass spectroscopy. Conjugates were released by incubation with β-glucuronidase.

The major metabolic products in the perfusate were the *trans*-5,6- (2) and *trans*-10,11-dihydrodiols (3) together with the *trans*-3,4-dihydrodiol (4), 7-hydroxymethyl-12-methylbenz[*a*]anthracene (5), 12-hydroxymethyl-7-methylbenz[*a*]anthracene (6), and 3-hydroxy- and 4-hydroxy-dimethylbenz[*a*]anthracenes (7) and (8). Following treatment with β-glucuronidase, the major metabolites in the bile were (5) and (6) with traces of (2), (3), and (4). Overall, livers from female rats were found to excrete metabolites of (1) at a higher rate in both bile and the perfusate than corresponding livers from male rats.

Reference

S. T. Vater, D. M. Baldwin, and D. Warshawsky, Hepatic metabolism of 7,12-dimethylbenz[a]anthracene in male, female, and ovariectomized Sprague–Dawley rats, *Cancer Res.*, 1991, **51**, 492.

7,12-Dimethylbenz[a]anthracene

Use/occurrence:	Model compound
Key functional groups:	Arylmethyl, polycyclic aromatic hydrocarbon
Test system:	Human and rat liver microsomes (50 μM substrate)

Structure and biotransformation pathway

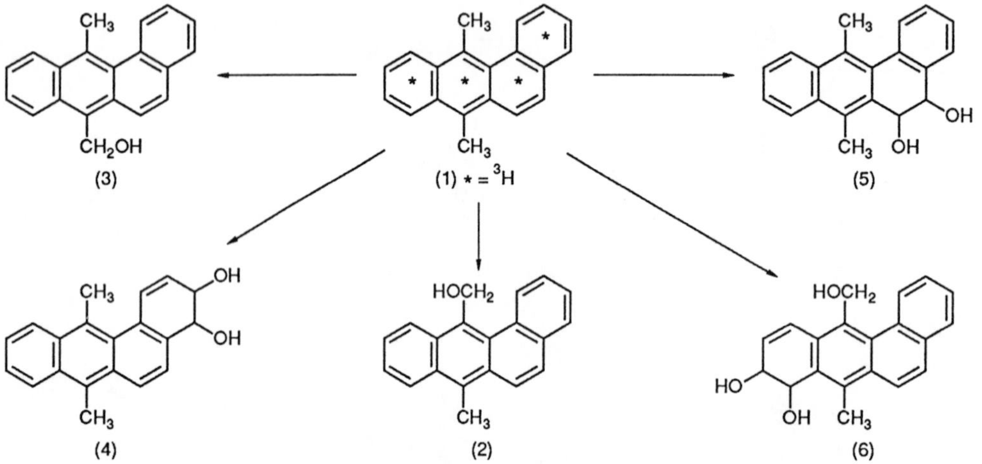

The metabolites of [³H]-(1) by human and rat liver microsomes were determined by HPLC co-chromatography with unlabelled standards. The major products of metabolism by human livers seemed to be mainly hydroxymethyl- [(2) and (3)] and dihydrodiol [(4)–(6)] metabolites. The rate of formation of these metabolites showed considerable interindividual variation. The rate of formation of the three dihydrodiols studied was shown to be linear with time.

Reference

V. M. Morrison, A. K. Burnett, L. M. Forrester, C. R. Wolf, and J. A. Craft, The contribution of specific cytochromes P-450 in the metabolism of 7,12-dimethylbenz[a]anthracene in rat and human liver microsomal membranes, *Chem. Biol. Interact.*, 1991, **79** 179.

7-Methylbenz[c]acridine

Use/occurrence:	Model compound
Key functional groups:	Polycyclic aromatic hydrocarbon
Test system:	Rat liver microsomes (40 μM substrate)

Structure and biotransformation pathway

The major microsomal metabolites of (1) were (2)–(5). The 8,9-dihydro-diol (4) was the most abundant metabolite in microsomal incubations from both control and 3-methylcholanthrene-pretreated rats. The dihydrodiols (3)–(5) formed by liver microsomes were separated as their bis-(+)-(1 R,2 S,4 S)-endo-1,4,5,6,7,7-hexachlorobicyclo[2.2.1]hept-5-ene-2-carbox-ylic acid esters. The enantiomeric purity was determined by HPLC, and quantitation was possible by both UV peak area and radioassay (comparable results were obtained by both methods). The 3,4-dihydrodiol (5) was formed with a high stereospecificity, the (3 R),(4 R)-enantiomer (> 90% purity) being predominantly formed with microsomes from control, phenobarbitone-induced and 3-methylcholanthrene-induced rats. The 8,9-dihydrodiol (4) was the predominant metabolite formed with microsomes from the 3-methyl-cholanthrene-induced rats (ca. 50% of total) and the (8 R),(9 R)-enantiomer was formed with a similar stereospecificity to the 3,4-diol (5). A lesser stereo-specificity occurred in the formation of this metabolite with microsomes from control and phenobarbitone-induced rats. The 5,6-diol (3) was the next most abundant metabolite (5–8% of total), and again the (5 R),(6 R)-enantiomer predominated (ca. 50% purity); with this metabolite similar results were obtained for all three microsomal preparations. Formation of the epoxide (2) was also stereospecific; however, a different specificity occurred with the different microsomal preparations.

Reference

C. C. Duke, T. W. Hambley, G. M. Holder, C. O. Navascues, S. Roberts-Thomson, and Y. Ye, Stereochemistry of the major rat liver microsomal metabolites of the carcinogen 7-methylbenz[*c*]acridine, *Chem. Res. Toxicol.*, 1991, **4**, 546.

Dibenz[*a,j*]acridine

Use/occurrence:	Chemical carcinogen
Key functional groups:	Polycyclic aromatic hydrocarbon, pyridine
Test system:	Isolated rat hepatocytes (substrate concentration 5–50 μM), rat (intraperitoneal or intravenous, 0.5 mg kg^{-1})

Structure and biotransformation pathway

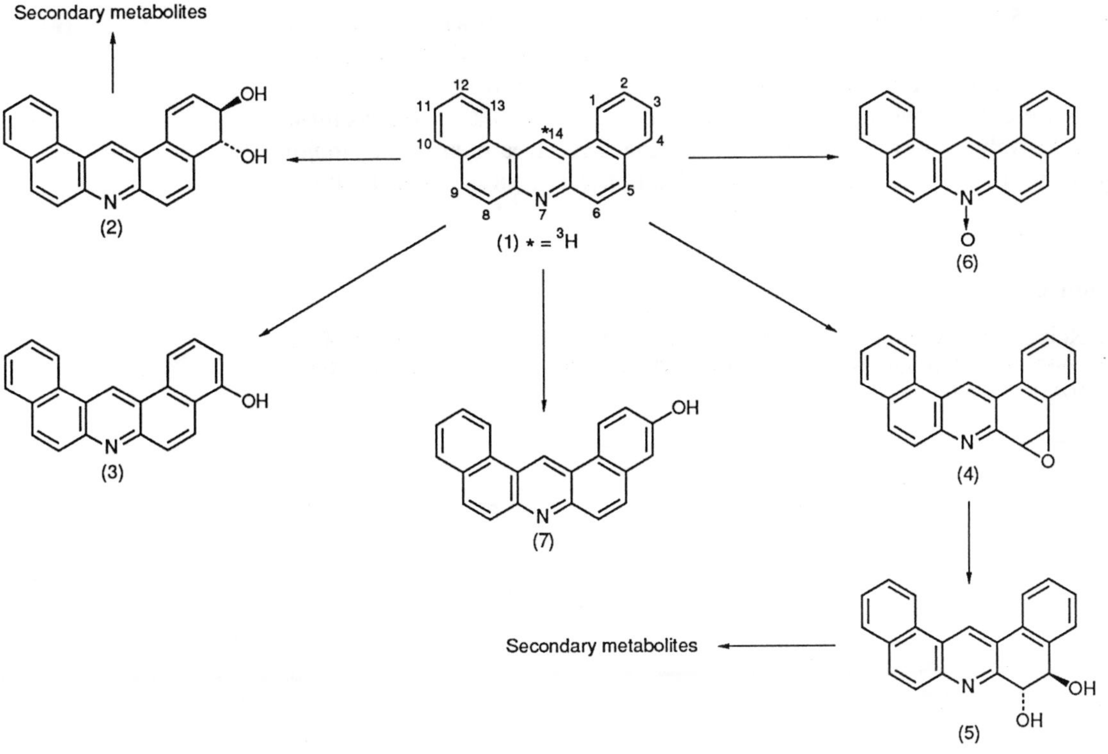

Metabolism of [14-^{3}H]dibenz[*a,j*]acridine (1) was maximal in hepatocyte suspensions prepared from untreated, phenobarbital- (PB), or 3-methyl-cholanthrene- (MC) treated rats at substrate concentrations of 30–50 μM. Following enzyme hydrolysis with β-glucuronidase/aryl sulfatase, the data suggested that only *ca.* 10% of the water-soluble metabolites were glucuronide/sulfate conjugates with the remainder being thioether (*i.e.* glutathione-derived) conjugates. Quantitative data indicated that the 3,4-dihydrodiol (2) was the principal metabolite of (1), accounting for about one-third (control and PB-treated hepatocytes) or two-thirds (MC-treated hepatocytes) of the solvent-extractable material from enzyme (β-glucuronidase/aryl sulfatase) hydrolysed preparations. [However, extensive secondary metabolism of (2), together with increased DNA and protein binding of ^{3}H-label, could be shown using high hepatocyte densities and prolonged incubation times]. The other major metabolite in control and PB-treated hepatocytes was the 4-hydroxy derivative (3), composing *ca.* 30% of the solvent-extractable radiolabel after hydrolytic enzyme treatment. Other metabolites identified in all hepatocyte incubations included the 5,6-oxide (4), the 5,6-dihydrodiol (5), the *N*-oxide (6), the 3-hydroxy metabolite (7), and secondary metabolites of (2) and (5).

Following intraperitoneal administration of (1) to male rats, *ca.* 8% of the radiolabel was recovered in the 7 day urine with the majority (74%) being excreted in the 7 day faeces. The major proportion of the ^{3}H was recovered within the first 48 hours. Biliary excretion of metabolites was indicated and 41–58% of the dose was recovered in bile collected up to 6 hours after an intravenous dose of (1) to control, PB-, and MC-treated rats.

About 25% of the biliary ^{3}H radiolabel was solvent-extractable following enzymic hydrolysis and contained secondary oxidation products of (2) and (5), *e.g.* diol epoxides. Other major metabolites in control and PB-treated animals were the 3-hydroxy (7) and 4-hydroxy (3) metabolites whereas these were relatively minor metabolites following MC treatment. The major hepatocyte metabolite, (2), the 3,4-dihydrodiol, was a minor metabolite in animals treated *in vivo*.

Reference

H. K. Robinson, C. C. Duke, G. M. Holder, and A. J. Ryan, The metabolism of the carcinogen dibenz[*a,j*]acridine in isolated rat hepatocytes and *in vivo* in rats, *Xenobiotica*, 1990, **20**, 457.

trans-1,2-Dihydrodiols of triphenylene

Use/occurrence:	Model compounds
Key functional groups:	Chiral carbon, dihydrodiol, polycyclic aromatic hydrocarbon
Test system:	Rat liver microsomes (0.05 mM substrate)

Structure and biotransformation pathway

(2)   (1) * = ^{3}H   (3)

Metabolism of both ^{3}H-labelled $(+)$-(S,S)-(1) and $(-)$-(R,R)-(1) was examined using rat liver microsomes and also using purified P-450 isozymes in a reconstituted mono-oxygenase system. Both enantiomers were metabolized at comparable rates; however, the distribution of metabolites between phenolic dihydrodiols (2) and bay region epoxide diastereoisomers (3) varied substantially with the different systems. Pretreatment of rats with either phenobarbital or 3-methylcholanthrene caused a slight reduction or less than two-fold increase, respectively, in the rate of total metabolism compared with control rats. Experiments with antibodies indicate that a large percentage of the metabolism by microsomes from control and phenobarbital-pretreated rats is catalysed by cytochrome P-450 IIIA1, and this may account for the differing selectivity seen in microsomes from 3-methylcholanthrene-pretreated rats (which contain predominantly P450 IA1). The diol epoxides (3) were hydrolysed to the corresponding tetraols before HPLC analysis. Metabolites were determined by radio-HPLC and using unlabelled standards to determine retention times.

Reference

D. R. Thakker, C. Boehlert, W. Levin, D. E. Ryan, P. E. Thomas, H. Yagi, and D. M. Jerina, Novel stereoselectivity of rat liver cytochrome(s) P-450 toward enantiomers of the *trans*-1,2-dihydrodiols of triphenylene, *Arch. Biochem. Biophys.*, 1991, **288**, 54.

1-Nitropyrene

Use/occurrence: Model compound (mutagen)

Key functional groups: Nitrophenyl, polycyclic aromatic hydrocarbon

Test system: Rat (intraperitoneal, *ca.* 5 mg kg^{-1})

Structure and biotransformation pathway

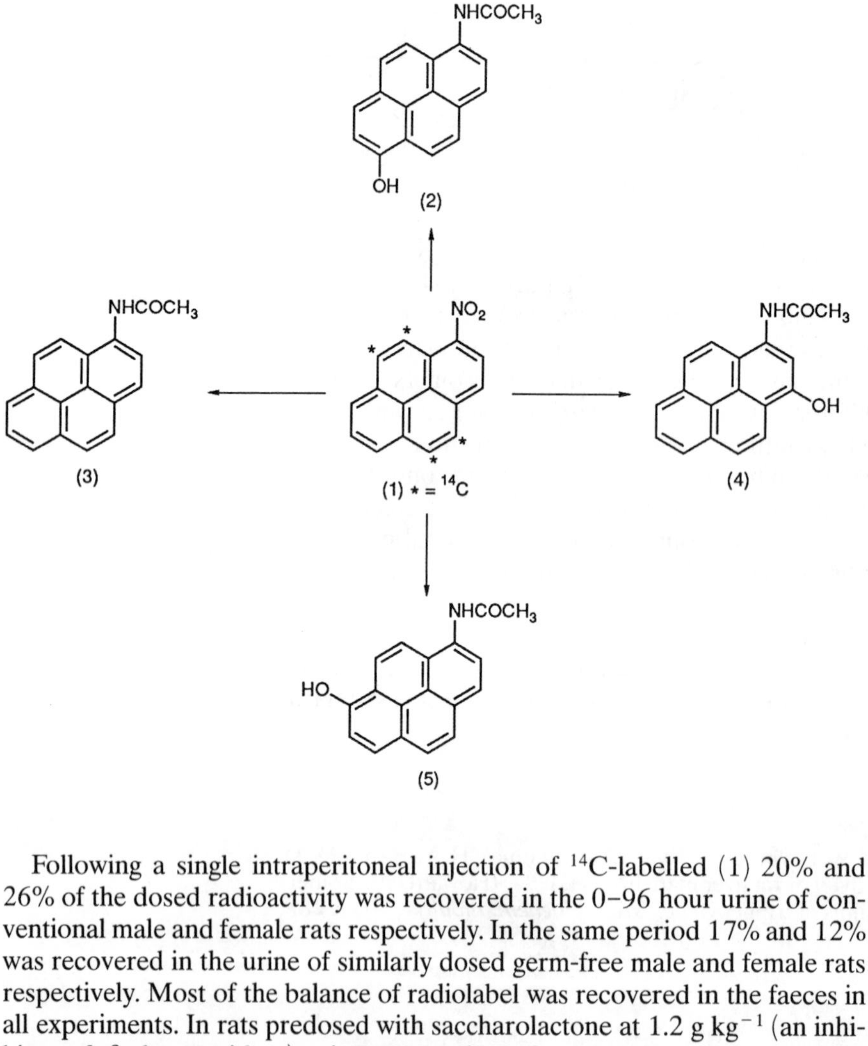

Following a single intraperitoneal injection of ^{14}C-labelled (1) 20% and 26% of the dosed radioactivity was recovered in the 0–96 hour urine of conventional male and female rats respectively. In the same period 17% and 12% was recovered in the urine of similarly dosed germ-free male and female rats respectively. Most of the balance of radiolabel was recovered in the faeces in all experiments. In rats predosed with saccharolactone at 1.2 g kg^{-1} (an inhibitor of β-glucuronidase) urinary excretion of total radioactivity was also slightly reduced. Analysis of urinary metabolites by HPLC showed that conventional rats excreted 10-fold more (2) than the germ-free rat; the amount of this metabolite excreted by conventional rats predosed with saccharolactone was also significantly reduced. Sufficient (2) was excreted in urine to allow the UV spectrum to be recorded, which compared closely with that of the synthetic standard. Other metabolites, (3)–(5), were identified by coelution with synthetic standards. Mutagenic activity of urine samples from rats dosed with 1-nitropyrene was correlated with the concentration of (2). A significant proportion of the urinary radioactivity from conventional rats was not associated with any of the reference markers, neither was the majority of the radiolabel in urine from germ-free rats. These results suggest that both bacterial β-glucuronidase and bacterial nitroreductases are important for the production of (2) in conventional rats.

Reference

L. M. Ball, J. J. Rafter, J. A. Gustafsson, B. E. Gustafsson, M. J. Kohan, and J. Lewtas, Formation of mutagenic urinary metabolites from 1-nitropyrene in germ-free and conventional rats: role of the gut flora, *Carcinogenesis*, 1991, **12**, 1.

1-Nitropyrene

Use/occurrence:	Environmental pollutant
Key functional groups:	Nitrophenyl, polycyclic aromatic hydrocarbon
Test system:	Human, mouse, rat, hamster, dog, and guinea-pig liver microsomes and cytosol, rat (oral, 30 μmol)

Structure and biotransformation pathway

Following incubation of [^{3}H]-1-nitropyrene (1) with microsomes from human, mouse, rat, hamster, dog, and guinea-pig liver the 4,5-oxide (2) and 9,10-oxide (3) were identified using HPLC. There were considerable inter-species differences, as well as inter-individual variations in the case of humans, in the ratio of (2):(3).

Further incubation of (2) and (3) with microsomes produced the corresponding dihydrodiols (4) and (5). Incubation of (2) and (3) with hepatic cytosol preparations led to the formation of the corresponding glutathione conjugates.

Analysis of bile by HPLC from rats dosed orally with (1) showed that the glutathione conjugates from (3) and (2) accounted for 12 and 2% respectively of the radioactivity in the 0–6 hour bile, and a novel metabolite, tentatively identified as a glucuronide or sulfate conjugate of (4), accounted for a further 10% of total metabolites in the bile.

Reference

K. Kataoka, T. Kinouchi, and Y. Ohnishi, Species differences in metabolic activation and inactivation of 1-nitropyrene in the liver, *Cancer Res.*, 1991, **51**, 3919.

7-Chlorobenz[*a*]anthracene, 7-Bromobenz[*a*]anthracene

Use/occurrence:	Model compounds
Key functional groups:	Polycyclic aromatic hydrocarbon
Test system:	Rat and mouse liver microsomes

Structure and biotransformation pathway

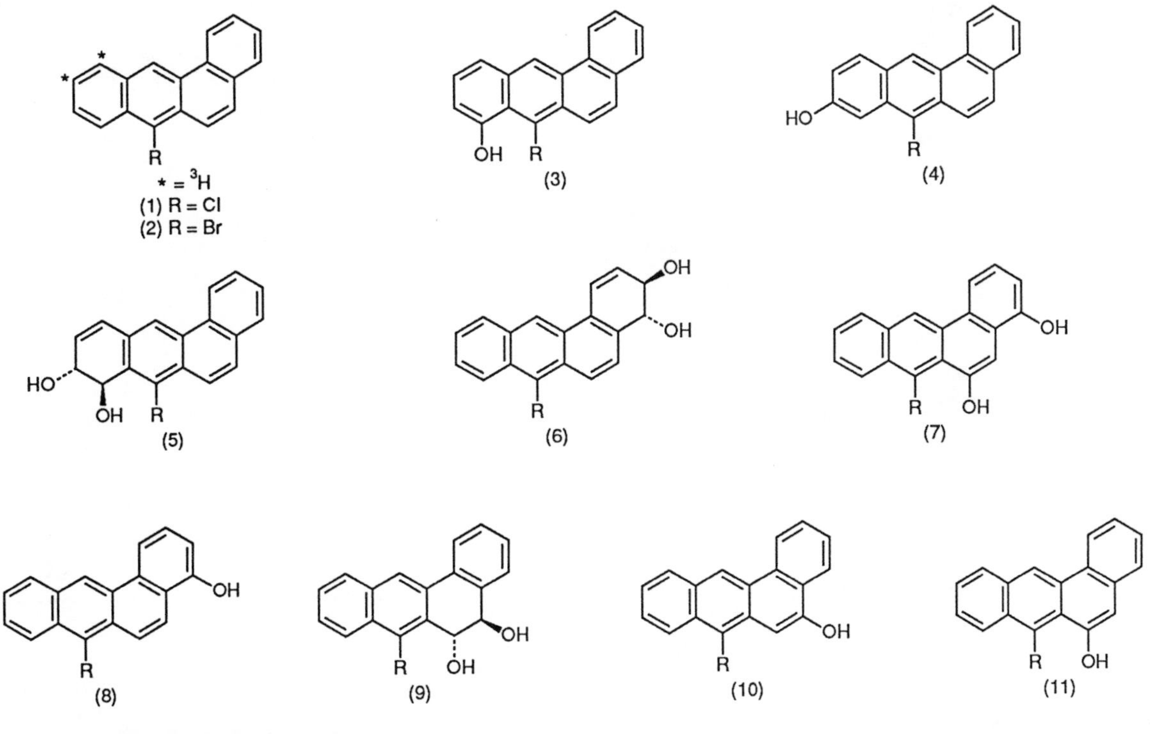

* = ³H
(1) R = Cl
(2) R = Br

Both rat and mouse liver microsomes preferentially metabolized (1) at the C-8 and C-9 aromatic double bond region producing metabolites (3)–(5), which together accounted for up to 50% of the total metabolites. The 3,4-dihydrodiol metabolite (6) of (1) formed 7–8% of the total metabolites with both microsomal preparations. The corresponding 3,4-dihydrodiol metabolite (6) of (2) was formed at yields of 16% and 9% respectively from mouse and rat liver microsomes. Both rat and mouse liver microsomes were found to metabolize both (1) and (2) in a stereoselective manner, preferentially producing *trans*-dihydrodiol metabolites with an *R,R* stereochemistry. The 3,4-*trans*-dihydrodiol metabolites (6) of both (1) and (2) exhibited enhanced mutagenicity when compared with the parent molecule. The other dihydrodiols (5) and (9) were essentially weakly mutagenic or inactive. Metabolites were measured by a radiochemical HPLC assay and the optical purity was determined by chiral chromatography.

Reference

P. P. Fu, L. S. V. Tungeln, L. E. Unruh, Y. Ni, and M. W. Chu, Comparative and regioselective metabolism of 7-chlorobenz[a]anthracene and 7-bromobenz[a]-anthracene by mouse and rat liver microsomes, *Carcinogenesis*, 1991, **12**, 371.

Alkenes, Halogenoalkanes, and Halogenoalkenes

2-Methylpropene (isobutene)

Use/occurrence:	Industrial chemical
Key functional groups:	Alkene
Test system:	Rat and mouse (inhalation exposure up to 10 000 p.p.m.) and mouse liver *in vitro* (9000 g supernatant with vapour exposure)

Structure and biotransformation pathway

(1)   (2)

Mouse liver homogenates were found to metabolize 2-methylpropene to the corresponding epoxide (2). The metabolite was detected following GC–MS analysis of the headspace. The reaction was sensitive to the presence of the P-450 inhibitors (metyrapone and SKF 525A), and to the absence of an NADPH generating system, demonstrating that the epoxidation was P-450 dependent.

The epoxide (2) was also detected in the expired air of rats exposed to (1) and mice were also presumed to metabolize (1) to (2) *in vivo*. Kinetic parameters for the elimination of (1) were determined in both rats and mice. In addition a scheme for the further metabolism and degradation of (2) was proposed.

References

G. A. Csanady, D. Friese, B. Denk, J. G Filser, M. Cornet, V. Rogiers, and R. J. Laib, Investigation of species differences in isobutene (2-methylpropene) metabolism between mice and rats, *Arch. Toxicol.*, 1991, **65**, 100.

M. Cornet, W. Sonck, A. Callaerts, G. Csanady, A. Vercruysse, and R. J. Laib, *In vitro* biotransformation 2-methylpropene (isobutene): epoxide formation in mouse liver, *Arch. Toxicol.*, 1991, **65**, 263.

Buta-1,3-diene

Use/occurrence:	Industrial chemical
Key functional groups:	Alkene
Test system:	Mouse liver microsomes

Structure and biotransformation pathway

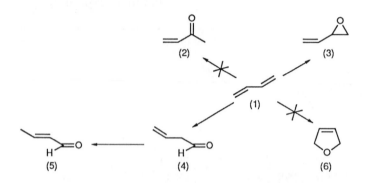

NADPH-dependent oxidation of buta-1,3-diene (1) by mouse liver microsomes or H_2O_2-dependent oxidation by chloroperoxidase produced both (3) and (5). The ketone (2) and dihydrofuran (6) were not detected. Metabolite (4) was a postulated intermediate in the formation of (5). Metabolites were determined by GC–MS.

Reference

A. A. Elfarra, R. J. Duescher, and C. M. Pasch, Mechanisms of 1,3-butadiene oxidations to butadiene monoxide and crotonaldehyde by mouse liver microsomes and chloroperoxidase, *Arch. Biochem. Biophys.*, 1991, **286**, 244.

4-Vinylcyclohexene

Use/occurrence:	Released during curing of rubber
Key functional groups:	Alkene, cyclohexene
Test system:	Human hepatic microsomes

Structure and biotransformation pathway

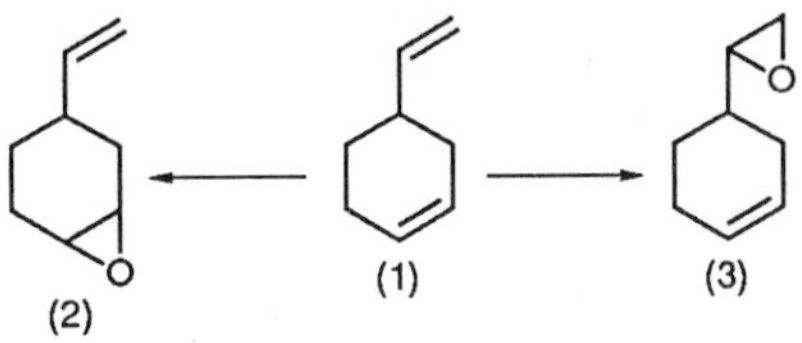

(2)   (1)   (3)

The rate of epoxidation of 4-vinylcyclohexene (1), an ovarian carcinogen in mice, has been studied using liver microsomes obtained from male and female humans, the metabolites being extracted with hexane and analysed by GC. The major product formed was the 1,2-epoxide (2) together with small amounts of the 7,8-epoxide (3). The rate of formation of (2) has been linked to the ovarian toxicity. The study showed that, in humans, formation of (2) was an order of magnitude slower than in the mouse, a susceptible species, and even slower than in the rat, a species resistant to 4-vinylcyclohexene-induced ovarian tumours.

Reference

B. J. Smith and I. G. Sipes, Epoxidation of 4-vinylcyclohexene by human hepatic microsomes, *Toxicol. Appl. Pharmacol.*, 1991, **109**, 367.

R-(+)-Pulegone, Menthofuran, Isopulegone

Use/occurrence:	Hepatotoxin
Key functional groups:	Alkene, allylic methyl, cyclohexanone
Test system:	Rat liver microsomes

Structure and biotransformation pathway

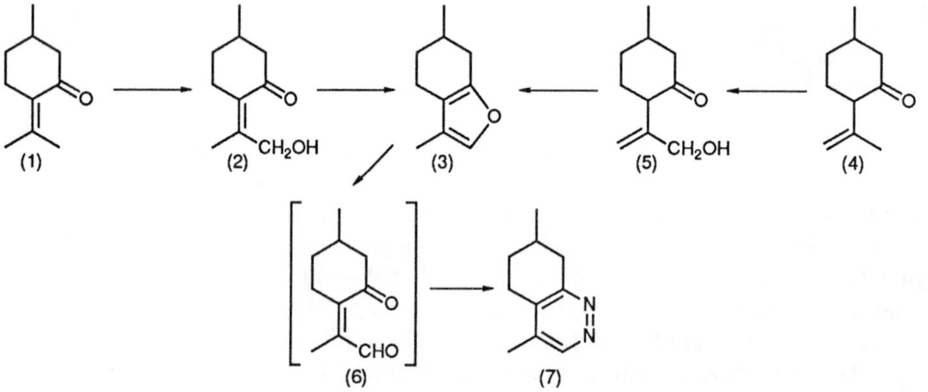

Incubation of pulegone (1) with liver microsomes showed formation of menthofuran (3) as the major metabolite. A minor metabolite was detected by GC–MS which gave a spectrum consistent with that of the allylic alcohol (2). Similar incubation of isopulegone (4) showed formation of the allylic alcohol (5) as a major metabolite and menthofuran as a minor metabolite. The allylic alcohol (5) was more stable than (2) and was only slowly converted into menthofuran. However, for both compounds microsomal oxidation of the methyl group *syn* to the ketone function appeared to predominate as the major primary pathway. After further incubation of menthofuran with liver microsomes formation of the aldehyde (6) was demonstrated by trapping it with semicarbazide as the cinnoline derivative (7). The aldehyde (6) has been proposed as the ultimate metabolite responsible for the observed hepato- and pneumo-toxicity in animals.

Reference

K. M. Madyastha and C. P. Raj, Biotransformations of *R*-(+)-pulegone and mentho-furan *in vitro*: chemical basis for toxicity, *Biochem. Biophys. Res. Commun.*, 1990, **173**, 1086.

Pulegone

Use/occurrence:	Natural product (hepatotoxin)
Key functional groups:	Alkene, allylic methyl, cyclohexanone
Test system:	Rat (intraperitoneal, 250 mg kg^{-1})

Structure and biotransformation pathway

(1) * = ^{14}C
† = ^{2}H

(2)

Pulegone (1), a hepatotoxic monoterpene, is a major constituent of the herbal abortifacient pennyroyal oil. The toxicity of pulegone is reported to be partially mitigated by glutathione. The objective of this study was to characterize Phase II metabolites, in particular glutathione conjugates, of (1) excreted in rat bile. Rats were administered an approximately equimolar mixture of deuterium- and ^{14}C-labelled (1) and *ca.* 35% dose was excreted in bile during 8.5 hours. Bile was treated with benzyl chloroformate and the metabolites and derivatives were separated by HPLC. Those components which could then be methylated with methanol–HCl were investigated using MS–MS. In this way ten conjugated metabolites (some of which were isomeric), were partially characterized. Although molecular formulae could be deduced, the regio- and stereo-chemistry of the metabolites could not be established. The most abundant biliary metabolites were glucuronides of hydroxylated and reduced hydroxylated Phase I metabolites of (1). Reduction could have occurred of either the exocyclic double bond or the ketone function. Three types of glutathione conjugate could be distinguished, containing the elements respectively of (1), reduced (1), and oxidized (1). The latter metabolite was tentatively identified as a glutathione conjugate of menthofuran (2), a known metabolite of (1) which is reported to contribute to its hepatotoxicity. The other two glutathione conjugates apparently underwent subsequent glucuronidation to yield novel glutathionyl glucuronide conjugates which contained non-hydroxylated exocon moieties.

Reference

D. Thomassen, P. G. Pearson, J. T. Slattery, and S. D. Nelson, Partial characterization of biliary metabolites of pulegone by tandem mass spectrometry, *Drug Metab. Dispos.*, 1991, **19**, 997.

Eugenol

Use/occurrence:	Model compound
Key functional groups:	Arylalkene, methoxyphenol, phenol
Test system:	Isolated rat hepatocytes (up to 1.5 mM substrate)

Structure and biotransformation pathway

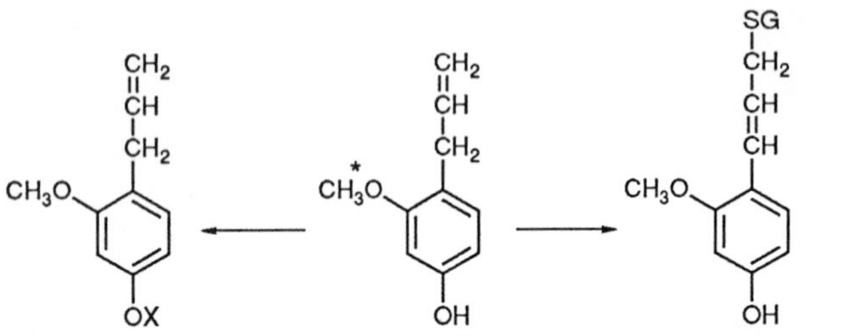

(2) X = sulfate
(3) X = glucuronyl

(1) * = ^{3}H

(4) SG = glutathionyl

Incubation of eugenol (1) with freshly isolated hepatocytes elicited a cytotoxic response which was concentration- and time-dependent. A concentration of 1 mM eugenol caused *ca.* 85% cell death over a five hour period compared with 35% cell death in control incubations. Analysis of eugenol metabolites in the incubation media revealed that three major metabolites (2)–(4) were being formed. Metabolites were analysed by HPLC and identified by comparison with standards.

Reference

D. C. Thompson, D. Constantin-Teodosiu, and P. Moldeus, Metabolism and cytotoxicity of eugenol in isolated rat hepatocytes, *Chem. Biol. Interact.*, 1991, **77**, 81.

Tamoxifen

Use/occurrence:	Anticancer agent
Key functional groups:	Arylalkene, alkyl aryl ether, dimethylaminoalkyl
Test system:	Rat (oral, 1 mg kg^{-1} daily), human (oral, 30–80 mg daily)

Structure and biotransformation pathway

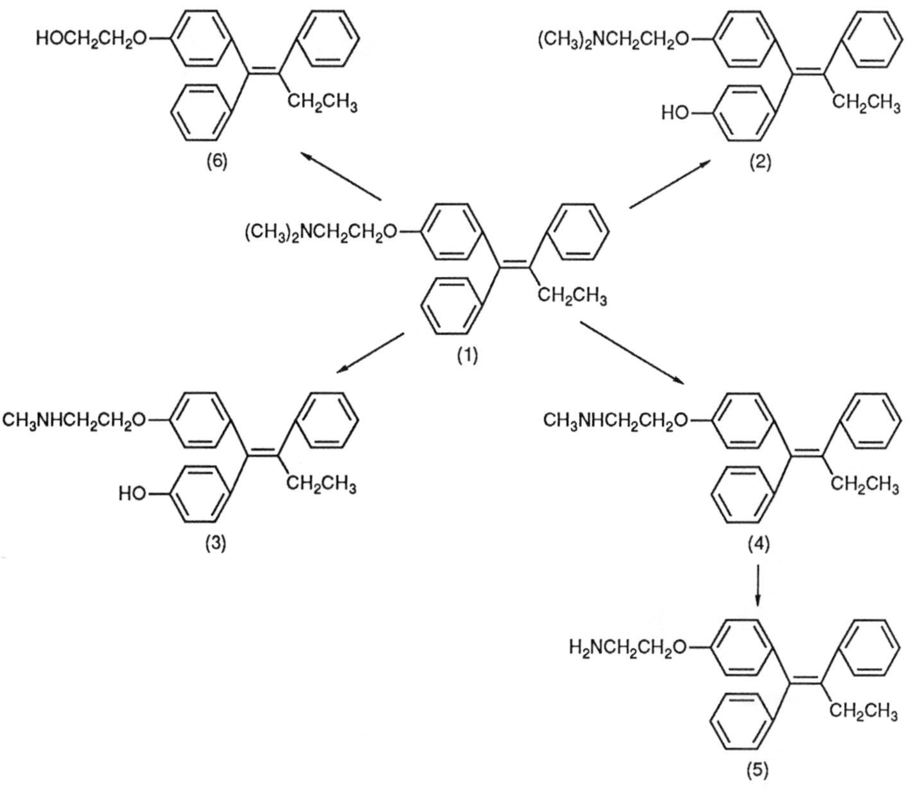

Rats were dosed with tamoxifen [*trans*-1-(4-β-dimethylaminoethoxy-phenyl)-1,2-diphenylbut-1-ene] (1) daily for either 3 or 14 days. Following necropsy tissues were extracted and analysed by HPLC–MS. Human tissues were also obtained from patients receiving (1) and were similarly processed.

In rats, the hydroxylated metabolites (2) and (3) and demethylated metabolite (4) together with unchanged (1) were abundant in all except fat tissues where unchanged (1) predominated. *N*-Desdimethyltamoxifen (5) was also present but in smaller amounts, though it was a more significant metabolite in the human tissues. The primary alcohol (6) was detected in serum, where it was present in trace amounts, but was not detectable in tissue extracts due to interference in the chromatograms.

Reference

E. A. Lien, E. Solheim, and P. M. Ueland, Distribution of tamoxifen and its metabolites in rats and human tissues during steady-state treatment, *Cancer Res.*, 1991, **51**, 4837.

4-Iodotamoxifen

Use/occurrence:	Anti-oestrogen
Key functional groups:	Arylalkene, dimethylaminoalkyl, iodophenyl
Test system:	Rat hepatocytes

Structure and biotransformation pathway

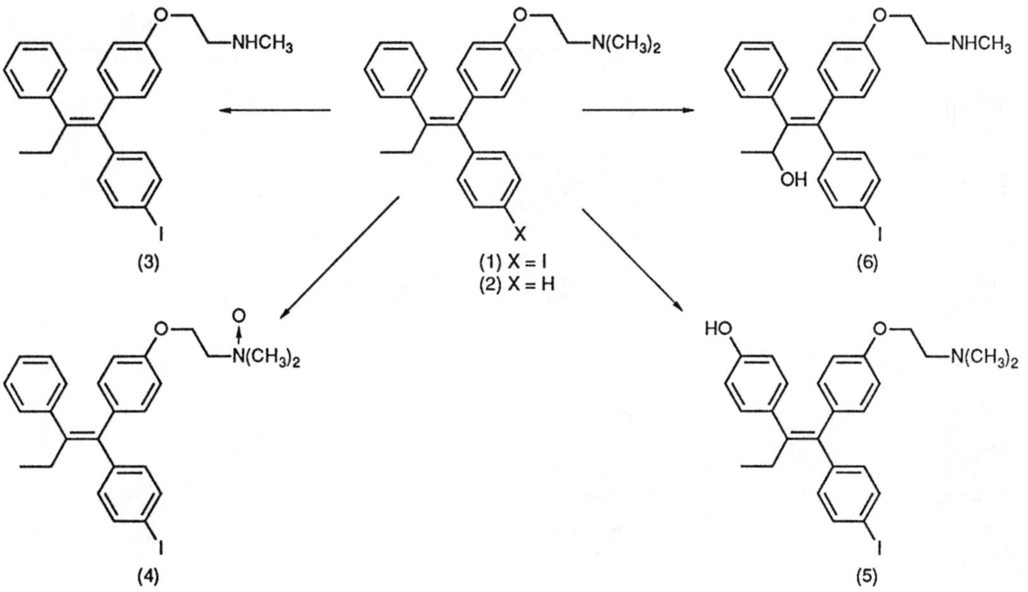

4-Iodotamoxifen (1) is a synthetic anti-oestrogen that has been found to have a greater affinity for cytosolic oestrogen receptors than tamoxifen (2) and a greater ability to inhibit the growth of the human breast cancer cell line MCF-7 *in vitro*. In this study 4-iodotamoxifen (1) was metabolized four times more slowly than tamoxifen (2) by freshly isolated rat liver hepatocytes. Four principal metabolites of (1) were isolated, (3)–(6). The identification of (3), (4), and (5) was confirmed by MS and by comparison with synthetic standards. The identification of (6) was revealed by ^{1}H NMR. The iodo-phenyl moiety was thus retained in all four metabolites. The iodine atom not only blocks metabolism in its vicinity but also apparently reduces the rate of side-chain demethylation and *N*-oxidation by three-fold. It was thus predicted that the presence of the iodine atom in (1) would lead to a greater duration of *in vivo* activity for this drug compared with (2).

Reference

R. McCague, I. B. Parr, and B. P. Haynes, Metabolism of the 4-iodo derivative of tamoxifen by isolated rat hepatocytes, *Biochem. Pharmacol.*, 1990, **40**, 2277.

61

Toremifene

Use/occurrence: Anticancer agent

Key functional groups: Arylalkene, chloroalkyl, dimethylaminoalkyl

Test system: Rat (oral, intravenous; 5, 10 mg kg^{-1})

Structure and biotransformation pathway

(1) R = CH_3
(2) R = H

(3) R = H
(7) R = OH

(4) R = CH_3
(5) R = H

(6)

(8)

(9) R = H
(10) R = OH

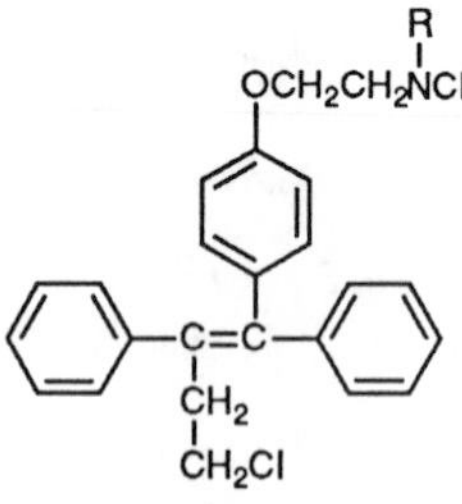

After oral or inravenous doses of [^{3}H]toremifene (1) to rats most of the radioactivity was excreted in the faeces. Faecal metabolites were extracted with methanol, isolated by preparative TLC, and purified by HPLC before mass spectral analysis. Metabolites were initially separated into three groups — amines, carboxylic acids, and alcohols. Acidic metabolites were esterified with HCl–methanol. The identified metabolites (2)–(10) were derived by degradation of the dimethylaminoalkyl side-chain and ring-hydroxylation. No metabolites resulting from biotransformation of the chloromethyl group were observed. Proportions of metabolites were measured but no detailed quantitative data were reported other than the relative amounts of the different groups of metabolites, which ranged from 20% to 50%.

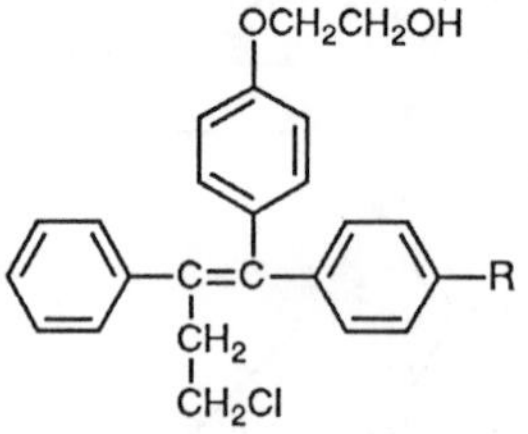

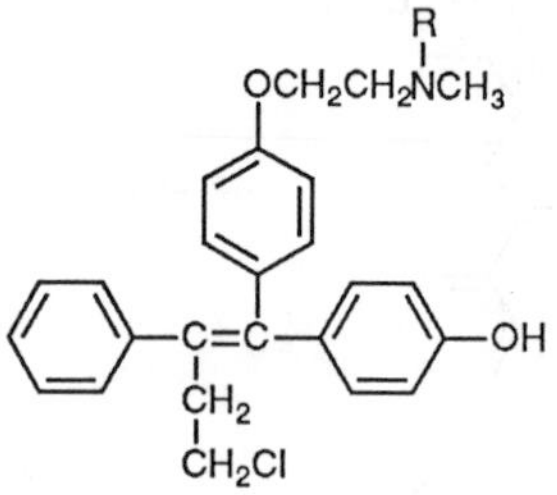

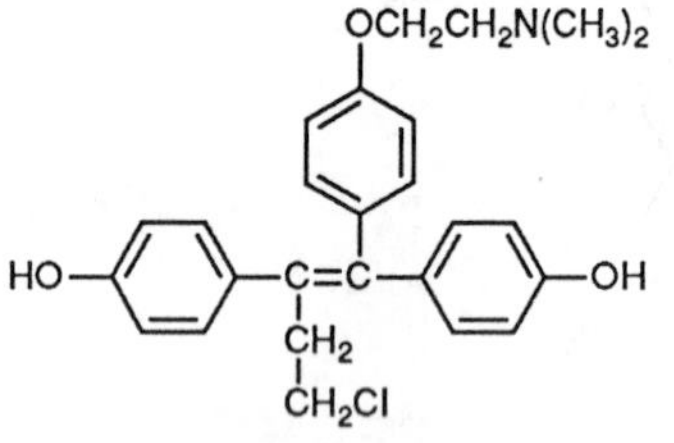

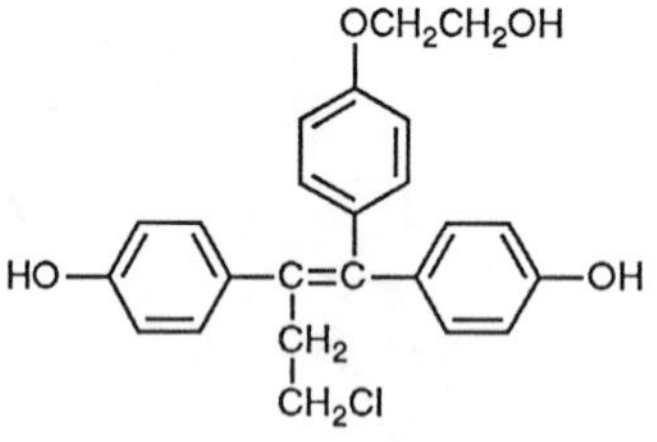

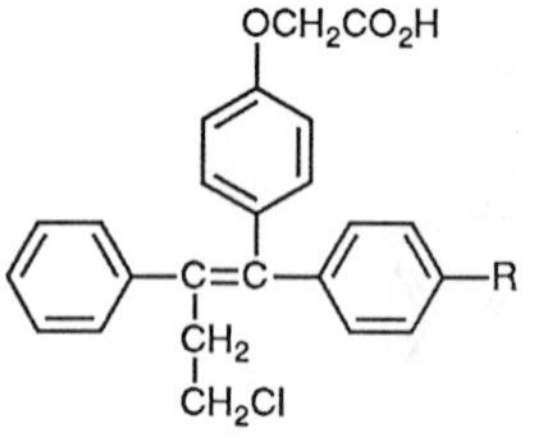

Reference

H. Sipila, L. Kangas, L. Vuorilehto, A. Kalapudar, M. Eloranta, M. Sodervall, R. Toivola, and M. Anttila, Metabolism of toremifene in the rat, *J. Steroid Biochem.*, 1990, **36**, 211.

1,2-Dichloropropane

Use/occurrence:	Industrial chemical
Key functional groups:	Chloroalkyl
Test system:	Rat (oral, 1 or 100 mg kg^{-1}; inhalation, 5 and 50 000 p.p.m.)

Structure and biotransformation pathway

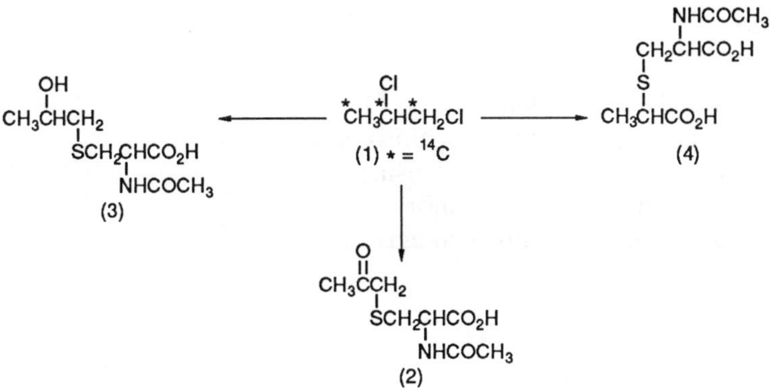

Following the oral administration of 1,2-dichloropropane (1) to rats at 1 or 100 mg kg^{-1} bodyweight, the dose was rapidly excreted mainly in the urine or expired air. Urinary metabolites were analysed by HPLC when four radioactive components were separated. The major component was further analysed using GC–MS and shown to consist of a mixture of three mercapturic acids (2)–(4). These accounted for 46–75% of the urinary metabolites. The three remaining components were not identified.

The major component in expired air was CO_2 (27–36% of the dose), though the amount of unchanged (1) rose 30-fold at the higher dose. The profile of metabolites following inhalation of (1) was similar to that following oral administration.

Reference

C. Timchalk, M. D. Dryzga, F. A. Smith, and M. J. Bartels, Disposition and metabolism of [^{14}C]-1,2-dichloropropane following oral and inhalation exposure to Fischer 344 rats, *Toxicology*, 1991, **68**, 291.

1,1,1-Trichloroethane

Use/occurrence:	Typewriter correction fluid, industrial degreasing solvent
Key functional groups:	Chloroalkyl
Test system:	Human (inhalation)

Structure and biotransformation pathway

$$Cl_3CCH_3 \longrightarrow Cl_3CCH_2OH \longrightarrow Cl_3CCO_2H$$

Following industrial exposure, the intake of 1,1,1-trichloroethane (1) was monitored in human urine by the determination of the total trichloro compounds, trichloroethanol (2), or trichloroacetic acid (3) present using GC. The results suggested that in man trichloroacetic acid was a minor metabolite, whereas trichloroethanol was a major metabolite. This is in agreement with previously observed data in the rat.

Reference

T. Kawai, K. Yamaoka, Y. Uchida, and M. Ikeda, Exposure of 1,1,1-trichloroethane and dose-related excretion of metabolites in urine of printing workers, *Toxicol. Lett.*, 1991, **55**, 39.

1,2,3-Trichloropropane

Use/occurrence:	Industrial chemical/solvent
Key functional groups:	Chloroalkyl
Test system:	Mouse (oral, 30 and 60 mg kg^{-1}), rat (oral, 30 mg kg^{-1})

Structure and biotransformation pathway

Following oral administration of 1,2,3-trichloro[2-^{14}C]propane (1) to male and female rats, *ca.* 50% of the dose was excreted in urine during 60 hours, 20% in the faeces, and a further 20% as $^{14}CO_2$ in the expired air. In treated mice, urinary excretion of radioactivity was slightly more extensive and faecal excretion less extensive. Radioactive components in the urine of intact rats and mice and in the bile of rats with cannulated bile ducts were separated by HPLC and the identities of the more important ones investigated by MS, NMR, and chromatographic comparison with synthetic standards. In rats, the major urinary metabolite was thus identified as the *N*-acetylcysteine conjugate (3). This compound accounted for *ca.* 40% of the 0–6 hour urinary radioactivity in male rats, but only 3% of that in mice. A minor rat urinary metabolite was similarly identified as the cysteine conjugate (2), but another important rat urinary metabolite, which was more polar than (2) or (3), was not identified. Three major rat biliary metabolites of (1) were separated by HPLC, none of which corresponded to parent compound. The most abundant of these, accounting for *ca.* 60% of the sample radioactivity, was identified as 2-(*S*-glutathionyl)malonic acid (4). Possible biotransformation pathways leading to the formation of CO_2 and the identified urinary and biliary metabolites are discussed in relation to the toxicity of (1) in laboratory animals.

Reference

N. A. Mahmood, D. Overstreet, and L. T. Burka, Comparative disposition and metabolism of 1,2,3-trichloropropane in rats and mice, *Drug Metab. Dispos.*, 1991, **19**, 411.

1,2-Dichloro-1,1-difluoroethane (HCFC-132b)

Use/occurrence:	Model compound
Key functional groups:	Chloroalkyl, fluoroalkyl
Test system:	Rat (intraperitoneal, 10 mmol kg^{-1})

Structure and biotransformation pathway

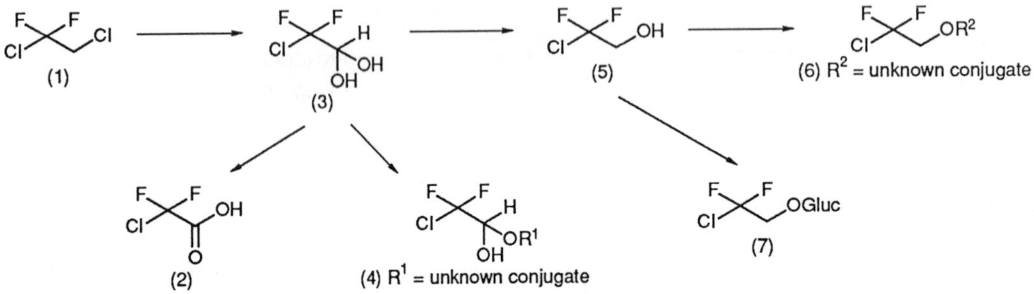

Urinary metabolites excreted in the first 24 hours following dosing accounted for *ca.* 1.8% of the dose as determined by ^{19}F NMR. During the first 6 hours following dosing identified urinary metabolites of (1) corresponded to (7), (2), and (3), the hydrated aldehyde. An unknown conjugate of (3) was also detected, together with another unknown conjugate which was believed to be an *O*-conjugate of (5). No unconjugated (5) was detected in the urine. Metabolite (3) was the only fluorine-containing metabolite detected in the microsomal incubations. Metabolites were identified by ^{19}F NMR both before and following enzyme and chemical hydrolysis. The *in vitro* metabolism of (1) was increased in microsomes from pyridine-treated rats, and was inhibited by *p*-nitrophenol. These data were taken to suggest that cytochrome P-450 IIE1 was largely responsible for the initial hydroxylation of (1).

Reference

J. W. Harris and M. W. Anders, Metabolism of the hydrofluorocarbon 1,2-dichloro-1,1-difluoroethane, *Chem. Res. Toxicol.*, 1991, **4**, 180.

1,1,1,2-Tetrafluoro-2-chloroethane

Use/occurrence:	Refrigerant
Key functional groups:	Chloroalkyl, fluoroalkyl
Test system:	Rat (inhalation, 10 000 p.p.m.), rat liver microsomes

Structure and biotransformation pathway

Urine was collected from rats during 24 hours following a two hour exposure to an atmosphere containing *ca.* 10 000 p.p.m. 1,1,1,2-tetrafluoro-2-chloroethane (1). ^{19}F NMR showed that the urine contained trifluoroacetic acid (4) and fluoride ion. No other fluorine-containing metabolites were detected in urine using this technique. Hepatic microsomes also produced (4) and fluoride ion together with an unidentified fluorine-containing metabolite. The rate of metabolism was described as low but pretreatment of rats with pyridine caused about a 20-fold increase in fluoride ion release, whereas the rate of defluorination was slightly decreased by phenobarbital administration. Microsomes from pyridine-treated rats also defluorinated (1) under conditions of reduced oxygen tension demonstrating that (1) could be reductively metabolized. The formation of trifluoroacetic acid and fluoride as products of oxidative metabolism of (1) led the authors to suggest a biotransformation pathway via the intermediates (2) and trifluoracetyl fluoride (3) shown in the scheme.

Reference

M. J. Olson, J. T. Johnson, J. F. O'Gara, and S. E. Surbrook Jr., Metabolism *in vivo* and *in vitro* of the refrigerant substitute 1,1,1,2-tetrafluoro-2-chloroethane, *Drug Metab. Dispos.*, 1991, **19**, 1004.

Triallate

Use/occurrence:	Herbicide
Key functional groups:	Chloroalkene, thiocarbamate
Test system:	Rat liver microsomes

Structure and biotransformation pathway

(1) C_3H_7 = Isopropyl
* = ^{14}C and ^{13}C

(2)

(3) GS = Glutathionyl

(4)

(5)

The main objective of this short communication was to describe the identification of the glutathione conjugate (2) using heteronuclear multiple quantum coherence (HMQC) NMR. This conjugate was formed when triallate (1) was incubated with rat liver microsomes fortified with NADPH and glutathione. A previously described bis-glutathione conjugate (3) was also formed under these conditions together with a small proportion of the 2-chloroacrylate (4). This latter metabolite (4) was a major product, together with the allylic alcohol (5), formed when (1) was incubated with microsomes fortified with NADPH only. Identification of these metabolites was based on HPLC–MS and ^1H NMR (4) or CI–MS and co-chromatography (5).

Reference

A. G. Hackett, J. J. Kotyk, H. Fujiwara, and E. W. Logusch, Microsomal hydroxylation of triallate: Identification of a 2-chloroacrylate glutathione conjugate using heteronuclear quantum coherence NMR spectroscopy, *Drug Metab. Dispos.*, 1991, **19**, 1163.

Hexachlorobuta-1,3-diene

Use/occurrence:	Environmental pollutant, nephrotoxin
Key functional groups:	Chloroalkene
Test system:	Isolated perfused rat liver

Structure and biotransformation pathway

$$G = HO_2CCHCH_2CH_2CONHCHCH_2-$$
$$NH_2 \qquad CONHCH_2CO_2H$$

Hexachlorobutadiene (1) is a selective nephrotoxin and nephrocarcinogen. Its organ-selective toxicity is believed to be based on a bioactivation mechanism initially involving hepatic conjugation with glutathione, but the mode of translocation to the kidneys is poorly understood. The objective of this study was to investigate the formation and excretion of the glutathione and cysteine conjugates, (2) and (3), respectively in the isolated perfused liver. Metabolite analysis was by HPLC.

Infusion of (1) into isolated perfused livers led to formation of both (2) and (3). At moderate infusion rates (2) was exclusively excreted into bile although increasing amounts of (2) were found in the perfusate as the rate of infusion of (1) was increased. The cysteine conjugate (3) was not detected in perfusate, but amounts found in bile correlated with the concentration of (2). The authors concluded that intestinal absorption of (2) or its metabolites is required for the induction of kidney damage *in vivo*.

Reference

Y. S. Gietl and M. W. Anders, Biosynthesis and biliary excretion of *S*-conjugates of hexachlorobuta-1,3-diene in the perfused rat liver, *Drug Metab. Dispos.*, 1991, **18**, 274.

Dichloroethyne

Use/occurrence:	Chemical by-product
Key functional groups:	Chloroalkyne
Test system:	Rat (inhalation, 20 and 40 p.p.m.)

Structure and biotransformation pathway

Dichloroethyne (1) has been shown to produce irreversible nerve damage in man following accidental exposure, but in rodents (1) is a potent and selective nephrotoxin.

Rats were exposed to [^{14}C]-(1) in an inhalation chamber such that only the noses were in contact with the exposure mixture. Immediately after treatment at two dose levels, the animals were transferred to metabolism cages and excreted ^{14}C was monitored. 15–18% of the dose was shown to be absorbed by the animals at both concentrations. During the 96 hour collection period, 60–68% of the absorbed radioactivity was eliminated in the urine and 27% in the faeces. 3.5% of the absorbed ^{14}C was associated with the organs/carcasses, but exhaled unchanged (1) was only a minor constituent of the recovered ^{14}C.

Urinary metabolites were separated by HPLC and identified by TSP–MS or by GC–MS following derivatization. The major metabolite in the urine was *N*-acetyl-*S*-(1,2-dichlorovinyl)-L-cysteine (2), which accounted for 61.8% of the urinary radioactivity. Other metabolites were dichloroethanol (3), 12.2% of urinary ^{14}C; dichloroacetic acid (4), 8.9%; chloroacetic acid (5), 4.7%; and oxalic acid (6), 8.3%. Metabolite (2) accounted for all the faecal radioactivity excreted. Following administration of [^{14}C]-(1) to bile duct-

70

cannulated rats, the only biliary metabolite identified was S-(1,2-dichloro-vinyl)glutathione (7). Biliary cannulation did not influence the renal excretion of (2), indicating that glutathione conjugate formation occurs in the kidney.

The results suggest that the major pathway of metabolism of (1) in the rat is the biosynthesis of toxic glutathione conjugates. The formation of 1,1-di-chloro compounds, produced after chlorine migration and mediated by cytochrome P-450, represents a minor pathway.

Reference

W. Kanhai, M. Koob, W. Dekant, and D. Henschler, Metabolism of [14C]dichloro-ethyne in rats, *Xenobiotica*, 1991, **21**, 905.

Acyclic Functional Compounds

Diethylene glycol

Use/occurrence:	Industrial chemical/solvent
Key functional groups:	Alkyl alcohol, dialkyl ether
Test system:	Rat (oral, 50 and 5000 mg kg^{-1}; intravenous, 50 mg kg^{-1}; dermal, 50 mg; drinking water, 0.3%, 1%, and 3%), dog (oral, 500 mg kg^{-1})

Structure and biotransformation pathway

(1) * = ^{14}C

(2)

Radioactivity excretion profiles following oral (intragastric gavage) and intravenous administration of [^{14}C]diethylene glycol (1) (50 mg kg^{-1}) to male rats were virtually identical. Thus, in each case *ca.* 85% of the dose was excreted in urine (mainly during 0–24 hours), 1% dose in the faeces, and 6% dose as $^{14}CO_2$ in the expired air, indicating that absorption of the oral dose of (1) was effectively complete in this species. After the 5000 mg kg^{-1} oral dose to rats, urinary excretion of radioactivity increased to 97% dose, and only 2% dose was eliminated as $^{14}CO_2$ in the expired air, suggesting that there was some saturation of metabolism at this dose level. Excretion profiles following administration of (1) to rats in the drinking water were similar to those observed after the gavage doses. Percutaneous absorption of (1) during 72 hours after application to the normal skin of rats was *ca.* 10%, and again most of the absorbed radioactivity was excreted in the urine. After single oral doses of (1) to female dogs, > 90% of the dose was excreted in urine, indicating extensive absorption in this species also. Radioactive components in the urine samples collected from rats and dogs during 0–6 hours after each dose were separated by HPLC using an ion exclusion column, and in every case only two such components were observed. The more abundant of these, which accounted for at least half of the dose, corresponded to unchanged (1), and the other was identified by ^{13}C NMR and HPLC comparison with the synthetic reference compound as the acid (2). In contrast to the results of some previous studies, no evidence was found for the presence of two-carbon metabolites resulting from ether cleavage; in particular, only traces of oxalic acid were detected in urine, suggesting that the toxic effects associated with diethylene glycol are not due to the formation of this acid metabolite.

Reference

J. M. Mathews, M. K. Parker, and H. B. Matthews, Metabolism and disposition of diethylene glycol in rat and dog, *Drug Metab. Dispos.*, 1991, **19**, 1066.

Valproic acid

Use/occurrence:	Antiepileptic drug
Key functional groups:	Isoalkyl carboxylic acid, fatty acid
Test system:	Rat liver mitochondrial preparations

Structure and biotransformation pathway

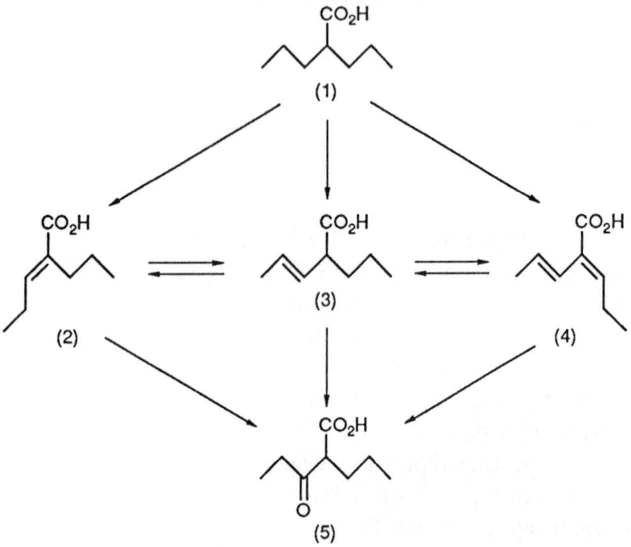

Incubation of valproic acid (1) with freshly isolated rat liver mitochondria led to the identification (by GC–MS) of the β-keto acid (5) and three unsaturated metabolites, (2)–(4). When incubations were carried out in a medium enriched with $H_2^{18}O$, metabolites were labelled in a manner consistent with the involvement of coenzyme A thioester intermediates. The β-keto acid (5) was the major metabolite. Separate incubations with synthetic metabolites showed that (2), (3), and (4) were interconverted and that all three served as precursors for (5). It was concluded that competition by (1) for enzymes or cofactors of the fatty acid β-oxidation complex might contribute to toxic effects observed *in vivo*.

Reference

S. M. Bjorge and T. A. Baillie, Studies on the β-oxidation of valproic acid in rat liver mitochondrial preparations, *Drug Metab. Dispos.*, 1991, **19**, 823.

Valproic acid (1), 2-Propylpent-4-enoic acid (2), (E)-2-Propylpenta-2,4-dienoic acid (3)

Use/occurrence:	(1) Antiepileptic agent; (2), (3) metabolites of (1)
Key functional groups:	Alkene, isoalkyl carboxylic acid, fatty acid
Test system:	Rat (intraperitoneal, 100 mg kg^{-1}), human pediatric patients

Structure and biotransformation pathway

(4) RS = Glutathionyl
(5) RS = N-Acetylcysteinyl

Rats were given intraperitoneal doses of valproic acid (1) or its unsaturated metabolites (2) and (3). After methylation, extracts of bile and urine were analysed by LC–MS–MS and GC–MS respectively. Identification of the conjugated metabolites was based on comparison with standards, the syntheses of which are described in the paper. The glutathione conjugate (4) was detected in the bile of rats treated with either (2) or (3). No mention was made of its detection in bile from animals treated with (1). The N-acetylcysteine conjugate (5) was stated to be a major (40% dose) metabolite in urine from rats receiving (3) and a prominent urinary metabolite of (2). Only very low levels of (5) could be detected in urine from animals treated with the parent compound (1). However, (5) was found to be a metabolite of (1) in the urine of 28 pediatric patients undergoing valproic acid therapy.

Reference

K. Kassahun, K. Farrell, and F. Abbott, Identification and characterization of the glutathione and N-acetylcysteine conjugates of (E)-2-propyl-2,4-pentadienoic acid, a toxic metabolite of valproic acid, in rats and humans, *Drug Metab. Dispos.*, 1991, **19**, 525.

Acrylonitrile

Use/occurrence:	Industrial chemical
Key functional groups:	Alkene, nitrile
Test system:	Rat and mouse (oral, 10 or 30 mg kg^{-1})

Structure and biotransformation pathway

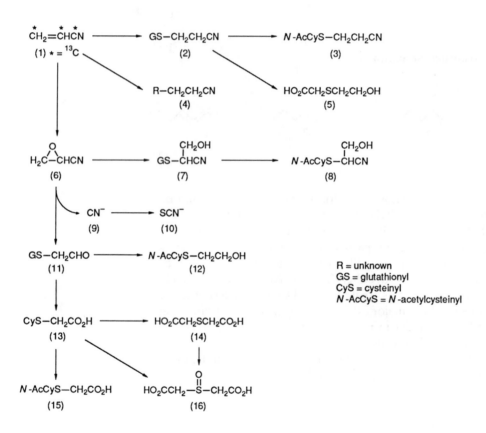

R = unknown
GS = glutathionyl
CyS = cysteinyl
N-AcCyS = N-acetylcysteinyl

Acrylonitrile (1) is a carcinogen in rats which is known to undergo extensive metabolism. In this study ^{13}C-enriched acrylonitrile was administered to both rats and mice to assist in the identification of urinary metabolites. The metabolites identified in both rat and mouse urine were (3), (8), (10), (12), (13) [or (15)], and (16). Metabolites were identified by a combination of ^{13}C and 2D NMR, and quantitated by integrating metabolic carbon resonances. The major metabolites were (14) and (13) or its N-acetyl derivative (15). Metabolites believed to be derived from the epoxide (6) accounted for ca. 60% of the products excreted in rat urine, compared with ca. 80% in mouse urine. It was postulated that the proportion of the dose metabolized via (6) may be an important determinant in the toxicity and carcinogenicity of (1).

Reference

T. R. Fennel, G. L. Kedderis, and S. C. J. Sumner, Urinary metabolites of [1,2,3-^{13}C]acrylonitrile in rats and mice detected by ^{13}C nuclear magnetic resonance spectroscopy, *Chem. Res. Toxicol.*, 1991, **4**, 678.

N,N'-Dimethylaminopropionitrile

Use/occurrence:	Manufacture of polyurethane foams
Key functional groups:	Alkyl nitrile, dimethylaminoalkyl
Test system:	Rat (oral, 525 mg kg^{-1}), rat liver, kidney, and bladder microsomes

Structure and biotransformation pathway

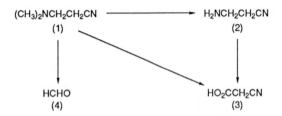

Following the oral administration of N,N'-dimethylaminopropionitrile (1) to rats at doses up to 525 mg kg^{-1} twice daily, *ca.* 44% of the dose was found to be excreted unchanged in the urine. The major metabolite product (2) was identified by TLC analysis, and metabolite (3) by a colorimetric assay.

Following incubation of (1) with microsomes from liver, kidney, and bladder, (3), formaldehyde (4), and cyanide were detected in the incubate. The presence of cyanide shows that (1) undergoes extensive oxidative degradation.

Reference

M. M. Mumtaz, M. Y. H. Farooqui, B. I. Ghanayem, and A. E. Ahmed, The urotoxic effects of N,N'-dimethylaminopropionitrile. 2. *In vivo* and *in vitro* metabolism, *Toxicol. Appl. Pharmacol.*, 1991, **110**, 61.

Benzyl selenocyanate, Dibenzyl diselenide

Use/occurrence:	Experimental antitumour agents
Key functional groups:	Benzyl, organoselenium
Test system:	Rat (oral, 12.5 mg kg^{-1}), rat liver 'S9' (0.25 mM substrate)

Structure and biotransformation pathway

Following a single oral dose of (1) or (2) to rats urine was the major excretory route. However, by 10 days following dosing only *ca.* 20% and 40% respectively of the dose was recovered in the urine. Faecal excretion of radioactivity in the same period accounted for less than 5% of the dose, and no radioactivity was detectable in expired air. Negligible amounts of radioactivity were recovered in excreta after 10 days. Following enzymic and/or acid hydrolysis of urine from rats dosed with (1), benzoic acid and hippuric acid were detected as the major extractable metabolites. From the enzymic hydrolysis results it was concluded that both sulfate and glucuronide conjugates of (7) and (8) were excreted in urine. Following acid hydrolysis of urine of rats dosed with (2), benzoic acid and hippuric acid were again detected as the major extractable metabolites suggesting that (1) and (2) shared a common fate. Further evidence suggesting a common fate was obtained following the *in vitro* incubations of (1), in which (2) was the major ethyl acetate soluble metabolite; (4) was also detected *in vitro* supporting the proposed metabolic scheme. The majority of the radioactivity following the *in vitro* incubations of (1) was recovered in the protein fraction. It was thus hypothesized that protein adducts containing Se—S or Se—Se bonds formed during the metabolism of (1) could explain the relatively low recovery of radioactivity *in vivo*. The formation of such protein adducts could play a key

81

role in the chemopreventative effects of (1). Metabolites were determined by HPLC assay.

Reference

K. El-Bayoumy, P. Upaddhyaya, V. Date, O. Sohn, E. S. Fiala, and B. Reddy, *Chem. Res. Toxicol.*, 1991, **4**, 546.

Rimantidine

Use/occurrence: Antiviral drug

Key functional groups: Alkyl amine, cycloalkyl

Test system: Rat (oral, 60 mg kg^{-1}; intravenous, 15 mg kg^{-1}), dog (oral, 5 and 10 mg kg^{-1}; intravenous, 5 mg kg^{-1})

Structure and biotransformation pathway

	R^1	R^2	R^3	R^4
(1)	NH$_2$	H	H	H
(2)	NH$_2$	OH	H	H
(3)	NH$_2$	H	H	OH
(4)	NH$_2$	H	OH	H
(5)	OH	OH	H	H
(6)	OH	H	H	OH
(7)	NO$_2$	H	H	OH

Following single oral or intravenous doses of [^{14}C]rimantidine [(1), as its hydrochloride salt] to rats and dogs, 81–87% of the dose was excreted in urine during 96 hours, indicating good absorption in both species. Radioactive components in urine were removed by solvent extraction at pH 11 before and after treatment with β-glucuronidase/sulfatase, then converted into their pentafluorobenzoyl derivatives for purification by HPLC and structural characterization by NMR, MS, and/or chromatographic comparison with reference compounds. In this manner, seven such components were identified [(1)–(7)]. Unchanged drug was excreted in small amounts as both the free and conjugated forms, demonstrating its extensive biotransformation before excretion, and its metabolism was essentially independent of the route of administration. The major metabolite (21–17% dose) in both species was unconjugated (2), but whereas rats also excreted a similar amount of (3) this metabolite was not found in dog urine; instead, dogs excreted (5) and (6), in which the amino group had been substituted by hydroxyl. In one hydroxylated metabolite, (7), the amino group had apparently been replaced by nitro, which was thought to have been formed by oxidation of the corresponding *N*-hydroxy compound during isolation. The main biotransformation pathways of (1) in these animal species were therefore similar to those reported previously for man, *viz.* *meta*- and *para*-hydroxylation, oxidation of the primary amino group, and *O*-glucuronidation/sulfation. Interestingly, whereas only *trans*-*p*-hydroxyrimantidine (3) was excreted by rats, humans excreted both *trans*- (3) and *cis*- (4) forms and dogs excreted neither.

Reference

A. C. Loh, A. J. Szuna, T. H. Williams, G. J. Sasso, and F.-J. Leinweber, The metabolism of [^{14}C]rimantidine hydrochloride in rats and dogs, *Drug Metab. Dispos.*, 1991, **19**, 381.

Methyl isocyanate

Use/occurrence:	Industrial chemical
Key functional groups:	Isocyanate
Test system:	Rat (intraperitoneal, 30 mg kg^{-1})

Structure and biotransformation pathway

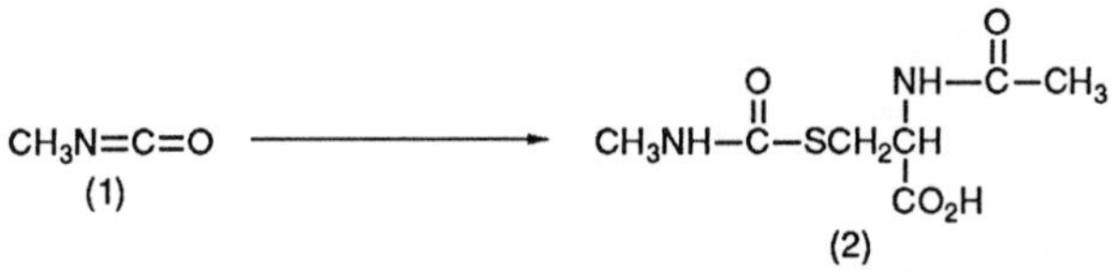

$$CH_3N=C=O \xrightarrow{\hspace{2cm}} CH_3NH-\underset{O}{\overset{O}{C}}-SCH_2\overset{NH-\overset{O}{C}-CH_3}{\underset{CO_2H}{CH}}$$

(1)   (2)

   Following a single intraperitoneal dose of methyl isocyanate (1) to rats, the 0–24 hour urinary excretion of the mercapturic acid metabolite (2) was shown to account for *ca.* 25% of the administered dose. Thus conjugation with glutathione and the subsequent conversion into the mercapturate (2) was shown to account for *ca.* 25% of the administered dose. Thus conjugation methyl isocyanate (1) in rats. The thermally labile mercapturate (2) was determined following methylation by LC–MS. The possible role of mercapturate formation in the long-term toxicity of (1) was discussed, with respect to the possible regeneration of free (1) from the mercapturate.

Reference

J. G. Slatter, M. S. Rashed, P. G. Pearson, D. Han, and T. A. Baillie, Biotransformation of methyl isocyanate in the rat. Evidence for glutathione conjugation as a major pathway of metabolism and implications for isocyanate-mediated toxicities, *Chem. Res. Toxicol.*, 1991, **4**, 157.

Ethyl carbamate

Use/occurrence: Model compound (carcinogen)

Key functional groups: Carbamate

Test system: Human liver microsomes (0.3 mM)

Structure and biotransformation pathway

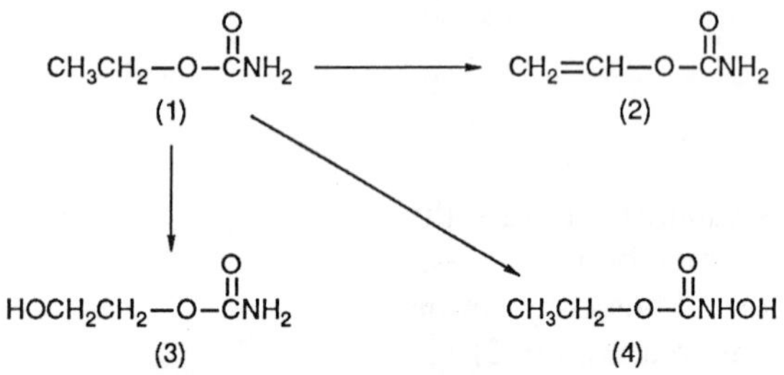

When human liver microsomes were incubated with (1) in the presence of NADPH the products (2)–(4) were detected by GC–MS or by HPLC co-chromatography. A large isotope effect was seen for the formation of (2) and (3) following incubation with the deuterated analogue of (1) with human microsomes.

Reference

F. P. Guengerich and D. Kim, Enzymic oxidation of ethyl carbamate to vinyl carbamate and its role as an intermediate in the formation of $1, N^6$-ethenoadenosine, *Chem. Res. Toxicol.*, 1991, **4**, 413.

Felbamate

Use/occurrence:	Anticonvulsant drug
Key functional groups:	Arylalkyl, carbamate
Test system:	Rat, rabbit, and dog (oral and intravenous, 1.6–300 mg kg^{-1})

Structure and biotransformation pathway

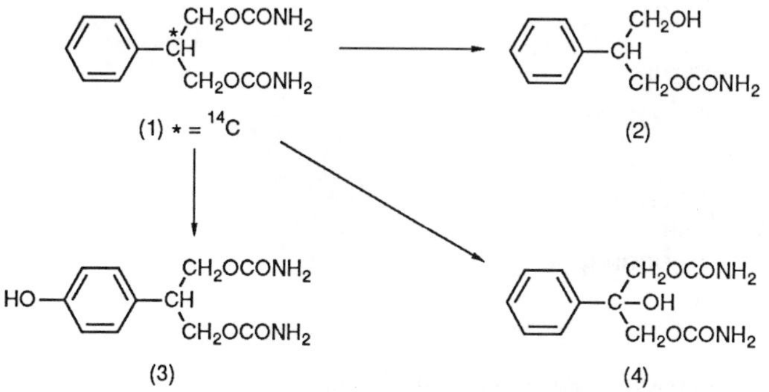

Animals dosed with [^{14}C]felbamate (1) excreted most (58–88%) of the radioactivity in the urine. The major Phase I pathways of metabolism in all three species were hydroxylation at the *para*-position of the aromatic ring (3) and hydroxylation at the benzylic carbon (4). The hydrolysis product (2) was a minor metabolite in the dog and rabbit and was not observed in the rat. Urine from all three species also contained a substantial proportion of unchanged (1). Metabolites were identified by EI- and CI-MS. Solvent extraction of urine before and after enzyme incubation indicated that a proportion of the urinary metabolites was excreted as conjugates, but these were not separately quantified or identified.

Reference

V. E. Adusumalli, J. T. Yang, K. K. Wong, N. Kucharczyk, and R. D. Sofin, Felbamate pharmacokinetics in rat, rabbit and dog, *Drug Metab. Dispos.*, 1991, **19**, 1116.
Felbamate metabolism in rat, rabbit and dog, *Drug Metab. Dispos.*, 1991, **19**, 1126.

Felbamate

Use/occurrence:	Anticonvulsant drug
Key functional groups:	Arylalkyl, carbamate
Test system:	Rat liver microsomes

Structure and biotransformation pathway

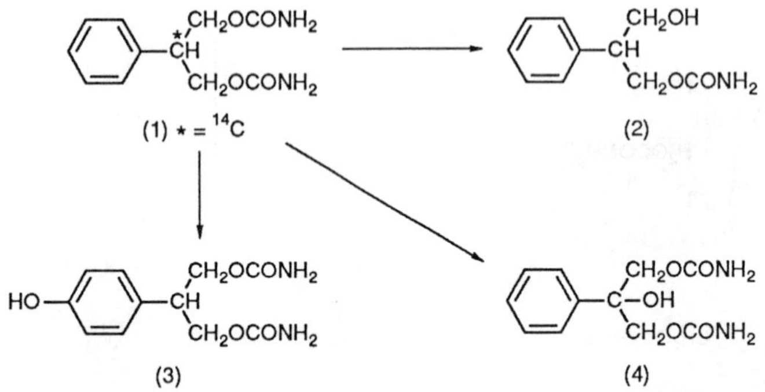

Incubation of [^{14}C]felbamate (1) with liver microsomes from untreated rats resulted in the formation of the hydroxylated metabolites (3) and (4) in an approximate ratio of 6:4. A small amount of the monocarbamate metabolite (2) was also formed. The use of microsomes from rats pretreated with (1) or phenobarbital led to an increase in the rate of metabolism of (1) which was mostly accounted for by increased hydroxylation at the benzylic carbon. Metabolites were assigned on the basis of HPLC comparison with reference standards.

Reference

V. E. Adusumalli, K. K. Wong, N. Kucharczyk, and R. D. Sofin, Felbamate *in vitro* metabolism by rat liver microsomes, *Drug Metab. Dispos.*, 1991, **19**, 1135.

Carbaryl

Use/occurrence:	Insecticide
Key functional groups:	Carbamate, naphthalene
Test system:	Rat liver and skin post-mitochondrial cell fractions, rat skin *in vitro*

Structure and biotransformation pathway

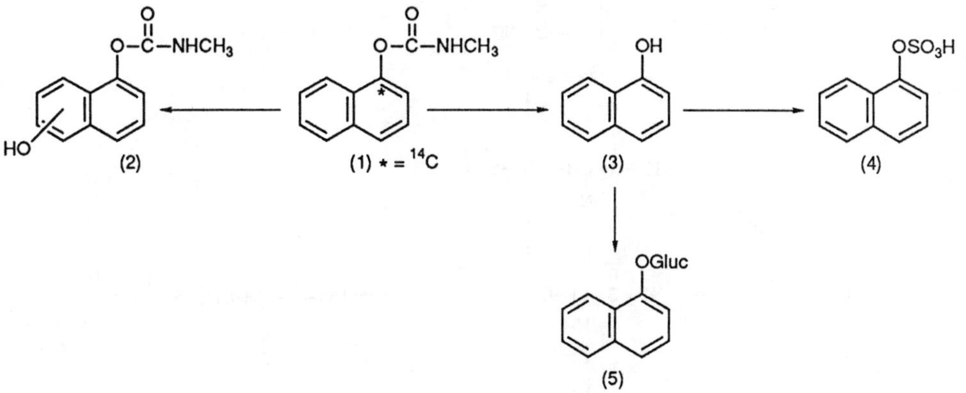

Both rat and skin post-mitochondrial fractions were shown to metabolize (1) by hydroxylation to produce (2), and by hydrolysis to (3) followed by conjugation to (4) and (5). Despite the capacity of the skin post-mitochondrial fraction to metabolize (1), no metabolism of (1) was detected when it was applied to whole skin in a passive diffusion cell model. The amount of skin present in the diffusion cell experiment was comparable to that used in the post-mitochondrial incubations, and it was anticipated that, had metabolism occurred, (3) or its conjugates would have been detected. The absence of any significant metabolism of (1) in intact skin could be attributed to its passage through skin via the intercellular route, since similar conditions had previously been shown to support the metabolism of aldrin to dieldrin during absorption. Metabolites were determined by a HPLC assay.

Reference

S. E. McPherson, R. C. Scott, and F. M. Williams, Fate of carbaryl in rat skin, *Arch. Toxicol.*, 1991, **65**, 594.

Sutan

Use/occurrence:	Herbicide
Key functional groups:	Dialkyl carbamate, thiocarbamate
Test system:	Rat (oral, 20 and 400 mg kg^{-1})

Structure and biotransformation pathway

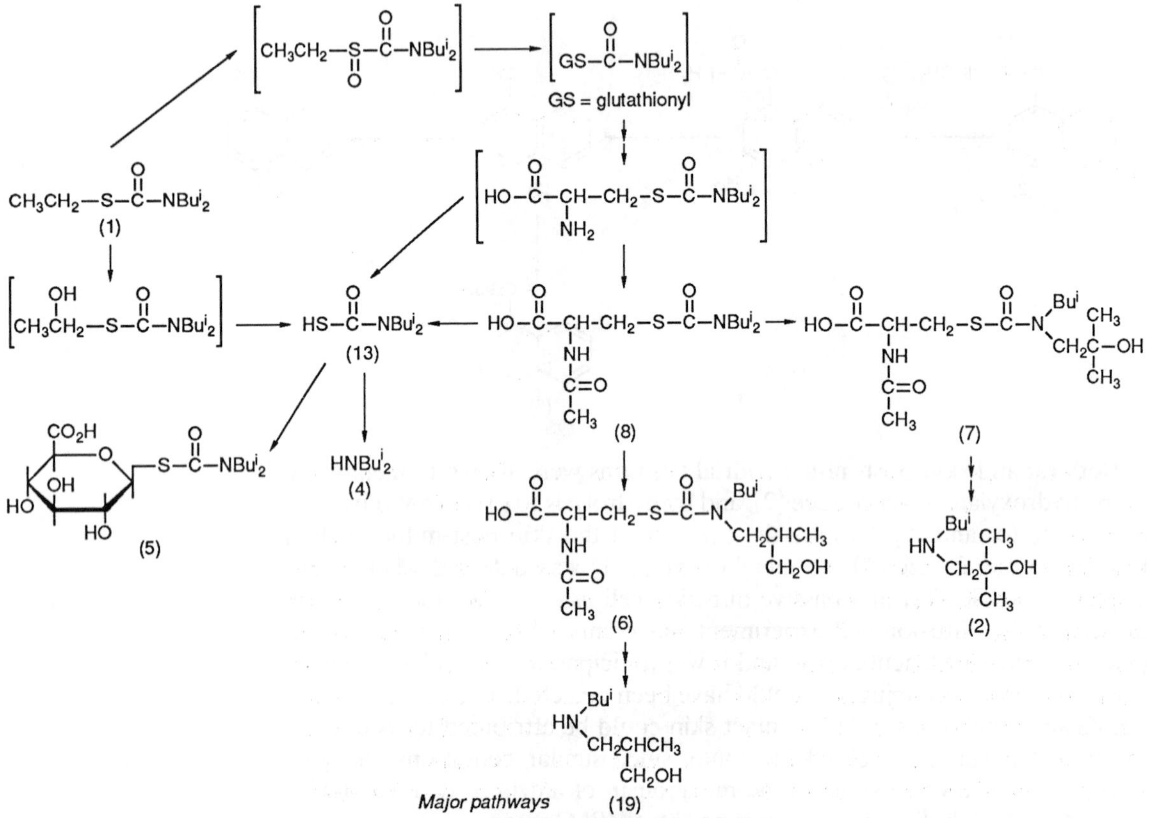

Major pathways

Sutan (*S*-ethyl-*N*,*N*-di-isobutyl thiocarbamate) (1) has been shown to be metabolized by sulfoxidation and subsequent glutathione conjugation in previous studies. However, the fate of the di-isobutylamine moiety and the identities of many minor metabolites have not been fully investigated. [^{14}C]-(1) was administered to male and female rats (and to male rats at a higher dose level) to elucidate the metabolic pathways for (1) in this species.

Following oral administration of [*isobutyl*-1-^{14}C]-(1), radioactivity was principally excreted in the urine in both male and female animals (79% of the dose in 72 hours). 5% of the dose (male) and 2% (female) was eliminated in the faeces and 3.9% (male) and 2.6% (female) was excreted as ^{14}CO$_2$.

Urinary metabolites were initially characterized by TSP–HPLC–MS and by co-chromatography with authentic standards on TLC. Subsequently acidic

Minor pathways

metabolites and *m*-toluyl derivatives of basic metabolites were further identified by GC–MS in the CI or EI mode.

At least 29 metabolites of greater than 0.2% of the 0–72 hour urinary
radioactivity were observed, 18 of which were identified. Identified metabolites represented 83.3–88.1% of the urinary radioactivity and the percentages
of total urinary ^{14}C which they represent are listed below. The major metabolite in the 0–72 hour urine in both male and female animals was di-isobutylamine (4) which accounted for 51% of the urinary ^{14}C. Other amine
metabolites observed were hydroxy compounds related to (4), the substituent
being present either on the primary carbon (19) (4–7%, male > female) or on
the tertiary carbon (2) (7–11%, male > female) of one isobutyl group. The
final amine metabolite was identified as isobutylamine (3) (1%). Metabolite
(5) (5–8%, female > male) was shown to be the *S*-glucuronyl conjugate of
(13), and metabolites (6)–(8) were identified as *N*-acetylcysteine conjugates
with either no hydroxyls (8) (1–3%, female > male) or one primary (6) (2%) or
tertiary (7) (4%) hydroxyl on an isobutyl group.

Minor metabolites of (1) included (9)–(12) and (18) which were cyclized
structures that resulted from an intramolecular reaction of a primary or
tertiary hydroxyl group with the carbonyl carbon of the thiocarbamate
moiety. Other minor metabolites identified were *N*,*N*-di-isobutylthio-
carbamic acid (13) and the oxidized metabolites (14)–(17). Each of these
metabolites represented 2% or less of the urinary radioactivity in both males
and females.

Reference

R. C. Peffer, D. D. Campbell, and J. C. Ritter, Metabolism of the thiocarbamate herbicide Sutan in rats, *Xenobiotica*, 1991, **21**, 1139.

91

IBI-P-01028

Use/occurrence:	Cytoprotective drug
Key functional groups:	Alkyl carboxylic acid, hydantoin
Test system:	Rat (oral, 150 mg kg^{-1}), dog (oral, 50 mg kg^{-1} day^{-1} for 7 days)

Structure and biotransformation pathway

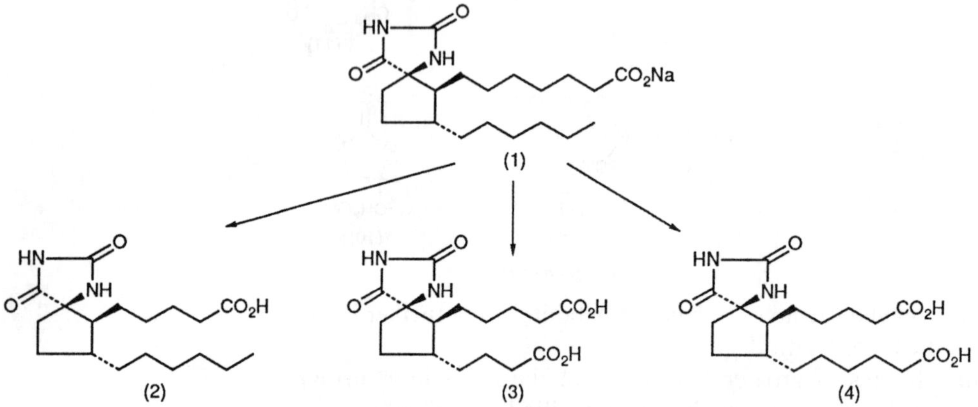

Drug-related components in solvent extracts of acidified urine and plasma from dogs treated with IBI-P-01028 (1) and of acidified urine from treated rats were analysed by GC–MS after suitable derivatization. Two such components were detected in dog plasma and urine, both of which were identified by GC and MS comparison with the corresponding reference compound. Thus, one corresponded to unchanged drug and the other to the side-chain-shortened metabolite (2). Two drug-related compounds were detected in rat urine, but neither corresponded to the parent drug or to the dog metabolite. Based on the MS data alone, these were identified as the side-chain-shortened dicarboxylic acid metabolites (3) and (4), and so demonstrate the different oxidative metabolic pathways in the two species.

Reference

E. Benfenati, R. Fanelli, E. Bosone, C. Biffi, R. Caponi, M. Cianetti, and P. Farina, Mass spectrometric identification of urinary and plasma metabolites of 6-(6′-carboxyhexyl)-7-n-hexyl-1,3-diazaspiro[4,4]nonan-2,4-dione, a new cytoprotective agent, *Drug Metab. Dispos.*, 1991, **19**, 913.

13-*cis* and all-*trans*-retinoic acids

Use/occurrence: Keratolytic drugs

Key functional groups: Alkene carboxylic acid, methylcyclohexene

Test system: Cynomolgus monkey (oral, 2 and 10 mg kg^{-1} day^{-1} for 10 days)

Structure and biotransformation pathway

This study was designed to obtain basic pharmacokinetic and metabolic information on 13-*cis*- and all-*trans*-retinoic acids [(1) and (2) respectively] in non-pregnant female monkeys in order to determine the suitability of this species for investigations into retinoid teratogenesis. To this end, plasma levels of each unchanged drug and its 4-keto [(3), (4)] and *O*-glucuronide [(5), (6)] metabolites were determined by HPLC on the first and last days of the 10 day dosing regime. Both compounds were administered at two dose levels, using the same four animals for each study. The identities of the glucuronide conjugates (5) and (6) in plasma were confirmed by thermospray LC–MS. Marked differences were found in both the plasma kinetics and metabolism of the two retinoic acids in monkeys. Thus, elimination of (2) was faster than that of (1), and the AUC for the former compound on day 10 was only one-third that on day 1, suggesting that (2) induces its own metabolism (in particular, its glucuronidation), whereas this effect was scarcely noticeable for (1). Furthermore, both (3) and (5) were major metabolites of (1), whereas the only important metabolite of (2) was (6); concentrations of both conjugates, however, increased during the treatment period. Low levels of isomerized metabolites were observed after administration of both retinoic acids. Serum samples were also obtained from a single female human subject undergoing treatment for acne with (1) $(0.7$ mg kg^{-1} day$^{-1})$. HPLC analysis showed that (3) was the most important metabolite therein, and that (5) was a new human metabolite; the isomerized compounds (2) and (4), which are possible teratogens, were also detected in human serum. Based on the results obtained from this study and others it is proposed that the mouse is the appropriate species for investigating the teratogenicity of (2), whereas for (1) the monkey may be of more use in this respect.

Reference

J. C. Kraft, W. Slikker, J. R. Bailey, L. G. Roberts, B. Fischer, W. Wittfoht, and H. Nau, Plasma pharmacokinetics and metabolism of 13-*cis*- and all-*trans*-retinoic acid in the cynomolgus monkey and the identification of 13-*cis*- and all-*trans*-retinoyl-β-glucuronides, *Drug Metab. Dispos.*, 1991, **19**, 317.

Sulprostone

Use/occurrence:	Prostaglandin E_2 analogue
Key functional groups:	sec-Alkyl alcohol, alkyl carboxamide, prostanoid, sulfonamide
Test system:	Female human volunteers (intravenous, 0.85 mg), guinea-pig (subcutaneous, 0.21 mg day^{-1}), isolated perfused guinea-pig liver

Structure and biotransformation pathway

The aim of this study was to identify the major metabolites of sulprostone (1) in human plasma. The guinea-pig was used to generate sufficient quantities of metabolites for characterization by GC–MS, ^{1}H NMR, and IR. The identified metabolites were then co-chromatographed with human plasma extracts. Four metabolites of sulprostone were thereby identified in human plasma, two major and two minor. One of the major metabolites (2) was the PGA analogue of the parent compound. The other (4) was probably formed by cyclization of the β-side-chain with the cyclopentenone ring, following reduction of the Δ^{13} double bond. The minor metabolites (3) and (5) were carboxylic acids derived from cleavage of the sulfonamide group from the parent compound or metabolite (2).

Reference

W. Kuhnz, G.-A. Hoyer, S. Backhus, and U. Jakobs, Identification of the major metabolites of the prostaglandin E_2-analogue sulprostone in human plasma, and isolation from urine (*in vivo*) and liver perfusate (*in vitro*) of female guinea pigs, *Drug Metab. Dispos.*, 1991, **19**, 920.

Tris(2-chloroethyl) phosphate

Use/occurrence:	Flame retardant
Key functional groups:	Alkyl phosphate, chloroalkyl
Test system:	Mouse, rat (oral, 175 mg kg^{-1})

Structure and biotransformation pathway

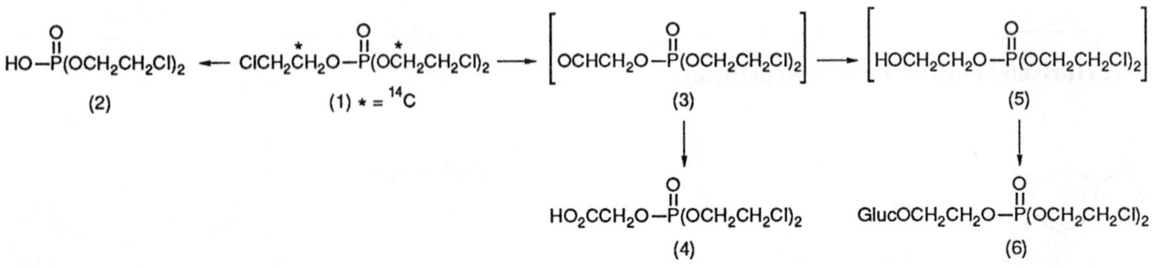

This study was designed to detect variations in the metabolism of tris(2-chloroethyl) phosphate (1) between male mice and male and female rats, which might explain observed species and sex differences in toxicity.

Following administration of [^{14}C]-(1), more than 75% dose was excreted in urine by both species within 24 hours, but the initial rate of excretion was faster in the male mouse (70% in 8 hours). Urinary metabolite profiles were qualitatively similar for the male mouse and both sexes of rat. Three metabolites were identified by ^{1}H NMR and FAB-MS. Three generally less important and more polar metabolites were unidentified. The carboxylic acid (4) was quantitatively by far the most important metabolite, particularly in the mouse, whilst the glucuronide conjugate (6) of the corresponding alcohol was the least important. The postulated aldehyde (3) and alcohol (5) intermediates were not observed. The metabolism of (1) was not induced or inhibited by nine daily 175 mg kg^{-1} doses.

Reference

L. T. Burka, J. M. Sanders, D. W. Herr, and H. B. Matthews, Metabolism of tris(2-chloroethyl) phosphate in rats and mice, *Drug Metab. Dispos.*, 1991, **19**, 443.

Cyclophosphamide, Phosphoramide mustard

Use/occurrence:	Anticancer drugs
Key functional groups:	N-Chloroethyl, phosphoramide
Test system:	Rabbit liver microsomes, cytosol and immobilized microsomal protein

Structure and biotransformation pathway

SG = glutathionyl

HPLC analysis of incubation mixtures of cyclophosphamide (1) and gluta-thione with rabbit liver cytosol, microsomes, or immobilized microsomal protein indicated the presence of a single reaction product. FAB-MS analysis of the isolated product showed that it corresponded to a monoglutathione conjugate containing one chlorine atom, *viz.* (3). Similarly, incubation of phosphoramide mustard (2) and glutathione with cytosol resulted in the formation of one major product which was identified as a diglutathione conjugate containing no chlorine atoms, *viz.* (4); in this case, however, no conjugate was formed during incubation with microsomes or immobilized microsomal protein, possibly due to its lipophobicity. The mechanism of formation of these two conjugates was investigated with the aid of suitable deuterium-labelled analogues of (1) and (2) and tandem MS. The results indicated that enzyme-mediated conversion of (1) into (3) involves direct replacement of the chlorine, as it does chemically, whereas both enzyme-catalysed and chemical (pH 10) transformation of (2) into (4) takes place via symmetrical aziridinium ions.

Reference

Z.-M. Yuan, P. B. Smith, R. B. Brundrett, M. Colvin, and C. Fenselau, Glutathione conjugation with phosphoramide mustard and cyclophosphamide, *Drug Metab. Dispos.*, 1991, **19**, 625.

Substituted Aromatic Compounds

Bromobenzene

Use/occurrence:	Model compound
Key functional groups:	Bromophenyl
Test system:	Rat (intraperitoneal, 2 mmol kg^{-1})

Structure and biotransformation pathway

Detailed analysis of the urinary metabolites from sodium phenobarbital-treated male Sprague–Dawley rats after the administration of bromobenzene (1) revealed that bromobenzene 2,3-oxide (2) is a discrete metabolite of (1) and not merely a hypothetical intermediate. This led to the isolation and characterization of *S*-(2-hydroxy-3-bromocyclohexa-3,5-dienyl)-*N*-acetyl-cysteine (3), but not its 2,6-regioisomer (4). Cytochrome P-448 catalysed oxidation of (1) afforded (2) and P-450 catalysed oxidation of (1) afforded (5). The epoxide (5) gave the regioisomeric premercapturic acids (6) and (7). Acid-catalysed dehydration of (6) and (7) (mixture) afforded only *para*-(8), whereas (3) gave *ortho*-(8) and *meta*-(8) which implies a sulfur atom rearrangement. Using HPLC, UV, 1-D and 2-D ^{1}H NMR, FAB-MS, and chemical degradation, the urinary premercapturic acid metabolites of (1) were isolated and characterized chemically and spectroscopically for the first time. However, it was not possible to determine whether the epoxides (2) and (5) were single diastereoisomers by the NMR techniques used, and the regioisomeric mercapturic acid metabolites (6) and (7) could not be separated by HPLC. Therefore, they were characterized following COSY-NMR and

101

parallel studies with specifically deuterated (1). The absence of the metabolite (4) in the urine of β-naphthoflavone pretreated rats, following intraperitoneal administration of (1) (3 mmol kg^{-1}), may imply a preference for glutathione S-transferase to conjugate at the less hindered C-3 position in bromobenzene 2,3-oxide (2). Some support for this proposal comes from the observation that 1,4-dihalogenobenzenes do not give rise to mercapturic acid metabolites in rats.

Reference

J. Zheng and R. P. Hanzlik, Premercapturic acid metabolites of bromobenzene derived via its 2,3-oxide and 3,4-oxide metabolites, *Xenobiotica*, 1991, **21**, 535.

1,2,4-Trichlorobenzene

Use/occurrence: Organic solvent

Key functional groups: Chlorophenyl

Test system: Rat liver microsomes

Structure and biotransformation pathway

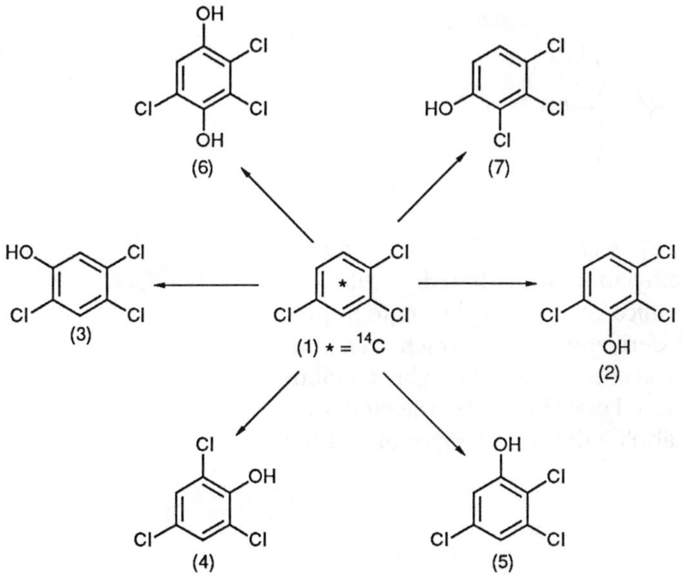

1,2,4-Trichlorobenzene (1) was incubated with microsomes prepared from rats pretreated with phenobarbital, 3-methylcholanthrene (MC), isosafrole, or dexamethasone. Control microsomes were also used. The metabolites formed were extracted and identified by HPLC comparison with standard compounds. The two major products were 2,3,6- and 2,4,5-trichlorophenol [(2) and (3)], together with 2,4,6- and 2,3,5-trichlorophenol [(4) and (5)] and trichlorohydroquinone (6), which were present in small amounts. 2,3,4-Tri-chlorophenol (7) was a significant metabolite only when microsomes from 3-MC-treated rats were used. The amount of (6) produced increased with increasing concentration of protein in the incubation as did the amount of covalently bound metabolites.

Reference

C. den Besten, M. C. C. Smink, E. J. de Vries, and P. J. van Bladeren, Metabolic activation of 1,2,4-trichlorobenzene and pentachlorobenzene by rat liver microsomes: a major role for quinone metabolites, *Toxicol. Appl. Pharmacol.*, 1991, **108**, 223.

3,4,5,3′,4′-Pentachlorobiphenyl

Use/occurrence:	Environmental contaminant
Key functional groups:	Chlorophenyl
Test system:	Rat (oral, 3 mg kg^{-1})

Structure and biotransformation pathway

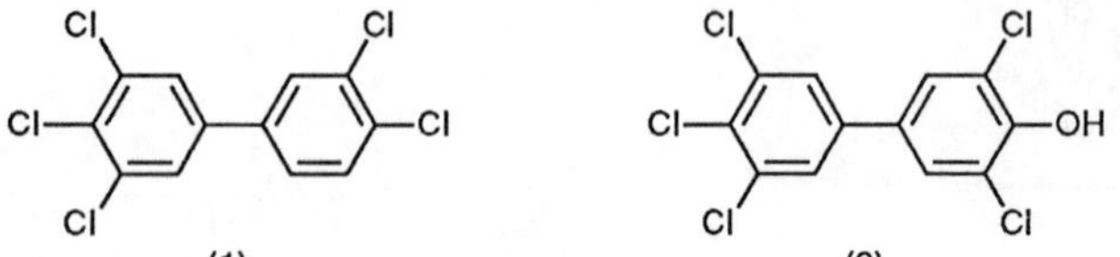

Metabolites of the pentachlorobiphenyl (1) were isolated from rat faeces by extraction with chloroform. After methylation with dimethyl sulfate a single metabolite was detected, purified by silica gel column chromatography, and identified by GC–MS as the methyl derivative of (2) which was confirmed by comparison with the authentic reference compound. This metabolite only represented *ca.* 1% of the administered dose in faeces collected for 5 days. Limited studies showed that the metabolite did not produce any of the toxic effects associated with PCB.

Reference

N. Koga, M. Beppu, and H. Yoshimura, Metabolism *in vivo* of 3,4,5,3′,4′-penta-chlorobiphenyl and toxicological assessment of the metabolite in rats, *J. Pharmacobio.-Dyn.*, 1990, **13**, 497.

Pentachlorophenol

Use/occurrence: Biocide

Key functional groups: Chlorophenyl, phenol

Test system: Rat (oral, 2.5 mg kg^{-1}; intravenous, 2.5 mg kg^{-1})

Structure and biotransformation pathway

(2)   (1)   (3)   (4)

Pentachlorophenol (1) intoxication has led to fatalities in humans after extensive exposure. Clinical signs of poisoning (fever, accelerated heart rate, profuse sweating) were consistent with the uncoupling of oxidative phosphorylation. The phenol (1) is carcinogenic in mice, but its widespread use makes it an environmental contaminant and occupational hazard. Therefore, the toxicokinetics were determined in male rats. The phenol (1) is conjugated with glucuronic acid (catalysed by UDP-glucuronyl transferase) to yield the ether glucuronide (2) or, after P-450-catalysed oxidative dechlorination to tetrachlorohydroquinone (3), to the corresponding mono-glucuronide (4). Following oral administration, absorption was complete (bioavailability 0.91–0.97), peak plasma concentration occurred at 1.5–2.0 hours, elimination half-life was 7.54 ± 0.44 hours, and clearance was essentially metabolic. There was some biliary excretion as 10% of the intravenous dose was recovered in the faeces. After an intravenous dose, metabolites (2) and (4) represented 27.4% and 30.5% dose respectively compared with 24.9% and 26.6% respectively after an oral dose (glucuronides hydrolysed with sulfuric acid). The metabolites were quantified using capillary GC.

Reference

B. G. Reigner, R. A. Gungon, M. K. Hoag, and T. N. Tozer, Pentachlorophenol toxico-kinetics after intravenous and oral administration to rats, *Xenobiotica*, 1991, **21**, 1547.

Phenol

Use/occurrence:	Industrial chemical
Key functional groups:	Phenol
Test system:	Lobster (intrapericardial, $0.02-3$ mg kg^{-1})

Structure and biotransformation pathway

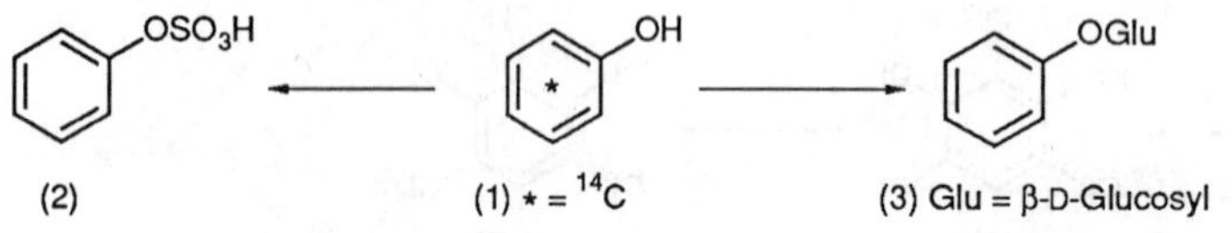

(2) (1) * = ^{14}C (3) Glu = β-D-Glucosyl

A proportion of the dose of [^{14}C]phenol (1), which increased to more than 90% of a 3 mg kg^{-1} dose, was excreted unchanged through the gills. Radioactivity was excreted in urine almost entirely as the sulfate conjugate (2), which was identified by FAB-MS, deconjugation experiments, and chromatographic comparison with a standard. A minor (0.1–0.3%) radioactive component in urine co-chromatographed with a standard of phenyl β-D-glucoside (3), and incubation of urine with β-glucosidase resulted in the formation of small amounts of phenol.

Reference

M. O. James, J. D. Schell, M. G. Barron, and C.-L. J. Li, Rapid dose-dependent elimination of phenol across the gills, and slow elimination of phenyl sulphate in the urine of phenol-dosed lobsters, *Homarus Americanus*, *Drug Metab. Dispos.*, 1991, **19**, 536.

o-, *m*-, and *p*-Hydroxybiphenyl

Use/occurrence:	Antifungal agents
Key functional groups:	Phenol
Test system:	Cytosol fractions from human liver, ileum, lung, colon, kidney, bladder, brain

Structure and biotransformation pathway

The glucuronidation and sulfation of hydroxybiphenyls have been studied in subcellular preparations of human liver and sulfation was shown to be the primary route of metabolism of (1) by this tissue. In contrast, glucuronidation and sulfation occurred at similar rates for (2) and (3).

In the present study, sulfotransferase activity using (1), (2), and (3) as substrates was studied in the human liver, ileum and colon mucosae, lung, kidney, urinary bladder, mucosa, and brain. The enzyme preparation used was the tissue cytosol and incubations were carried out in the presence of ^{35}S-labelled PAPS. Sulfotransferase activity was detectable in all the tissues studied. In a single tissue, the V_{max} values were similar for the three substrates but the K_m values were in the order of $(2) < (1) < (3)$. The V_{max}/K_m ratio, a measure of the optimal metabolizing capacity of the tissue, showed that the liver was the most efficient tissue in the sulfation of (1), (2), and (3). The V_{max}/K_m ratios for this reaction were liver > ileum > lung > colon > kidney > bladder > brain.

Reference

G. M. Pacifici, L. Vannucci, C. Bencini, G. Tusini, and F. Mosca, Sulphation of hydroxybiphenyls in human tissues, *Xenobiotica*, 1991, **21**, 1113.

Harmol, 1-Naphthol

Use/occurrence: Model compounds

Key functional groups: Phenol

Test system: Isolated rat intestinal loop *in situ*

Structure and biotransformation pathway

(1) R = H
(2) R = Gluc
(3) R = SO₃H

(4) R = H; * = ¹⁴C
(5) R = Gluc
(6) R = SO₃H

Harmol (1) (2, 20, and 200 μmol) and 1-naphthol (4) (0.1, 1, and 10 μmol) were separately administered to rats intraluminally into the isolated intestinal loop preparation *in situ*. The mesenteric venous (portal) blood was collected continuously for 1 hour and monitored for (1)–(3) by TLC with fluorimetric detection or for (4)–(6) by HPLC with UV absorbance or radioactivity detection. The results showed that both substrates were rapidly and extensively metabolized in the intestine before absorption. Formation of the glucuronide conjugates (2) and (5) was rapid and represented the predominant intestinal biotransformation pathway in each case, but evidence was found for saturation at the high dose level. Cumulative formation of these conjugates was similar for all doses of (1) and for the lower doses of (4). On the other hand, the extent of formation of both sulfate conjugates (3) and (6) increased with dose level and equalled that of glucuronidation at the high dose level. The results therefore demonstrate dose-dependent intestinal biotransformation of the two phenols in the rat *in situ*. The nature of this dependency is similar to that reported previously for paracetamol and both contrast to the normal results found for Phase II metabolism of phenolic substrates in the liver, where dose dependency is characterized by the predominance of sulfation at low dosages and the extent of glucuronidation increasing with dose level.

Reference

D. Goon and C. D. Klaassen, Intestinal biotransformation of harmol and 1-naphthol in the rat, *Drug Metab. Dispos.*, 1991, **19**, 340.

Propofol

Use/occurrence:	Anaesthetic
Key functional groups:	Arylalkyl, phenol
Test system:	Rat (intravenous, 10 mg kg^{-1}), dog (intravenous, 10 mg kg^{-1}), rabbit (intravenous, 7.2 mg kg^{-1})

Structure and biotransformation pathway

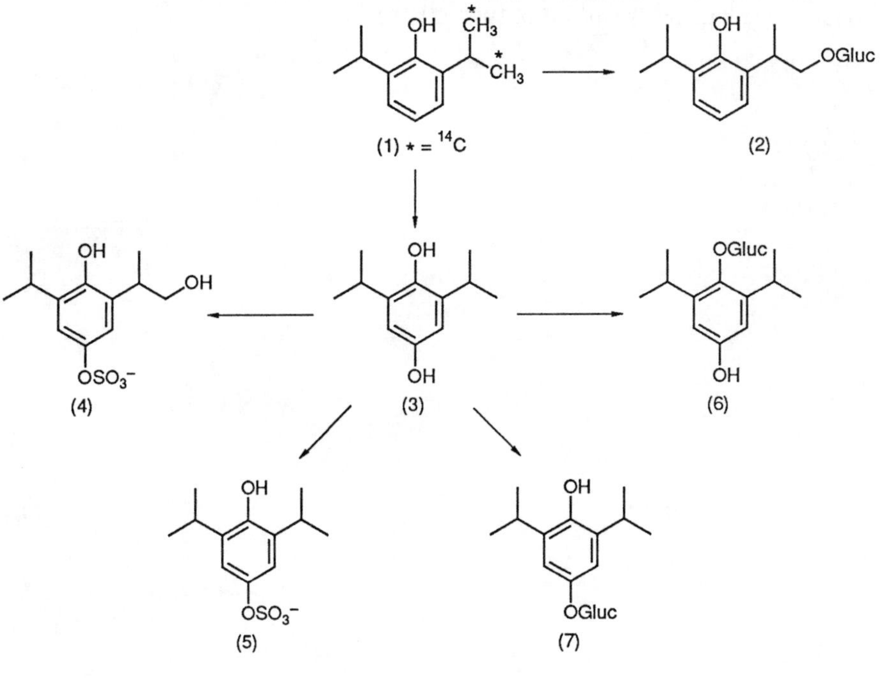

Propofol (1), a lipophilic, sterically hindered alkylphenol, is metabolized primarily by glucuronide and sulfate Phase II conjugation processes, by aromatic ring hydroxylation, and by alkyl group oxidation to primary alcohols. Bolus intravenous doses to rat, dog, or rabbit were eliminated primarily in the urine (60–95% of the dose); faecal elimination occurred in rat and dog (13–31%), but this was minimal in rabbit (<2%). Propofol (1) formed a sulfate in rabbit (4% of the dose in 24 hour urine), and a glucuronide in dog (2%) and in rabbit (12%). Phase I oxidation to the primary alcohol, followed by glucuronidation, afforded (2) in rabbit (6%). *para*-Hydroxylation to the quinol metabolite (3), which was not itself detected, afforded the sulfate (4) in rat (16%) and rabbit (14%), the sulfate (5) in rat (30%), dog (20%), and rabbit (5%), and the regioisomeric glucuronides (6) and (7) [dog (2%) and rabbit (4%); rat (9%), dog (20%), and rabbit (29%)

109

respectively]. No unchanged (1) was detected in the urine, and only traces were detected in the bile. The biliary metabolites were almost entirely glucuronides and not sulfates, which is consistent with the molecular weight threshold for anion excretion in rats of 325 Da. Analytical methods included liquid scintillation counting, HPLC, TLC, GC–MS, NMR, and acid and enzyme hydrolysis.

References

P. J. Simons, I. D. Cockshott, E. J. Douglas, E. A. Gordon, S. Knott, and R. J. Ruane, Distribution in female rats of an anaesthetic intravenous dose of ^{14}C-propofol, *Xenobiotica*, 1991, **21**, 1325.

P. J. Simons, I. D. Cockshott, E. J. Douglas, E. A. Gordon, S. Knott, and R. J. Ruane, Species differences in blood profiles, metabolism and excretion of ^{14}C-propofol after intravenous dosing to rat, dog and rabbit, *Xenobiotica*, 1991, **21**, 1243.

3-t-Butyl-4-hydroxyanisole, 2-t-Butylhydroquinone

Use/occurrence:	Antioxidant in food
Key functional groups:	Aryl tert-alkyl, methoxyphenyl, phenol
Test system:	Rat (intraperitoneal, 400 mg kg^{-1}), rat liver microsomes

Structure and biotransformation pathway

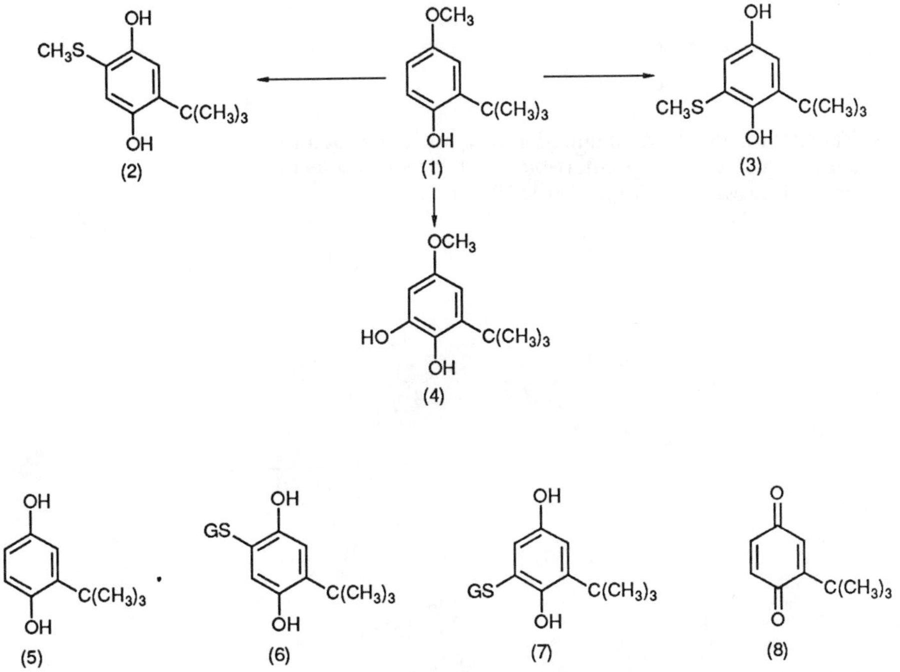

GS = glutathionyl

The purpose of the present study was to investigate the possible formation of sulfur-containing metabolites of 3-t-butyl-4-hydroxyanisole (1), since this compound is known to undergo extensive *O*-demethylation *in vivo*, and the resulting hydroquinone would be potentially capable of being further oxidized to the quinone, which, in turn, may react with sulfhydryl and other nucleophiles. GC–MS analysis of solvent extracts of acidified urine from rats treated with (1) showed the presence of two products, and these were identified as the methylthio metabolites (2) and (3) by GC and MS comparison with the synthetic reference compounds. Similarly, solvent extracts of urine previously incubated with *β*-glucuronidase/sulfatase were shown to contain three such products, which were identified as (2)–(4). Metabolites (2) and (3)

111

were also found in the urine of rats dosed with 2-t-butylhydroquinone (5), but none of the above metabolites accounted for more than 1% of the dose of (1) or (5) in 0–48 hour urine. Incubation of (5) with rat liver microsomes in the presence of an NADPH-generating system and GSH (reduced glutathione) was shown by HPLC to result in the formation of two main metabolites; these were isolated and further purified by HPLC, and identified as the glutathione conjugates (6) and (7) by ^{1}H and ^{13}C NMR and FAB-MS. The formation of these metabolites was shown to be dependent on the presence of NADPH, molecular oxygen, and GSH, and to be inhibited by the presence of SKF 525-A and metyrapone. These results suggest that (5) is first converted by cytochrome P-450-mediated mono-oxygenases into the reactive quinone intermediate (8), which then conjugates with GSH. The use of cytosol rather than microsomes yielded essentially no GSH conjugates, and addition of cytosol fraction to the microsomes did not noticeably enhance their formation, so GSH S-transferases were apparently not involved in the conjugation process.

Reference

K. Tajima, M. Hashizaki, K. Yamamoto, and T. Mizutani, Identification and structure characterization of S-containing metabolites of 3-tert-butyl-4-hydroxyanisole in rat urine and liver microsomes, *Drug Metab. Dispos.*, 1991, **19**, 1028.

Quercetin, Catechin

Use/occurrence:	Plant flavonoids
Key functional groups:	Catechol, phenol
Test system:	Isolated perfused rat liver

Structure and biotransformation pathway

(1)   (2)

The metabolism of the plant flavonoids quercetin (1) and (+)-catechin (2) have been studied using the isolated perfused rat liver in the presence of ^{35}S-labelled Na_2SO_4. The ^{35}S-labelled sulfates were located (following separation by TLC) by autoradiography and quantified by removal of appropriate areas from the thin layer plate followed by scintillation counting. This system produced three sulfated metabolites from each of (1) and (2): these were secreted into the bile and the perfusate. Two double conjugates, containing sulfate and glucuronic acid, were formed from (1) by the perfused rat liver along with a metabolite containing sulfate alone; catechin gave one such double conjugate and two sulfates. The glucuronide conjugates were identified following β-glucuronidase digestion.

The results of the perfusion studies show that (1) and (2) behave as typical polyhydroxy compounds in that they form not only sulfate esters but also doubly conjugated molecules containing both sulfate and glucuronide. It was possible that (2) also formed a bis-sulfate. Simple glucuronide conjugates must also have been formed, but these would not have been detected in the present study.

Reference

N. A. Shali, C. G. Curtis, G. M. Powell, and A. B. Roy, Sulphation of the flavonoids quercetin and catechin by rat liver, *Xenobiotica*, 1991, **21**, 881.

(±)-*trans*-2-(3'-Methoxy-5'-methylsulfonyl-4'-propoxyphenyl)-5-(3'',4'',5''-trimethoxyphenyl)-tetrahydrofuran (L-659,989)

Use/occurrence:	PAF (platelet activating factor) antagonist
Key functional groups:	Alkoxyphenyl, methoxyphenyl, tetrahydrofuran
Test system:	Rhesus monkey (oral and intravenous, 10 mg kg^{-1})

Structure and biotransformation pathway

The major routes of metabolism and excretion of [^{3}H]-L-659,989 (1) were studied in male and female rhesus monkeys following either intravenous or oral administration. The drug was well absorbed in both sexes based on plasma radiolabel measurements but the bioavailability was low (≤ 10%), indicative of extensive metabolism. Approximately 80% of the radiolabelled dose was excreted in the urine and faeces collected up to 96 hours (*ca.* 40% of the dose excreted by each route), with the largest percentage of the tritiated dose (31%) in the 0–24 hour urine.

The drug was rapidly metabolized in the monkey, predominantly at the C-4'-propoxy side-chain. The metabolites were identified by comparison with authentic chemical standard [metabolite (2)] or by ^{1}H NMR [metabolites (3) and (4)] following separation by HPLC. The two major plasma metabolites were identified as the *trans*-4'-(2-hydroxypropoxy) derivative (2) and the 4'-hydroxy metabolite (3) which was isolated as a 2:1 mixture of (±)-*trans*:(±)-*cis*. However, it is unknown whether the isomerization of (3) occurred after excretion or *in vivo*. The major metabolites in the excreta were

114

(3) and the glucuronide conjugate of the *trans*-isomer of (3), metabolite (4). Metabolite (4) accounted for 21% and 4% of the dose excreted in the urine following intravenous and oral administration respectively. Metabolite (3) was excreted in both urine and faeces, and accounted for $\leq 0.1\%$ and 7.4% of the dose in intravenous and orally dosed monkeys, respectively.

The identification of the C-4'-propoxy side-chain as a primary site of metabolism should assist future synthetic efforts to improve the metabolic stability of this class of PAF antagonists.

Reference

K. L. Thomson, M. N. Chang, R. L. Bugianesi, M. M. Ponpipom, B. H. Arison, H. B. Hucker, B. M. Sweeney, S. D. White, and J. C. Chabala, Metabolism of the platelet activating factor antagonist ($\pm$)-*trans*-2-(3'-methoxy-5'-methylsulphonyl-4'-propoxyphenyl)-5-(3'',4'',5''-trimethoxyphenyl)tetrahydrofuran (L-659,989) in rhesus monkeys, *Xenobiotica*, 1991, **21**, 613.

Pentachlorothioanisole (PCTA), 1,4-Bis(methylthio)tetrachlorobenzene (MTTCB), Terbutryn

Use/occurrence:	Model compounds
Key functional groups:	Chlorophenyl, methyl thioether
Test system:	Immobilized rat liver microsomal enzymes, rat (oral, PCTA 5 mg, MTTCB 5 mg, terbutryn 2 mg)

Structure and biotransformation pathway

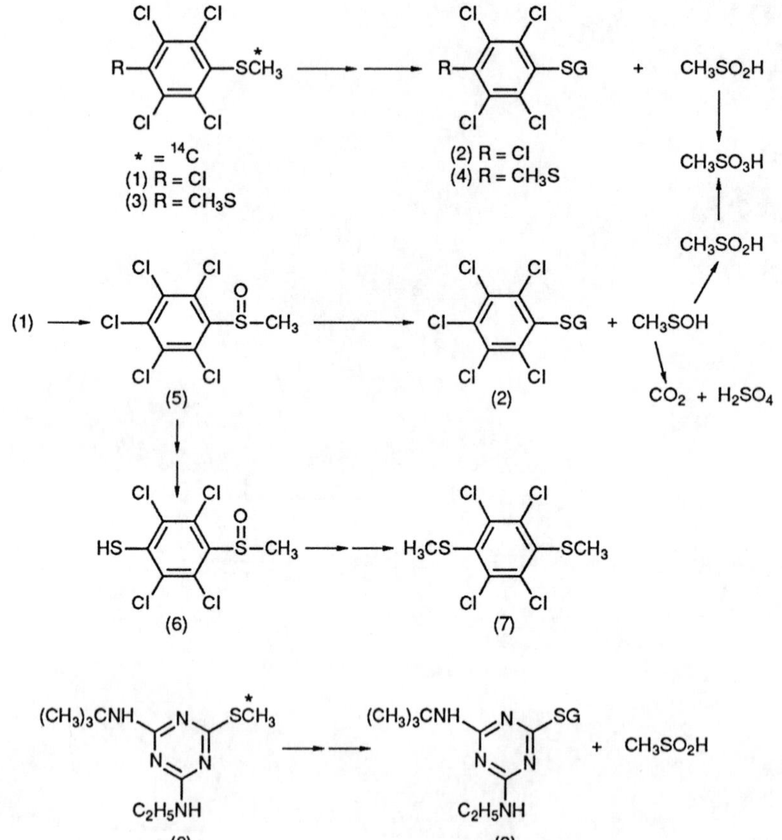

Pentachlorothioanisole (PCTA) (1) and 1,4-bis(methylthio)tetrachlorobenzene (MTTCB) (3) are biotransformed by oxidation (32% of the dose) and by substitution with glutathione (GSH) into the corresponding tripeptide conjugates (2) and (4) (7%) respectively, together with methanesulfinic and ultimately methanesulfonic acids. If the oxidative product derived from (1) is

116

the sulfoxide (5), rather than the sulfone, a different metabolic pathway may be followed: the sulfoxide may be displaced as methanesulfenic acid, the lowest oxidation congener, which is oxidized to sulfate and carbon dioxide as well as methanesulfinic acid. However, these important displacement pathways compete in (5) with conjugation at C-4 by GSH. If (1) undergoes oxidation to (5) and GSH conjugation at C-4, peptidase processing, and then C—S bond lysis, the sulfoxide-thiophenol (6) may be subsequently methylated and reduced to the bisthioether (7) (2%) where only one thiomethyl group was present in the starting material and the other has been derived from cysteine (S atom) and presumably from methionine (methyl group). Terbutryn (8) gave an analogous GSH conjugate (9) and methanesulfinic acid via oxidation (59%) and a substitution reaction. Full TLC, positive ion FAB-MS, and ^{1}H and ^{13}C NMR data are given, together with information on the synthesis of reference compounds and enzyme immobilization experimental details. The *in vitro/in vivo* comparisons were analysed and the displacement products were shown to be qualitatively similar although quantitatively different. Several of these GSH conjugation reactions occur spontaneously *in vitro* and presumably *in vivo* as well.

Reference

J. K. Huwe, V. J. Feil, J. E. Bakke, and D. J. Mulford, Studies on the displacement of methylthio groups by glutathione, *Xenobiotica*, 1991, **21**, 179.

Pentachlorothioanisole

Use/occurrence:	Model compound
Key functional groups:	Chlorophenyl, methyl thioether
Test system:	Rat (oral, 5 mg)

Structure and biotransformation pathway

$$\overset{*}{C}H_3\overset{\dagger}{S}-\overset{\ddagger}{C}_6Cl_5$$

$$(1)\ * = {}^{13}C$$
$$\dagger = {}^{35}S$$
$$\ddagger = {}^{14}C$$

$$\left[\begin{array}{c}\overset{*}{C}H_3\overset{\dagger}{S}-\overset{\ddagger}{C}_6Cl_4-SCH_3 \\ \text{and} \\ CH_3S-\overset{\ddagger}{C}_6Cl_4-SCH_3\end{array}\right]$$

$$(2)$$

$$\left[\begin{array}{c}\overset{*}{C}H_3-\overset{\dagger}{S}O_2H\ \ (3) \\ \overset{\dagger}{S}O_4{}^{2-} \\ \overset{*}{C}O_2\end{array}\right]$$

$$\overset{*}{C}H_3\overset{\dagger}{S}-\overset{\ddagger}{C}_6Cl_5 \longrightarrow CH_3\overset{O}{\overset{\|}{S}}-C_6Cl_5 \longrightarrow CH_3\overset{O}{\underset{O}{\overset{\|}{\underset{\|}{S}}}}-C_6Cl_5$$

$$(1)$$

$$CH_3S-C_6Cl_4-SG \qquad CH_3\overset{O}{\overset{\|}{S}}-C_6Cl_4-SG \qquad CH_3\overset{O}{\underset{O}{\overset{\|}{\underset{\|}{S}}}}-C_6Cl_4-SG$$

GS = glutathionyl

$$\overset{*}{C}H_3-\overset{\dagger}{S}O_2H\ \ (3)$$
$$\overset{\dagger}{S}O_4{}^{2-}$$
$$\overset{*}{C}O_2$$

$$\left[\begin{array}{c}\overset{*}{C}H_3\overset{\dagger}{S}-\overset{\ddagger}{C}_6Cl_4-SCH_3 \\ \text{and} \\ CH_3S-\overset{\ddagger}{C}_6Cl_4-SCH_3\end{array}\right]\ \left[\begin{array}{c}\overset{*}{C}H_3\overset{\dagger}{S}-\overset{\ddagger}{C}_6Cl_4-H \\ \\ CH_3S-\overset{\ddagger}{C}_6Cl_4-H\end{array}\right]\ +\ CH_3-\overset{\ddagger}{C}_6Cl_5$$

$$(2) \qquad\qquad\qquad\qquad\qquad\qquad\qquad (1)$$

Pentachlorothioanisole (1) is a metabolite of hexachlorobenzene and pentachloronitrobenzene in the rat. It has also been shown to undergo further metabolism to bis(methylthio)tetrachlorobenzene (2) in this species. The persistence of (2) in the environment is thought to be due to its unique metabolism which involves the turnover of a methylthio group.

The conversion of (1) into (2) was studied following oral administration of ^{13}C/^{14}C/^{35}S-labelled (1) to male rats. Metabolites were separated by TLC, GC, and HPLC and identified by FAB-MS. Metabolites of (1) identified in the urine were methanesulfinic acid (3), inorganic sulfate, and carbon dioxide in expired air.

The data suggest that although (1) is metabolized to (2) in part via displacement of a chlorine by glutathione, the major pathway may be oxidation at the sulfur atom of (1). The corresponding sulfoxide and sulfone may then undergo glutathione conjugation, subsequent hydrolysis to the cysteine conjugate, cleavage by cysteine conjugate β-lyase, S-methylation, and reduction, yielding metabolite (2).

Reference

D. J. Mulford, J. E. Bakke, and V. J. Feil, Glutathione-S-transferase-mediated methyl-thio displacement and turnover in the metabolism of pentachlorothioanisole by rats, *Xenobiotica*, 1991, **21**, 597.

3-[2-(*N*,*N*-Dimethylaminomethyl)phenyl-thio]phenol

Use/occurrence:	Anticholinergic agent
Key functional groups:	Diaryl thioether, dimethylaminoalkyl
Test system:	Rat (oral, 100 mg kg^{-1}), mouse, rabbit, dog (oral, 50 mg kg^{-1})

Structure and biotransformation pathway

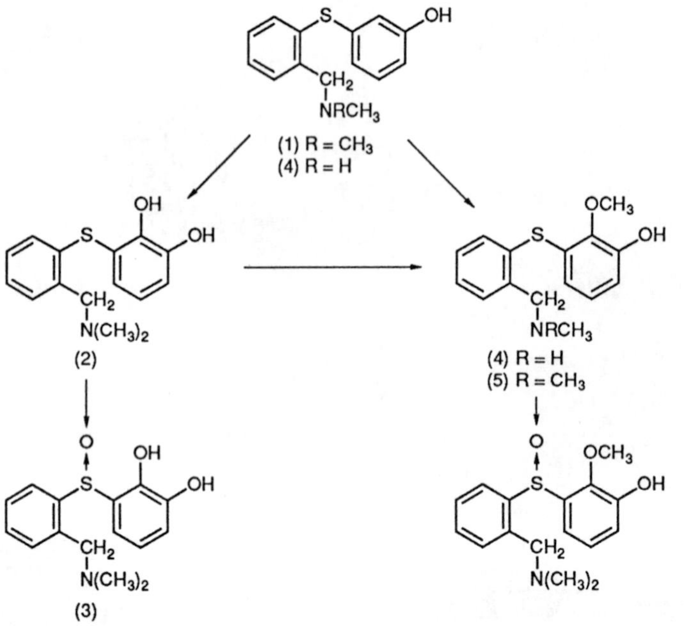

Urine samples collected after administration of oral doses of (1) to animals were extracted with benzene and ethyl acetate at pH 7 and 9. Metabolites were separated by TLC and detected by comparison with control urine extracts. Similar extracts were prepared after treatment of aqueous phases with β-glucuronidase. Isolated metabolites were analysed by MS. Up to six metabolites (2)–(6) were identified, derived by *N*-demethylation, ring-hydroxylation, and *S*-oxidation. All these metabolites except (3) were present in rat urine. Components (1) and (6) were present in mouse urine, (1), (3), (4), and (6) in dog urine, and (1), (6), and (7) in rabbit urine. The metabolites were also reported to be present as conjugates but no quantitative data were presented except a statement that the highest amounts were detected in rat urine.

Reference

M. Ryska, I. Koruna, L. Polakova, and E. Svatek, Mass spectrometric structure determination of metabolites of 3-[2-(*N,N*-dimethylaminomethyl)phenylthio]phenol, *J. Chromatogr.*, 1991, **562**, 289.

Pentamidine

Use/occurrence:	Antiparasitic drug
Key functional groups:	Alkyl aryl ether, amidine
Test system:	Rat liver microsomes

Structure and biotransformation pathway

Pentamidine [1,5-bis(4-amidinophenoxy)pentane (1)], has been used for the treatment of several parasitic diseases including *Pneumocystis carinii* pneumonia, an opportunistic infection seen in a number of AIDS victims.

Following incubation of (1) with rat liver microsomes, seven putative metabolites were separated by HPLC. Metabolites were identified by co-elution on HPLC with synthetic standards and comparison of putative metabolites and standards using LSIMS (liquid secondary ion MS). The two major metabolites identified were the pentan-2-ol and pentan-3-ol analogues of (1), metabolites (2) and (3) respectively. The other metabolites were present in much smaller amounts than (2) and (3). Two of the minor metabolites were identified as the mono- and di-hydroxy metabolites of (1), metabolites (4) and (5) respectively. The possibility of hydroxylation on the α-carbon of the alkane chain was investigated and 4-hydroxybenzamidine (thought to be formed following hydrolysis of the resultant hemiacetal compound) was identified as metabolite (6). Two other minor peaks in the incubation mixture were unidentified.

Experiments conducted using MS–MS could not identify metabolites of (1) unequivocally, since (2), (3), and (4) all gave the same fragmentation patterns. In addition standard (6) gave the same fragmentation pattern as part

of the standard (5). It would thus be desirable to identify the metabolites of
(1) by additional means.

Reference

B. J. Berger, V. V. Reddy, S. T. Le, R. J. Lombardy, J. E. Hall, and R. R. Tidwell,
  Hydroxylation of pentamidine by rat liver microsomes, *J. Pharmacol. Exptl.
  Therap.*, 1991, **256**, 883.

1,3-Dinitrobenzene

Use/occurrence:	Industrial chemical
Key functional groups:	Nitrophenyl
Test system:	Rat, hamster (intraperitoneal, 25 mg kg^{-1})

Structure and biotransformation pathway

(1) $* = {}^{14}C$ (2) (3) (4) (5) (6) (7)

The objective of this study was to test the hypothesis that species differences in susceptibility to testicular toxicity and methaemoglobinemia induced by 1,3-dinitrobenzene (1) could be explained by differences in its metabolism and pharmacokinetics. HPLC was used for analysis of metabolites in blood and urine and GC–MS for identification.

Rat and hamster each excreted *ca.* 80% of an intraperitoneal dose of [^{14}C]-(1) in urine during 48 hours. All of the metabolites in the scheme except (4) were detected in enzyme-hydrolysed urine from both species. Metabolite (4) was detected in blood, together with (2) and (6). Biotransformation of (1) involved reduction of the nitro groups, followed by acetylation, and aromatic hydroxylation. The formation of (3) was considered to involve deamination of an aniline species by intestinal microflora. Analysis of untreated urine indicated that nitroaniline (2) was excreted both as a free metabolite and as a conjugate, whereas the phenolic metabolites were excreted only as conjugates. Identified metabolites represented *ca.* 74% urinary radioactivity in both species and urinary metabolite profiles were qualitatively similar. Quantitative differences (percentage urinary radioactivity) concerned metabolite (6), which was more important in the rat (26%) than in the hamster (10%), and metabolite (5) which was more important in hamster (21%) than in the rat (9%). Metabolites (2) (14–15%) and (7) (23–26%) were of similar importance in both species while (3) was a minor metabolite (2–3%). Rat blood contained higher levels of nitroaniline than hamster blood and the authors tentatively concluded that the most likely metabolic explanation for species differences

124

in toxicity was the apparently greater ability of the rat to reduce the first nitro group of 1,3-dinitrobenzene.

Reference

S. F. Mceuen and M. G. Miller, Metabolism and pharmacokinetics of 1,3-dinitrobenzene in the rat and the hamster, *Drug Metab. Dispos.*, 1991, **19**, 661.

4-Nitrophenol, 4-Nitroanisole

Use/occurrence: Model compounds

Key functional groups: Nitrophenyl, phenol

Test system: Perfused rabbit ear
(0.1–200 μM at 150 μl
$\min^{-1} g^{-1}$)

Structure and biotransformation pathway

(2)   (1) * = ^{14}C (3)   (4)

4-Nitrophenol (1) and 4-nitroanisole (2) were appied either dermally or arterially to isolated perfused rabbit ears. The anisole (2) yielded only the Phase II conjugate metabolites of the phenol (1). The apparent V_{max} values for the glucuronidation and sulfation of (1) were 20 and 10 pmol $\min^{-1} cm^{-2}$ respectively. The analytical methods included UV detection, either directly at 400 nm or after separation of the metabolites by HPLC, and comparison with authentic standards after incubation with β-glucuronidase/arylsulfatase. As no Phase I metabolite of (2) [e.g. (1) or reduction of the nitro functional group to 4-anisidine or 4-aminophenol] was detected, it was concluded that the conjugation of (1) was rapid. Only the glucuronide (3) and the sulfate (4) were detected. The substrate concentration for the Phase I O-demethylation of (2) to (1) was sub-optimal, as the solubility of (2) was limited to 150 μM in the buffer.

Reference

B. M. Henrikus, W. Breuer, and H. G. Kampffmeyer, Dermal metabolism of 4-nitrophenol and 4-nitroanisole in single-pass perfused rabbit ears, *Xenobiotica*, 1991, **21**, 1229.

p-Chloronitrobenzene

Use/occurrence:	Model compound
Key functional groups:	Chlorophenyl, nitrophenyl
Test system:	Rat (intraperitoneal, 100 mg kg^{-1})

Structure and biotransformation pathway

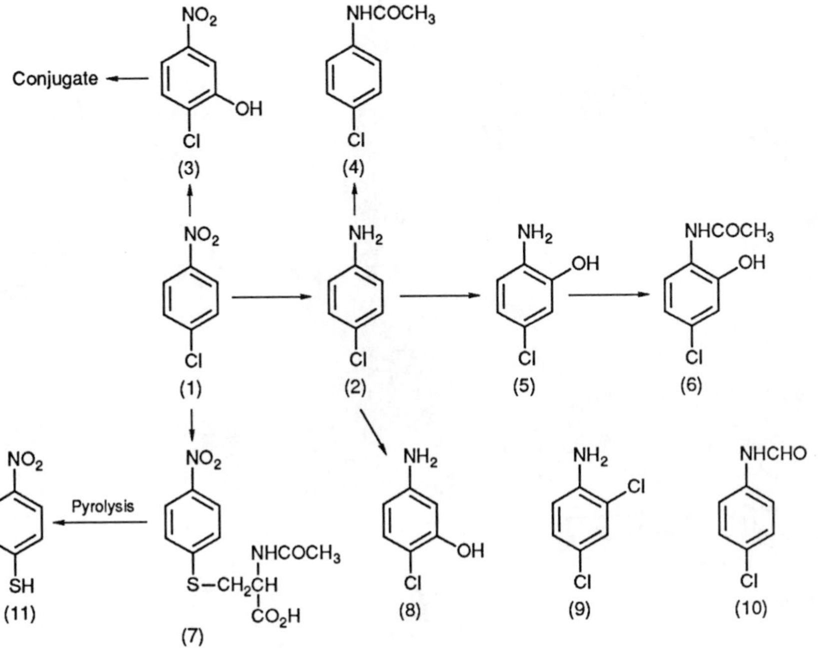

Urine was collected from 8 to 24 hours following administration of a single intraperitoneal dose of *p*-chloronitrobenzene (1). Urinary metabolites of (1) were identified by GC–MS and by comparison with authentic standards. Urinary metabolites were extracted with diethyl ether at pH 1.0 and pH 10.0 from urine samples hydrolysed with either acid or base and from untreated urines. Six metabolites [(2)–(6) and (8)] were found in urine. Unchanged (1), together with three probable artefacts, *p*-nitrothiophenyl (11), 2,4-dichloro-aniline (9), and *p*-chloroformanilide (10), were also detected in urine samples. The detection of *p*-nitrothiophenyl (11) may have been due to the mercapturate (7) which was found to be pyrolysed in the injection port of the GC. The detection of *p*-chloroformanilide (10) may have been due to another metabolite which was pyrolysed in the injection port, whereas 2,4-dichloroaniline (9) was probably formed on sample preparation. 2-Chloro-5-nitrophenol (3) was

127

believed to be mainly excreted as a conjugate. No absolute quantitation of any of the metabolites was given.

Reference

T. Yoshida, K. Andoh, and T. Tabuchi, Identification of urinary metabolites in rats treated with *p*-chloronitrobenzene, *Arch. Toxicol.*, 1991, **65**, 52.

5-t-Butyl-2,4,6-trinitroxylene (musk xylene)

Use/occurrence:	Fragrance chemical
Key functional groups:	Alkylphenyl, arylmethyl, nitrophenyl
Test system:	Rat (oral, 70 mg kg^{-1})

Structure and biotransformation pathway

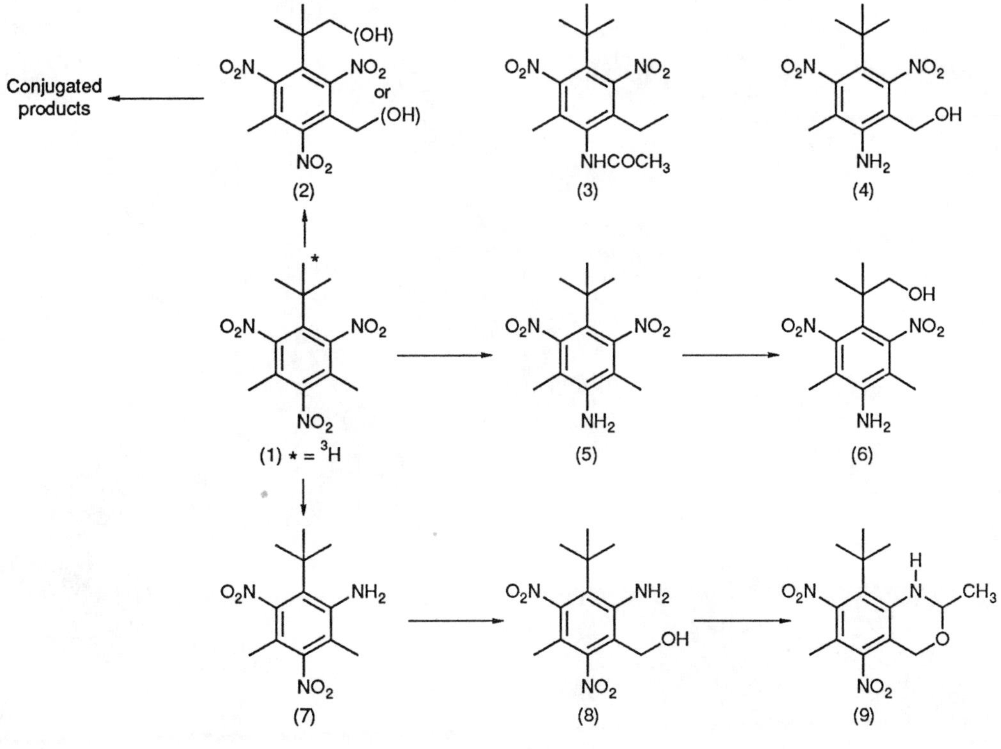

Following the single oral administration of [^{3}H]-(1), 10% and 70% of the dosed radioactivity was recovered in urine and faeces respectively by 7 days, and the total residue in tissues accounted for less than 2% of the dose. The highest concentration of radioactivity was in adipose tissue, and the next highest concentration was found in the liver.

After purification by TLC or column chromatography some of the metabolites were identified by GC–MS and NMR; other metabolites were presumably identified following TLC co-chromatography. Unchanged (1) and the metabolites (3)–(6) were found in faeces, bile, and urine. Metabolites (7) and (9) were present in both urine and faeces, metabolite (8) was present in urine, and metabolite (2) was present in bile. The major route of excretion of (1) and its metabolites was the faeces via bile. The reduction of the 2-nitro group of (1)–(5) was a key step in the metabolism of (1).

Reference

K. Minegishi, S. Nambaru, M. Fukuoka, A. Tanaka, and T. Nishimaki-Mogami, Distribution metabolism and excretion of musk xylene in rats, *Arch. Toxicol.*, 1991, **65**, 273.

Nitecapone

Use/occurrence:	Catechol-O-methyl transferase inhibitor
Key functional groups:	Alkenyl ketone, nitrophenyl, phenol
Test system:	Human volunteers (oral, 100 mg)

Structure and biotransformation pathway

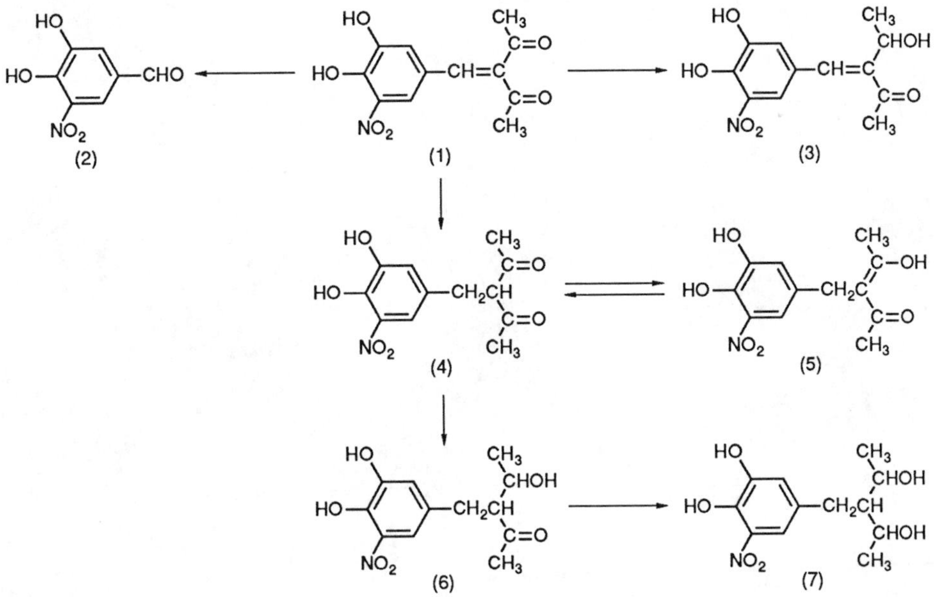

Metabolites of nitecapone (1) were isolated from urine both as intact conjugates and as free aglycones after hydrolysis with β-glucuronidase. The Phase I metabolites were identified by EI-MS, ^{1}H NMR, and IR. The majority of the Phase I metabolites were formed by reduction of the alkene bond or the carbonyl groups. The exception, metabolite (2), was thought to be an artefact formed on storage of urine. Two separate metabolites were assigned structure (6) and were assumed to be diastereoisomers. All the Phase I metabolites were present in urine mainly as their glucuronide conjugates. Conjugates were also isolated and were identified by FAB-MS. Only one glucuronide was observed for each Phase I metabolite implying regioselectivity of the conjugation reaction and stereoselectivity of the reduction reactions. Quantitatively the glucuronide of unchanged nitecapone was the most important urinary metabolite, representing an estimated 60–65% of all metabolites found. The absence of metabolites resulting from nitro group

131

reduction, catechol methylation, or sulfate conjugation was considered to be unusual.

Reference

J. Taskinen, T. Wikberg, P. Ottoila, L. Kanner, T. Lotta, A. Pippuri, and R. Bäckström, Identification of major metabolites of the catechol-*O*-methyltransferase-inhibitor nitecapone in human urine, *Drug Metab. Dispos.*, 1991, **19**, 178.

Ibuprofen

Use/occurrence:	Anti-inflammatory drug
Key functional groups:	Arylalkyl, aryl propionic acid, chiral carbon
Test system:	Rat (oral, 15 mg kg^{-1})

Structure and biotransformation pathway

The purpose of this study was to investigate further the process by which R-ibuprofen (1) undergoes metabolic chiral inversion *in vivo* to its pharmacologically more active S-enantiomer (2) in the rat. In particular, deuterium-labelled analogues and quantitative stereoselective GC–MS analysis of rat plasma were used to show that removal of the metabolically labile C-2 hydrogen of the propionic acid side-chain is not the rate-limiting step in the chiral inversion process, and that this process proceeds reversibly via a symmetrical intermediate. After oral administration of a 1:1 mixture of (1) and R-[^{2}H$_5$]ibuprofen (3) to rats, plasma levels of the S-isomers exceeded those of the R-isomers at all times after 0.3 hours. Furthermore, the S-ibuprofen consisted of equal proportions of the unlabelled (2) and tetradeuterated (5) species only, thereby demonstrating that inversion involved the loss of a single deuterium atom which was not subject to a measurable kinetic isotope effect and so was not the rate-limiting step. Confirmation that this deuterium atom was located at C-2 of the acid side-chain was obtained from a separate experiment involving the administration of RS-[*ring*-^{2}H$_4$]ibuprofen to rats and analysing the plasma as before, when it was found that all four ring deuteriums were retained. The R-ibuprofen in the plasma of rats treated with the mixture of (1) and (3) consisted not only of unlabelled and pentadeuterated species, but also an appreciable proportion of the tetradeuterated species (4), which indirectly confirms the involvement of a symmetrical intermediate in the inversion process. The mechanism proposed for the chiral inversion of (1) to (2) *in vivo* is shown, the key stages of which invoke stereoselective formation of the coenzyme A thioester metabolite of *R*-ibuprofen (6) and its subsequent reversible enzyme-catalysed transformation into the symmetrical enolate tautomer (7).

Reference

S. M. Sanins, W. J. Adams, D. G. Kaiser, G. W. Halstead, J. Hosley, H. Barnes, and T. A. Baillie, Mechanistic studies on the metabolic chiral inversion of *R*-ibuprofen in the rat, *Drug Metab. Dispos.*, 1991, **19**, 405.

Naproxen

Use/occurrence:	Anti-inflammatory drug
Key functional groups:	Aryl propionic acid, chiral carbon
Test system:	Isolated cells from guinea-pig liver, stomach, intestine, and colon

Structure and biotransformation pathway

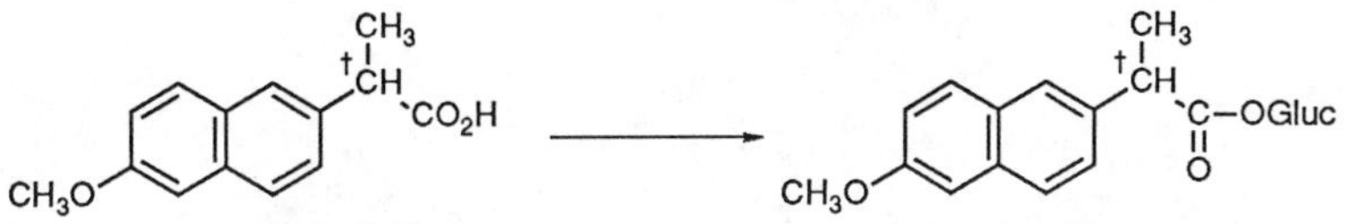

(1) † = chiral carbon   (2)

The objective of this study was to investigate the comparative stereoselectivity of the glucuronidation of racemic naproxen (1) by gastrointestinal cells. Metabolite analysis was by HPLC.

The rate of the conjugation reaction in liver cells was about thirty times greater than in intestinal cells. In all three types of gastrointestinal cell the rate of formation of S-naproxen glucuronide was greater, with gastric cells showing the highest S/R ratio (7.5) and the small intestine the lowest (1.7). In contrast liver cells preferentially formed the R-naproxen glucuronide, with an S/R ratio of 0.5.

Reference

M. El Mouehli and M. Schwenk, Stereoselective glucuronidation of naproxen in isolated cells from liver, stomach, intestine and colon of the guinea pig, *Drug Metab. Dispos.*, 1991, **19**, 844.

Carprofen, Flunoxaprofen, Naproxen

Use/occurrence:	Anti-inflammatory drugs
Key functional groups:	Aryl propionic acid
Test system:	Rat (intravenous, 11 μmol kg^{-1})

Structure and biotransformation pathway

(1) R =

(2) R =

(3) R =

These studies were designed to investigate the stereoselective disposition of the non-steroidal anti-inflammatory drugs carprofen (1), flunoxaprofen (2), and naproxen (3) following intravenous administration of each racemate and enantiomer to intact or bile duct-cannulated rats. Concentrations of the enantiomers of (1)–(3) in plasma, urine, and bile were determined by quantitative HPLC using S-($-$)-α-methylbenzylamine as a pre-column derivatizing agent. Concentrations of the corresponding acyl glucuronide conjugate of each compound in urine and bile were determined from the difference in drug concentration in untreated and base-hydrolysed samples. The pharmacokinetic results showed that the total clearances of the R-enantiomers of (1) and (2) were significantly greater than those of the corresponding S-antipodes, whereas the clearance of R-(3) was similar to that of S-(3), but there were no appreciable differences in the volume of distribution at steady-state between the R- and S-enantiomers for any of the three compounds investigated. Interestingly, the R- to S-enantiomer inversion ratio for (2) was 0.54 in rats, but only 0.003 and 0.02 for (1) and (3) respectively. Excretion of (1)–(3) in urine as the parent enantiomer or glucuronide conjugate during 5 or 8 hours post-administration was less than 0.5% of the dose. Excretion in the bile, on the other hand, was quite extensive and stereoselective in each case. Thus, biliary excretions of R-(1) and its glucuronide conju-

136

gate were greater than those of the corresponding *S*-enantiomers, whereas the reverse situation obtained for the optical isomers of (2), (3), and their conjugates. The results illustrate well, therefore, the wide variability in the extent of metabolic chiral inversion of the 2-arylpropionic acid NSAIDs (and the difficulty in correlating this with chemical structure), and in the stereoselective disposition of their acyl glucuronide conjugates.

Reference

S. Iwakawa, H. Spahn, L. Z. Benet, and E. T. Lin, Stereoselective disposition of carprofen, flunoxaprofen and naproxen in rats, *Drug Metab. Dispos.*, 1991, **19**, 853.

R-Flamprop-methyl

Use/occurrence:	Herbicide
Key functional groups:	Aryl carboxamide, chiral carbon, methyl ester
Test system:	Rat (oral, 4 mg kg^{-1})

Structure and biotransformation pathway

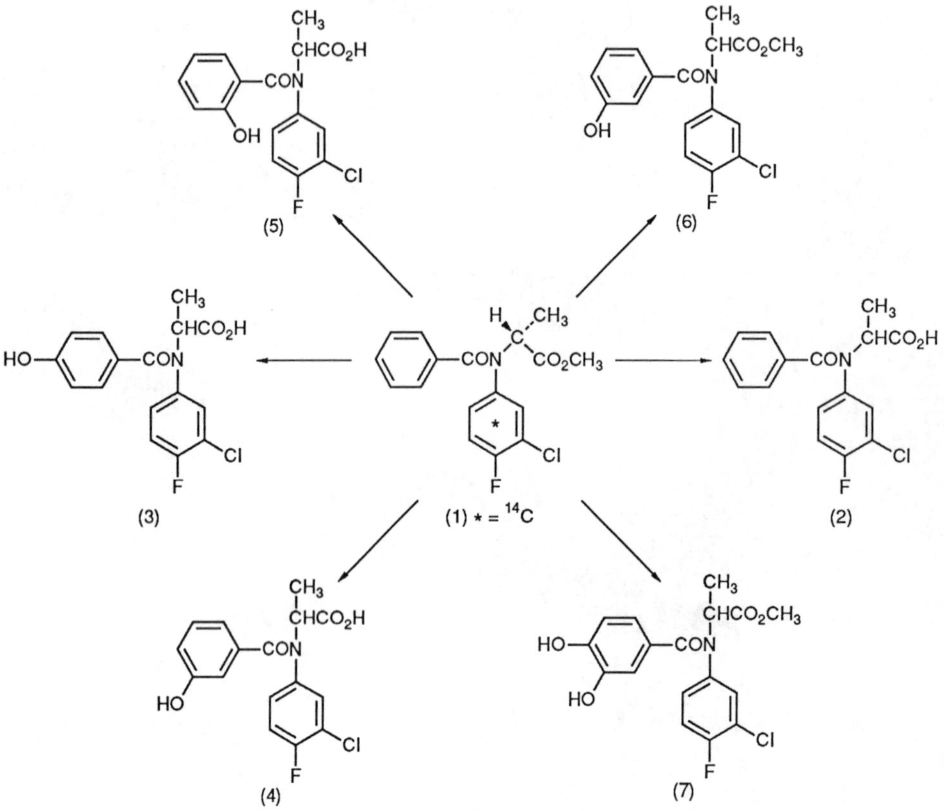

(5)

(6)

(3)

(1) * = ^{14}C

(2)

(4)

(7)

Following the oral administration of R-flamprop-methyl (1) to rats the dose was rapidly eliminated with the faeces being the major route of excretion. The acid (2) was the principal metabolic product excreted mainly in the faeces of males and the urine of females. The other metabolites which were identified by TLC and HPLC were the 4-, 3-, and 2-hydroxy acids (3)–(5), 3,4-dihydroxyflamprop-methyl (7), 3-hydroxyflamprop-methyl (6), and unchanged (1).

Examination of the enantiomer ratio of (2) excreted following dosing with the R-enantiomer of (1) showed that metabolic inversion had taken place as (2) contained *ca.* 13% of the S-enantiomer.

138

Reference

D. H. Hutson, C. J. Logan, and B. Taylor, The metabolism in the rat of the herbicid-ally active isomer (*R*)-flamprop-methyl in comparison with the racemic form, *Pestic. Sci.*, 1991, **31**, 151.

Ethyl 2-carbamoyloxybenzoate

Use/occurrence: Anti-inflammatory analgesic (pro-drug)

Key functional groups: Alkyl ester, aryl carboxylate, carbamate

Test system: Hepatic post mitochondrial supernatant from rat, rabbit, and dog

Structure and biotransformation pathway

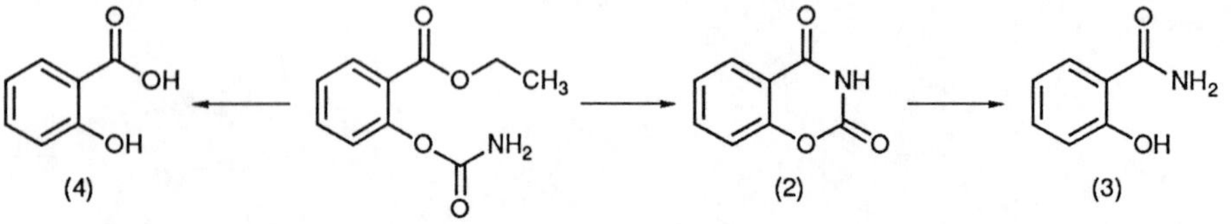

Ethyl 2-carbamoyloxybenzoate (1) is a pro-drug of salicylic acid (4), carsalam (2), and salicylamide (3). The metabolism of (1), to the three metabolites (2)–(4) was studied using rat, rabbit, and dog hepatic post mitochondrial supernatants. The addition of potassium fluoride to the incubations inhibited the formation of (3). Metabolites were determined by HPLC and confirmed by NMR and MS.

Reference

A. Kamal, Metabolism of ethyl 2-carbamoyloxybenzoate (4003/2), a prodrug of salicylic acid, carsalam and salicylamide, *Biochem. Pharmacol.*, 1990, **40**, 1669.

t-Butyl bicyclo-orthobenzoate

Use/occurrence: Insecticide

Key functional groups: Orthoester

Test system: Rat (intravenous, intragastric, 33–187 ng kg^{-1})

Structure and biotransformation pathway

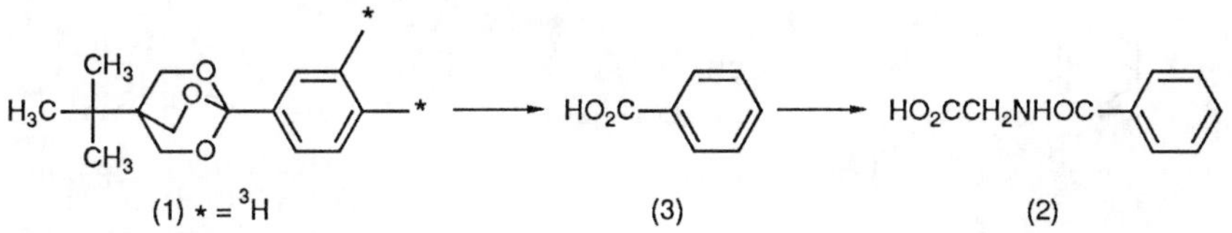

(1) ∗ = ^{3}H   (3)   (2)

Bicyclo-orthocarboxylates such as (1) are neurotoxicants which antagonize GABA-mediated inhibitory neurotransmission. The metabolism of [^{3}H]-(1) was investigated in the rat following intravenous (i.v.) or intragastric (i.g.) administration. The bulk of the metabolites from i.g. administration (*ca.* 85% of the dose) was excreted in the urine within 24 hours. Metabolites were separated by TLC. The most polar fraction in the urine was a mixture which was tentatively identified as hippuric acid (2) and benzoic acid (3): these metabolites also predominated in the faeces. Components (2) and (3) were the major metabolites following either i.v. or i.g. administration, typically accounting for 91–99% of the radioactivity present in either urine or faeces.

Reference

R. Zierer and J. Seifert, *Tert*-Butylbicycloortho[^{3}H]benzoate (^{3}H-TBOB) toxico-kinetics and disposition in rats, *Xenobiotica*, 1991, **21**, 839.

Kawain

Use/occurrence:	Analgesic and muscle-relaxant drug
Key functional groups:	Arylalkene, lactone
Test system:	Human (oral, 200 mg)

Structure and biotransformation pathway

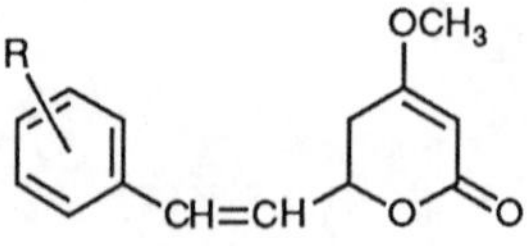

(1) R = H
(4) R = OH

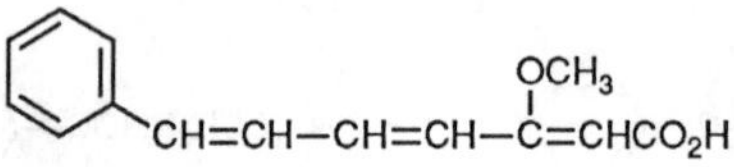

(2)

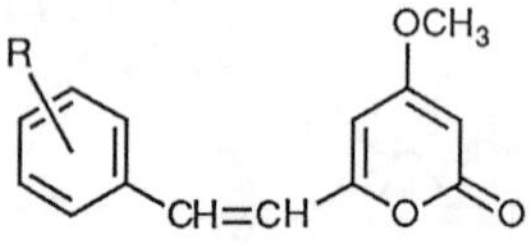

(3) R = H
(6) R = OH

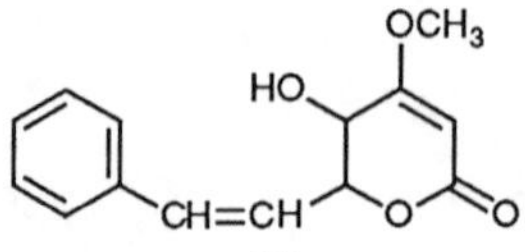

(5)

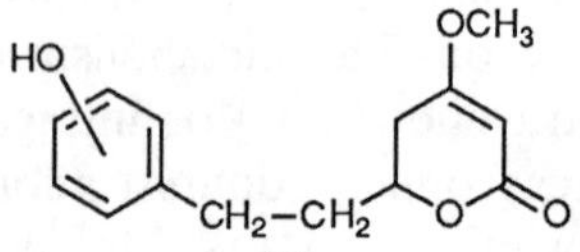

(7)

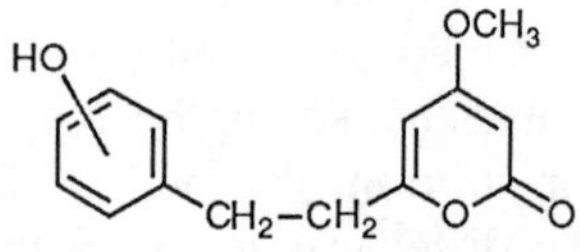

(8)

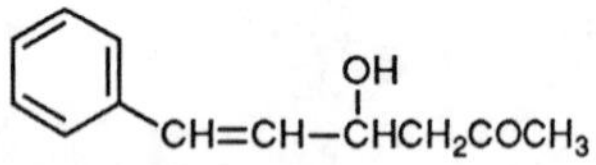

(9)

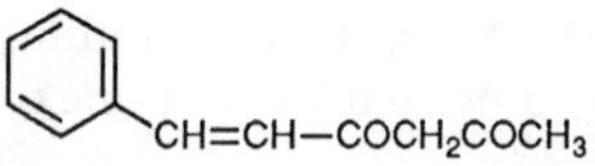

(10)

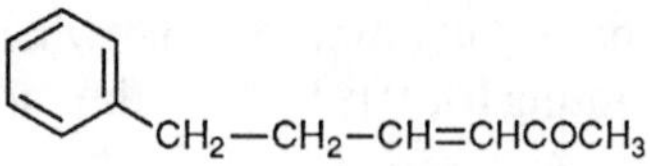

(11)

Urine samples collected for 24 hours after administration of kawain (1) to human volunteers were extracted with ether before and after incubation with β-glucuronidase. Extracts were analysed by GC–MS directly and after methylation with diazomethane. Eight metabolites [(3)–(7), (9), and (11)] were identified by direct analysis of urine. Five metabolites [(2), (4), (6), (7), and (8)] were identified after methylation. The major metabolite was indicated as (4) formed by ring-hydroxylation. Most metabolites were reported to be excreted as conjugates. The quantitative importance of each metabolite is unknown.

Reference

C. Koppel and J. Tenczer, Mass spectral characterisation of urinary metabolites of D,L-kawain, *J. Chromatogr.*, 1991, **562**, 207.

Coumarin

Use/occurrence:	Plant oil
Key functional groups:	Benzopyranone, lactone
Test system:	Rat (oral, 10, 100, 150 mg kg^{-1})

Structure and biotransformation pathway

(2)   (1) * = ^{14}C   (4)

(3)   (5)   (6)

(7)   (8)   (9)

In the high-dose experiment urinary metabolites excreted in the first 24 hours following dosing accounted for *ca.* 47% of the dose. The major urinary metabolite was identified as (8), accounting for *ca.* 20% of the dose. More than 10% of the dose was excreted as polar conjugates which were characterized following acid hydrolysis to the aglycones (5), (7), and (8). A novel metabolite (6) was identified as a mercapturic acid, accounting for *ca.* 6% of the dose. Metabolites (2)–(5), (7), and (9) together accounted for most of the remaining urinary radioactivity. The presence of the mercapturic acid metabolite (6) suggested the intermediate formation of the 3,4-epoxide, which in turn reacts with glutathione to produce a glutathione adduct. Previous studies have shown that coumarin itself does not react with glutathione nor is it a substrate for glutathione transferases. The formation of a reactive epoxide may account for the observed hepatotoxicity of (1) in rats and dogs, which has been shown to be due to its bioactivation by P-450-dependent enzymes. The mercapturate metabolite was identified by FAB-MS, by ^{1}H NMR, and by comparison with a synthetic standard of the 4-isomer. Other metabolites were identified following HPLC co-chromatography with reference standards, although not all of the standards were

143

separated. Following the 10 mg kg^{-1} dose the mercapturate metabolite accounted for a lesser proportion of the administered dose (2.6% *cf.* 7.5% and 5.8% respectively at 100 and 150 mg kg^{-1} doses).

Reference

T. Huwer, H. Altmann, W. Grunow, S. Lenhardt, M. Przybylski, and G. Eisenbrand, Coumarin mercapturic acid isolated from rat urine indicates metabolic formation of coumarin 3,4-epoxide, *Chem. Res. Toxicol.*, 1991, **4**, 586.

Coumarin

Use/occurrence:	Natural product, food, and pharmaceutical flavourant
Key functional groups:	Benzopyranone, lactone
Test system:	Rat liver microsomes and purified cytochromes P-450 IA1 and P-450 IIB1

Structure and biotransformation pathway

(1) * = ^{14}C   (2)   (3)

(4)

   Coumarin (1,2-benzopyrone) (1), a widely distributed natural product, is used in various cosmetic and detergent preparations. It produces hepatic centrilobular necrosis in the rat and, following chronic administration, it results in bile duct lesions. The toxicity may arise from covalent binding of coumarin 3,4-epoxide (2), an intermediate formed by cytochrome P-450 oxidation of (1). The epoxide (2) may rearrange to 3-hydroxycoumarin (3) or undergo further metabolism to coumarin 3,4-dihydrodiol (4). Detoxification of the epoxide (2) is believed to be by glutathione conjugation. The 3-hydroxy metabolite (3) can undergo lactone ring-opening and further metabolism by oxidative decarboxylation to *o*-hydroxyphenylacetic acid, although, in humans, coumarin (1) is extensively metabolized to 7-hydroxycoumarin. The induction of several mixed-function oxidase enzymes was demonstrated to affect the metabolism of (1), isoenzymes P-450 IA and P-450 IIB subfamilies playing significant roles in the metabolism. HPLC profiles showed a total of twelve polar metabolic products from (1).

Reference

M. M. C. G. Peters, D. G. Walters, B. van Ommen, P. J. van Bladeren, and B. G. Lake, Effect of inducers of cytochrome P-450 on the metabolism of [3-^{14}C]coumarin by rat hepatic microsomes, *Xenobiotica*, 1991, **21**, 499.

7-Ethoxycoumarin

Use/occurrence:	Model substrate
Key functional groups:	Alkyl aryl ether
Test system:	Isolated perfused rat intestinal loops

Structure and biotransformation pathway

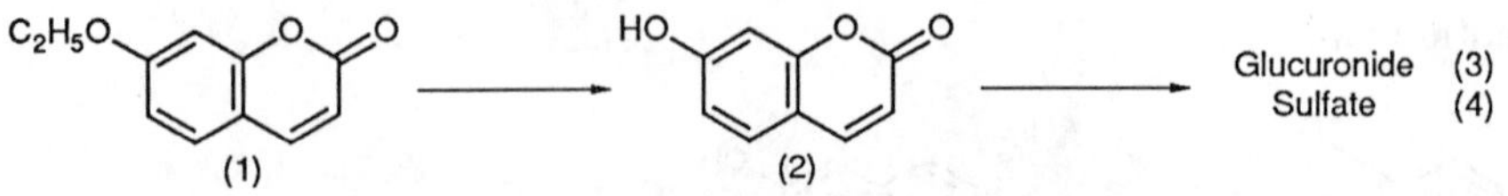

7-Ethoxycoumarin (1) was infused into cannulated, perfused loops of rat intestine, and the level of the de-ethylated metabolite (2) in the blood or lumen perfusion media was determined by extraction and spectrofluorimetry. The major products were found to be the glucuronide or sulfate conjugates [(3) and (4)] of (2) which were deconjugated using β-glucuronidase or aryl sulfatase. The phenol (2) was then measured as above.

The sulfate conjugate (4) was the major metabolite present in the lumen, whereas the glucuronide (3) predominated in the blood. Unchanged (2) was found to be minor component (2%) of the total metabolites in both blood and lumen perfusion media.

Reference

R. Albers and D. W. Rosenberg, *In vivo* intestinal metabolism of 7-ethoxycoumarin in the rat: production and distribution of phase I and II metabolites in the isolated, perfused intestinal loop, *Toxicol. Appl. Pharmacol.*, 1991, **109**, 507.

Warfarin, Acenocoumarol (4′-nitro-warfarin)

Use/occurrence:	Anticoagulants
Key functional groups:	Alkyl ketone, coumarin, nitrophenyl
Test system:	Rat liver microsomes

Structure and biotransformation pathway

Warfarin (1) R = H
Acenocoumarol (2) R = NO$_2$

R = H	R = NO$_2$
(3) 6-hydroxy	(9) 6-hydroxy
(4) 7-hydroxy	(10) 7-hydroxy
(5) 8-hydroxy	(11) 8-hydroxy
(6) 9-hydroxy	(12) 9-hydroxy
(benzylhydroxy)	(benzylhydroxy)
(7) 10-hydroxy	(13) 10-hydroxy
(8) Dehydro	(14) Dehydro
(C-9=C-10 double bond)	(C-9=C-10 double bond)
(15) 4′-hydroxy	

The oral anticoagulants warfarin (1) and acenocoumarol [4′-nitrowarfarin (2)] are administered clinically as racemates. The aim of the present study was to elucidate the mechanism behind the remarkable differences seen *in vivo* between the clearances of *R*- and *S*-warfarin and *R*- and *S*-acenocoumarol.

Following incubation of either (1) or (2) with rat hepatic microsomes in the presence of a NADPH-regenerating system at substrate concentrations of 0.02–1.6 mM, a range of hydroxylated metabolites were detected. Metabolites were separated by HPLC and identified by UV spectra (diode array detector analysis) and retention times. The *in vitro* metabolism of (1) and (2) involved the same metabolic reactions in that hydroxylation occurred at the coumarin moiety at the 6-, 7-, and 8-positions [metabolites (3)–(5) and (9)–(11)], and at C-9 [benzylhydroxy, (6) and (12)] and C-10 [(7) and (13)]. The 6- and 7-hydroxy metabolites were the major metabolites of both (1) and (2). However, the enantiomers of (2) appeared to be the preferred substrate for the 6- and 7-hydroxylation, with apparent K_m values being 2–19-fold

147

lower. In addition (1) but not (2) was metabolized to a 4'-hydroxy metabolite by a non-stereoselective pathway.

Formation of the 6-, 7-, and 8-hydroxy metabolites of (1) was stereoselective for the *R*-enantiomer whereas formation of the same metabolites of (2) showed stereoselectivity for the *S*-enantiomer. These observations reflect the differences in clearance seen *in vivo* in the rat for the enantiomers of the two compounds. Further experiments using induced hepatic microsomes from rats pretreated with phenobarbitone or 3-methylcholanthrene or inhibition experiments with cimetidine were also conducted. Taken together, these data strongly indicated that some of the differences in the metabolism of (1) and (2) could be attributed to catalysis by different isozymes of cytochrome P-450.

Reference

J. J. R. Hermans and H. H. W. Thijssen, Comparison of the rat liver microsomal metabolism of the enantiomers of warfarin and 4'-nitrowarfarin (Acenocoumarol), *Xenobiotica*, 1991, **21**, 295.

N-Methyl-*N*-alkyl-*p*-chloroamides

Use/occurrence:	Model compounds
Key functional groups:	Aryl carboxamide, dialkyl amide
Test system:	Rat liver microsomes

Structure and biotransformation pathway

Compound	R	Metabolites
(1)	CH_3	(8), (9)
(2)	C_2H_5	(8), (10)
(3)	$n\text{-}C_3H_7$	(8), (11)
(4)	$n\text{-}C_4H_9$	(8), (12)
(5)	$C_6H_5CH_2$	(8), (13)
(6)	$iso\text{-}C_3H_7$	(8), (14)
(7)	$cyclo\text{-}C_3H_7$	(8), (15)

Compounds (1)–(7) were metabolized by phenobarbitone-induced rat liver microsomes in the presence of NADPH exclusively to mono-*N*-dealkylated products; with each compound both possible routes of metabolism, dealkylation [to metabolite (8)], and demethylation [to metabolites (9)–(15)] occurred. Didealkylation (to chlorobenzamide), *p*-chlorobenzoic acid, or ring-hydroxylated products were not detected in this metabolizing system with any of the substrates (1)–(7). *N*-Alkyl-*N*-(α-hydroxyalkyl) products (which are known stable intermediates in amide *N*-dealkylation) were not observed in the present study because base was added to decompose them before extraction. Metabolites were separated by HPLC and were identified by co-chromatography, EI-MS, and ^{1}H NMR of authentic standards.

The demethylation/dealkylation ratio varied from 0.3 to 2.2 among the primary alkyl groups [1.0 for compound (1); 2.2 (2); 1.9 (3); 0.9 (4); and 0.3 (5)] but was *ca.* 40 when the alkyl group was isopropyl (6) or cyclopropyl (7). Dealkylation of the benzyl substituent [from (5)] was 2–3 times more favourable than for any other primary alkyl group. Despite the wide variation in the demethylation/dealkylation ratios for (1)–(7), the rates of total oxidation (demethylation plus dealkylation) were similar across the series.

Reference

L. R. Hall and R. P. Hanzlik, *N*-Dealkylation of tertiary amides by cytochrome P-450, *Xenobiotica*, 1991, **21**, 1127.

N,N-Diethylphenylacetamide

Use/occurrence: Insect repellant

Key functional groups: Acetamide, alkyl amide

Test system: Rat (oral, 851 or 70 mg kg^{-1})

Structure and biotransformation pathway

The metabolism of N,N-diethylphenylacetamide (1) has been investigated in the rat following oral dosing. Urine samples were extracted and analysed by comparing retention times of metabolites and synthetic standards on GC. Unchanged (1) was found to be excreted, together with ethylphenylacetamide (2), phenylacetamide (3), and phenylacetic acid (4).

Following administration of the higher dose, blood, liver, lung, and kidney contained unchanged (1) together with smaller amounts of (2).

Reference

S. S. Rao, D. K. Jaiswal, and P. K. Ramachandran, Distribution and metabolism of the insect repellant N,N-diethylphenylacetamide on oral exposure in rats, *Toxicol. Lett.*, 1991, **55**, 243.

CI-976

Use/occurrence:	Anti-atherosclerotic agent, plasma lipid regulator
Key functional groups:	Alkyl carboxamide, aryl amide
Test system:	Rat (intravenous, 10 mg kg^{-1}; oral, 50 mg kg^{-1}), Cynomolgus monkey (intravenous, 2 mg kg^{-1}; oral, 50 mg kg^{-1})

Structure and biotransformation pathway

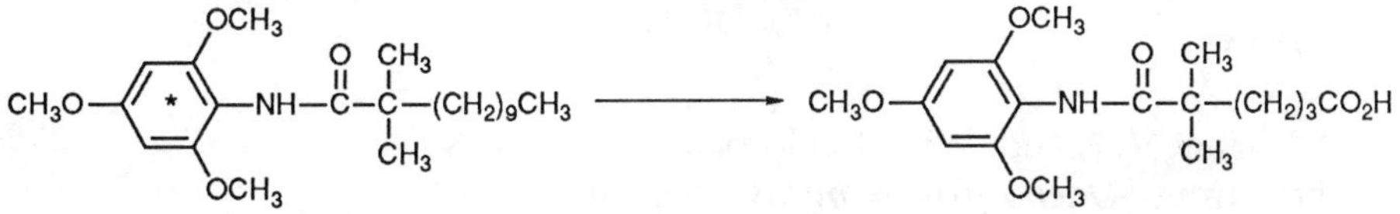

(1) * = ^{14}C

(2)

In the rat, excretion of radioactivity following oral or intravenous administration of [^{14}C]-CI-976 (1) was evenly divided between urine and faeces. Most of the urinary radioactivity was associated with a single metabolite, which was identified (GC–MS, LC–MS, and ^{1}H NMR) as the six-carbon chain-shortened carboxylic acid (2). This metabolite was also the major radioactive component excreted in bile. In the monkey *ca.* 70% of an intravenous dose was excreted in urine and *ca.* 43% was associated with metabolite (2). No mention was made of the nature of the remaining urinary radioactivity but the authors reported that they found no evidence of intermediate chain-shortened carboxylic acids or *O*-demethylated metabolites.

Reference

T. F. Woolf, S. M. Bjorge, A. E. Black, A. Holmes, and T. Chang, Metabolism of the acyl-CoA: cholesterol acyltransferase inhibitor 2,2-dimethyl-*N*-(2,4,6-trimethoxy-phenyl)dodecanamide in rat and monkey, *Drug Metab. Dispos.*, 1991, **19**, 696.

Aniline, 4-Chloroaniline

Use/occurrence:	Model compounds
Key functional groups:	Aryl amine, chlorophenyl
Test system:	Rainbow trout liver microsomes

Structure and biotransformation pathway

(1) R = H
(2) R = Cl

(3) R = H
(4) R = Cl

N-Hydroxylation of [^{14}C]aniline (1) and [^{14}C]-4-chloroaniline (2) was quantified in juvenile rainbow trout (*Onchorhyncus mykiss*) hepatic microsomes by HPLC and liquid scintillation counting. The products phenylhydroxylamine (3) and its 4-chloro analogue (4) were identified by co-elution with authentic synthetic standards. Quantification of the *N*-hydroxy metabolites using reverse-phase gradient HPLC was achieved without derivatization or extraction procedures. The pH optima for *N*-hydroxylation of (1) and (2) were different; different isoenzymes are possibly contributing to the product formation. Although primary aromatic amine *N*-hydroxylation is well characterized in mammals, it has not been previously observed in fish. Further oxidation of (3) and (4) was not observed, although small amounts of *p*-hydroxyaniline were obtained from (1). Other Phase I oxidative metabolites of (1), *e.g.* 2-aminophenol, azobenzene, and azoxybenzene, were not detected. At 25 °C, the V_{max} for *N*-hydroxylation of (1) and (2) was 33.8 and 22.0 pmol min^{-1} mg^{-1} respectively, with K_m values of 1.0 and 0.8 mM respectively. These activities were not induced by treatment of the trout with Aroclor 1254. At 11 °C, the physiological temperature of rainbow trout in this study, V_{max} for (2) was 6.4 pmol min^{-1} mg^{-1} with K_m 0.5 mM.

Reference

J. M. Dady, S. P. Bradbury, A. D. Hoffman, M. M. Voit, and D. L. Olson, Hepatic microsomal *N*-hydroxylation of aniline and 4-chloroaniline by rainbow trout (*Onchorhyncus mykiss*), *Xenobiotica*, 1991, **21**, 1605.

4-Aminobiphenyl

Use/occurrence:	Carcinogen in cancer research also used in the detection of sulfate functional groups
Key functional groups:	Aryl amine, biphenyl
Test system:	Rat (intraperitoneal and oral, 100 mg kg^{-1})

Structure and biotransformation pathway

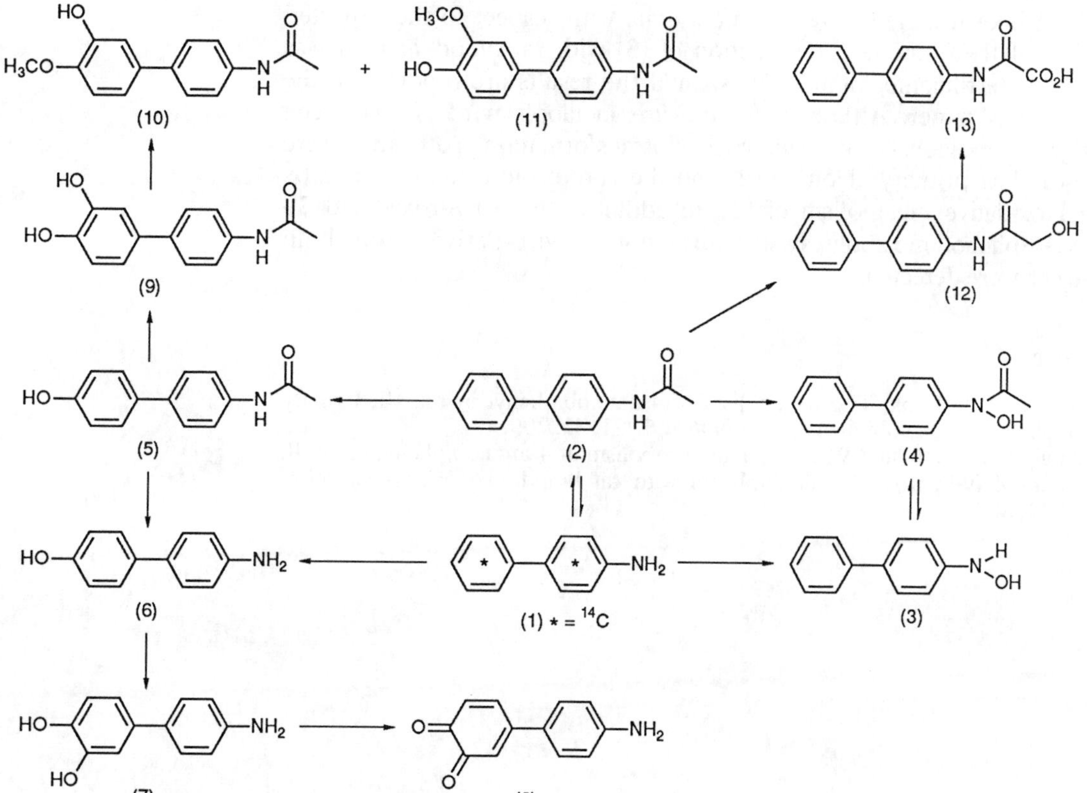

Urine from rats administered uniformly labelled [^{14}C]-4-aminobiphenyl (1) (orally or intraperitoneally) showed mainly hydroxylated metabolites (up to eight) on TLC analysis after treatment with urease and β-glucuronidase-aryl-sulfatase. TLC, UV, MS, and NMR data were identical with those from synthetic standards and the synthetic experimental details are given in full. Several-fold chromatographic purification of the metabolites was required for an unambiguous identification of the regioisomers, and the study is there-fore a qualitative analysis. The urinary metabolites from rats dosed with (1)

were detected as up to 18 discrete bands on TLC when the chromatogram was developed ten times. In addition to the nine hydroxylated metabolites, further minor phenolic compounds remained unidentified owing to their trace quantities. No nitrosobiphenyls were detected, but the acetanilide (2) was rapidly and reversibly formed together with its N-hydroxy analogue (3) which was in equilibrium with N-hydroxy-4-acetylaminobiphenyl (4). Unconjugated (4) was not detected, but (4) can also arise from (1) via (2). p-Hydroxylation of (2) gave (5) which was hydrolysed to 4'-hydroxy-4-aminobiphenyl (6), which is also the product of para-hydroxylation of (1). Subsequent metabolism of (5) afforded the catechol 3',4'-dihydroxy- (9) and the regioisomeric O-methyl ethers 3'-hydroxy-4'-methoxy- (10) and 4'-hydroxy-3'-methoxy-4-acetylaminobiphenyl (11). These phenols are analogous to the metabolites of biphenyl detected in rats. Subsequent metabolism of (6) afforded the catechol (7) which was biotransformed into the 1,2-benzoquinone (8), characterized by MS and UV with the quinone displaying $(M+2)$, a known increase by two mass units with respect to the expected mass ion of the quinone. The reaction of (3) with rat blood in vitro was assayed for ferrihaemoglobin synthesis and the results are reported in the accompanying paper. Although the in vitro metabolism of (1) has been investigated previously, the in vivo biotransformation pathways were unknown. The primary alcohol (12) and the carboxylic acid (13) arise by Phase I oxidative metabolism of (2). In addition, the 3-hydroxy- and 2'-hydroxy-o-phenol metabolites (structures not shown), derived from both (1) and (2), were detected.

References

S. Karreth and W. Lenk, The metabolism of 4-aminobiphenyl in rat. III. Urinary metabolites of 4-aminobiphenyl, *Xenobiotica*, 1991, **21**, 709.

R. Heilmair, S. Karreth, and W. Lenk, The metabolism of 4-aminobiphenyl in rat. II. Reaction of N-hydroxy-4-aminobiphenyl with rat blood *in vitro*, *Xenobiotica*, 1991, **21**, 805.

4-Aminobiphenyl

Use/occurrence:	Environmental carcinogen
Key functional groups:	Aryl amine, biphenyl
Test system:	Dogs (oral, 5 mg kg^{-1})

Structure and biotransformation pathway

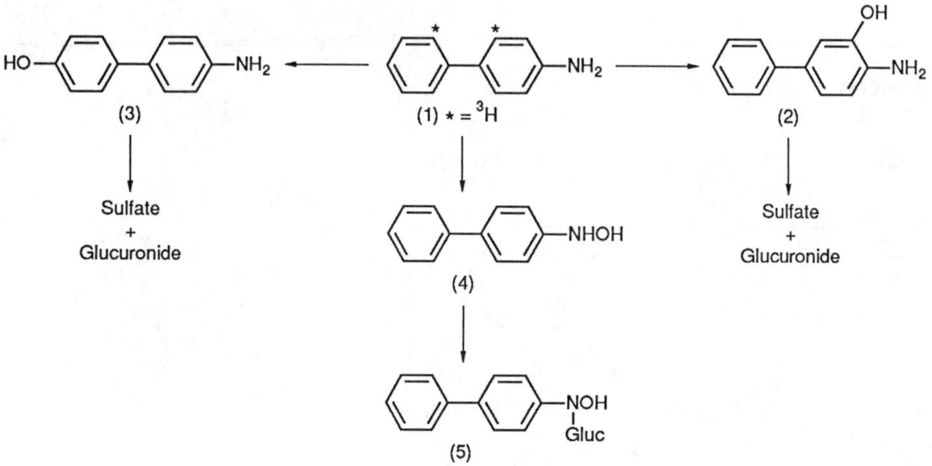

The major urinary metabolites of 4-aminobiphenyl (1) in the dog following oral administration were analysed by HPLC. The sulfate and glucuronide conjugates of (2) and (3) were the major urinary products together with unchanged (1) and the hydroxylamine (4). There were also small amounts of the N-glucuronides of (4) and (1) excreted. The ratio of (4) to (5) depended on the voiding interval of the urine. The presence of unconjugated (4) in the circulating blood is the probable source of the blood-haemoglobin adducts.

Reference

F. F. Kadlubar, K. L. Dooley, C. H. Teitel, D. W. Roberts, R. W. Benson, M. A. Butler, J. R. Bailey, J. F. Young, P. W. Skipper, and S. R. Tannenbaum, Frequency of urination and its effect on metabolism, pharmacokinetics, blood haemoglobin adduct formation, and liver and urinary bladder DNA adduct levels in beagle dogs given the carcinogen 4-aminobiphenyl, *Cancer Res.*, 1991, **51**, 4371.

2-Fluoroaniline

Use/occurrence:	Chemical intermediate
Key functional groups:	Aryl amine, fluorophenyl
Test system:	Rat (oral, 50 mg kg^{-1}), rat liver microsomes, rat hepatocytes, isolated liver perfusion, substrate concentration 3 mM

Structure and biotransformation pathway

Gluc = glucuronyl

2-Fluoroaniline is a chemical used in the manufacture of herbicides and plant growth regulators. The metabolism of 2-fluoroaniline (1) was investigated both after *in vivo* exposure of rats following oral administration and in *in vitro* model systems. The pattern of metabolism was elucidated using ^{19}F NMR, enabling metabolites of (1) to be identified on the basis of their ^{19}F NMR resonances.

Following an oral dose of (1) at a level of 50 mg kg^{-1}, 90% of the dose was eliminated in the urine over a 24 hour period. About nine major metabolites were seen in the ^{19}F NMR spectrum of urine samples, seven of which were identified. The major urinary metabolite was 4-amino-3-fluorophenyl sulfate (3), which accounted for 53.5% of the total ^{19}F NMR signals. The other metabolites were 4-acetamido-3-fluorophenol (6), 0.5%; the corresponding sulfate (7), 18.1%; and glucuronide conjugate (8), 9.9%; and the glucuronide conjugate (4) of 4-amino-3-fluorophenol, 3.3%. Unchanged (1) accounted for 4.9% of total ^{19}F intensity and F$^-$ ion for 1.6%.

The major metabolite of (1) in rat hepatic microsomal incubations fortified with NADPH was (2), 4-amino-3-fluorophenol. Fluoride was a minor metabolite in these incubations. In contrast, incubation of isolated rat hepatocytes in Krebs–Ringer buffer in the presence of (1) produced a number of metabolites. N-Acetylation was the major route of metabolism yielding metabolite (5), 2-fluoroacetanilide, although *para*-hydroxylation to (2) also occurred to a significant extent as did F$^-$ production. Minor metabolites seen were (3), (4), and (6)–(8). When hepatocytes were incubated with (1) in Williams Medium E (a sulfate-rich medium), all phenolic metabolites appeared to be conjugated with sulfate, yielding metabolite (3) exclusively. The isolated perfused rat liver preparation gave results similar to those for rat hepatocytes using (1) as substrate. However, the main difference was that glucuronide conjugates were not detected using the perfused liver system even when perfusate sulfate concentrations were limiting, in contrast to hepatocyte incubations.

In summary, the major biotransformation route for (1) is via *para*-hydroxylation. No *ortho*-hydroxylated metabolites, typically seen with other aniline derivatives, were observed as major excretion products. In addition, sulfation is the predominant Phase II reaction for metabolites of (1).

Reference

J. Vervoort, P. A. de Jager, J. Steenbergen, and I. M. C. M. Rietjens, Development of a ^{19}F-NMR method for studies on the *in vivo* and *in vitro* metabolism of 2-fluoro-aniline, *Xenobiotica*, 1990, **20**, 657.

Aniline, 4-Fluoroaniline, Pentafluoroaniline, 2,3,5,6-Tetrafluoroaniline

Use/occurrence:	Model compounds
Key functional groups:	Arylamine, fluorophenyl
Test system:	Rat liver microsomes (3 mM substrate) and, for 2,3,5,6-tetrafluoroaniline and pentafluoroaniline, rat (oral, 25 mg kg^{-1})

Structure and biotransformation pathway

Following a single oral administration of (7) or (8) to rats 62% and 90% of the dose respectively was recovered in the urine by 48 hours (based on recovery of fluorine-containing metabolites). No fluorine-containing metabolites were detected in the faeces of rats dosed with (7), suggesting that some of the metabolites are withheld in the body. The major urinary metabolites of (7) were fluoride ion (6) and the phenyl sulfate (11). Fluoride ion was not a significant metabolite of (8) *in vivo*. *In vitro* experiments with (7) and (8) using NADPH as the electron donor demonstrated the formation of (10) from both compounds, and as expected with (7) formation of fluoride ion (6). Experiments using alternative electron donors suggested that (10) was formed as a primary metabolite of (8), whereas the formation of (10) from (7) involved the intermediacy of the quinoneimine metabolite (9).

Similarly *in vitro* experiments with (1) and (3) using NADPH as the electron donor demonstrated the formation of (2) from both compounds, and as expected with (3) formation of fluoride ion. The intermediacy of a quinone-imine was proposed in the metabolism of (3) to (2). The production of (2) following microsomal incubations of both (1) and (3) was studied (using a spectrophotometric method). Other metabolites of (3), (7), and (8), including fluoride ion (6), were determined by ^{19}F NMR. Metabolites of (3) included (4), (5), and (6). No 3-fluoro-4-hydroxyaniline, which would result from *para*-hydroxylation of (3) followed by an NIH shift, was detected.

Reference

I. M. C. M. Rietjens and J. Vervoort, Bioactivation of 4-fluorinated anilines to benzo-quinoneimines as primary reaction products, *Chem. Biol. Interact.*, 1991, 77, 263.

Benzocaine hydrochloride

Use/occurrence: Anaesthetic

Key functional groups: Alkyl ester, aryl amine, aryl carboxylate

Test system: Rainbow trout (via dorsal aortic cannula, 20.8 μg kg^{-1})

Structure and biotransformation pathway

(1) $\star$ = ^{14}C (2) (4) (3)

Benzocaine (ethyl p-aminobenzoate) (1) metabolism has not been well characterized in fish. Biotransformation of (1) is by N-acetylation of the aniline to the amide (2) (67% of the aromatics detected in effluent water and 8% in urine after one hour), hydrolysis of the ethyl ester to the benzoic acid (3) (1% in urine after 20 hours), and conversion of (2) and (3) into the corresponding benzoic acid acetanilide (4) (97% in urine after 20 hours). Benzocaine (1) and the hydrophobic N-acetylated metabolite (2) were eliminated in effluent water primarily through the gills; renal and biliary pathways are less significant. The more polar metabolites (3) and (4) were retained and excreted more slowly in the urine. Effluent water and urine were analysed by HPLC; metabolites were separated and detected by radiochromatography using TLC and HPLC with UV detection at 286 nm. Radioactive residues were identified by comparing the retention times of peaks on radiochromatograms with authentic standards. In fish, biliary excretion of charged compounds with molecular weights under 600 is limited. Although benzoic acid forms free amino acid conjugates in fish, no amino acid conjugates were detected in the urine of rainbow trout (*Oncorhynchus mykiss*) exposed to (1). Benzocaine, a local anaesthetic in human and veterinary medicine, is an effective and safe fish anaesthetic.

Reference

J. R. Meinertz, W. H. Gingerich, and J. L. Allen, Metabolism and elimination of benzocaine by rainbow trout, *Oncorhynchus mykiss*, *Xenobiotica*, 1991, **21**, 525.

Arsanilic acid

Use/occurrence:	Growth promoter in animal feeds. Control of swine dysentry
Key functional groups:	Aryl amine, aryl arsenic acid
Test system:	Pig (oral, 2–3 mg kg^{-1}), chicken (oral, 10 mg kg^{-1})

Structure and biotransformation pathway

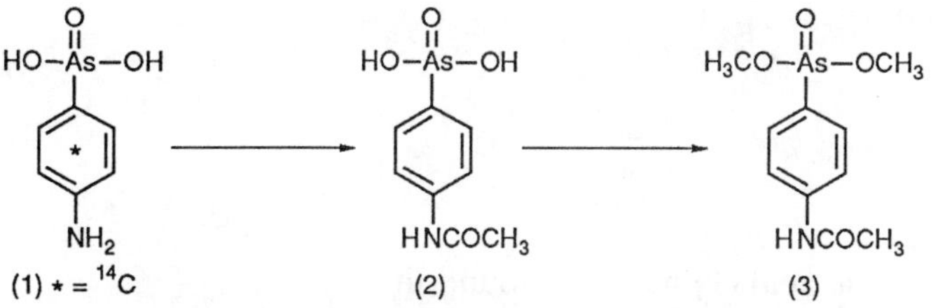

$$\text{(1)} \quad \star = {}^{14}C \qquad \text{(2)} \qquad \text{(3)}$$

Following oral dosing to pigs, 35–40% of the dose was found in the 0–12 hour urine. Overall, 59 ± 8% of the dose was excreted in the urine during 96 hours, with 4.7% in the bile. Tissue residues 96 hours after dosing were highest in the liver (1.3 p.p.m.) and kidney (0.2 p.p.m.). The urinary metabolites were isolated by low-pressure (Porapak Q) and high-pressure liquid chromatography before being identified using FAB-MS and NMR.

Unchanged arsanilic acid (1) accounted for 17–39% of the urinary metabolites, *N*-acetylarsanilic acid (2) for 15–29%, and metabolite (3), tentatively identified as (4-acetamidophenyl)dimethylarsine oxide, was present in low amounts (1.5–4%).

Only unchanged arsanilic acid was isolated from the urine of roosters which had been colostomized.

Reference

P. W. Aschbacher and V. J. Feil, Fate of [^{14}C]Arsanilic acid in pigs and chickens, *J. Agric. Food Chem.*, 1991, **39**, 146.

5-Aminosalicylic acid

Use/occurrence:	Treatment of inflammatory bowel disease
Key functional groups:	Aryl amine
Test system:	Mini-pig (oral, 2 g kg^{-1}), human (intravenous, 250 mg), rat liver homogenate

Structure and biotransformation pathway

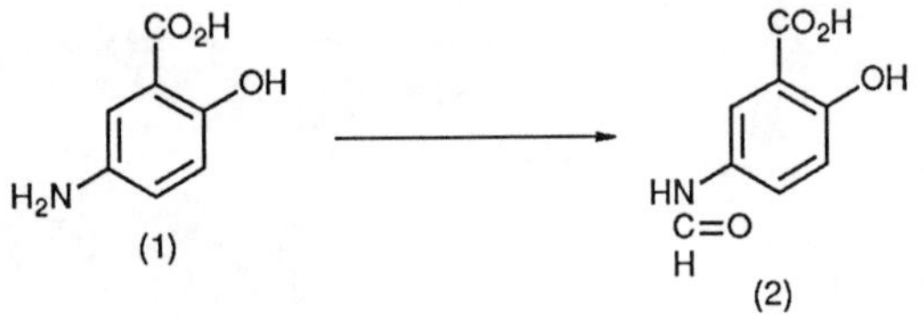

A new metabolite of the drug 5-aminosalicylic acid (1) has been found in the urine from pigs following oral administration. The metabolite, 5-formamidosalicylic acid (2), was also found in human plasma following an intravenous dose. Metabolite (2) was isolated from pig urine on an XAD-2 column and purified using preparative HPLC. The N-formyl metabolite (2) was identified using ^{1}H and ^{13}C NMR and thermospray MS–MS and the structure was confirmed by chemical synthesis. The metabolite was also readily formed by rat liver homogenate in the presence of N-formyl-L-kynurenine and (1). The authors have suggested that (2) may be formed *in vivo* by the action of formamidase.

Reference

J. Tjørnelund, S. H. Hansen, and C. Cornett, New metabolites of the drug 5-amino-salicylic acid. II. N-Formyl-5-aminosalicylic acid, *Xenobiotica*, 1991, **21**, 605.

p-Aminophenol, Phenacetin, Acetaminophen (paracetamol)

Use/occurrence:	Analgesic drugs
Key functional groups:	*N*-Acetyl aryl amine, alkyl aryl ether, phenol
Test system:	Rat liver and kidney microsomes

Structure and biotransformation pathway

The major metabolite of phenacetin (1) is acetaminophen, the analgesic drug paracetamol (2). Large quantities of (2) induce hepatic necrosis and acute renal failure in humans via metabolic oxidation to the *N*-acetyl-*p*-benzoquinoneimine (3) or epoxidation. The quinoneimine (3) is detoxified by glutathione conjugation or it binds to cellular macromolecules. *p*-Aminophenol (4) is a minor metabolite of both (1) and (2) and it is similarly oxidized to the Michael acceptor (5). The binding site of the glutathione conjugate derived from (5), in the presence of horseradish peroxidase, was therefore elucidated. Analysis was by HPLC and TLC, positive ion FAB-MS and ^{1}H NMR; full spectroscopic details are given. The metabolite (4) is a potent nephrotoxin. Even in small doses it produces a site-specific necrosis of proximal convoluted tubules in rats. It is detoxified by glutathione conjugation *ortho* to the amine, affording (6). It is not known whether this regiospecific Michael addition reflects the greater Michael acceptor potency of the enone over the unsaturated imine only, or whether it also reflects the enzymatic specificity of glutathione *S*-transferase. The haemoglobin-SH functional

163

group (cys β93), in human and canine erythrocytes, forms adducts with p-aminophenol (4).

Reference

R. Eyanagi, Y. Hisanari, and H. Shigematsu, Studies of paracetamol/phenacetin toxicity: isolation and characterization of p-aminophenol-glutathione conjugate, *Xenobiotica*, 1991, **21**, 793.

Propanil (3,4-dichloropropionanilide)

Use/occurrence:	Herbicide
Key functional groups:	Aryl amide, chlorophenyl
Test system:	Rat (intraperitoneal, 1.0 mmol kg^{-1})

Structure and biotransformation pathway

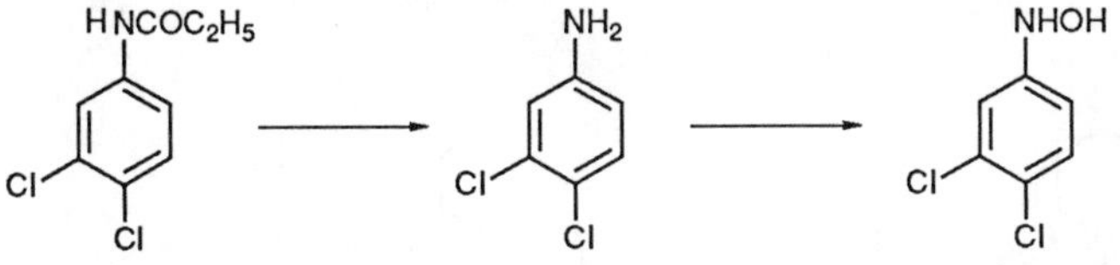

(1)   (2)   (3)

Following intraperitoneal dosing of propanil (1), 3,4-dichloroaniline (2) and the N-hydroxylated metabolite (3), previously shown *in vivo* to mediate the haemolitic anaemia induced by propanil, were detected in the blood of rats, by extraction and HPLC using electrochemical detection.

Reference

D. C. McMillan, T. P. Bradshaw, J. A. Hinson, and D. J. Jollow, Role of metabolites in propanil-induced haemolytic anaemia, *Toxicol. Appl. Pharmacol.*, 1991, **110**, 70.

Encainide

Use/occurrence: Antiarrythmic drug

Key functional groups: Chiral carbon, methoxyphenyl, N-methylpiperidine

Test system: Rat liver microsomes (10–2000 μM substrate)

Structure and biotransformation pathway

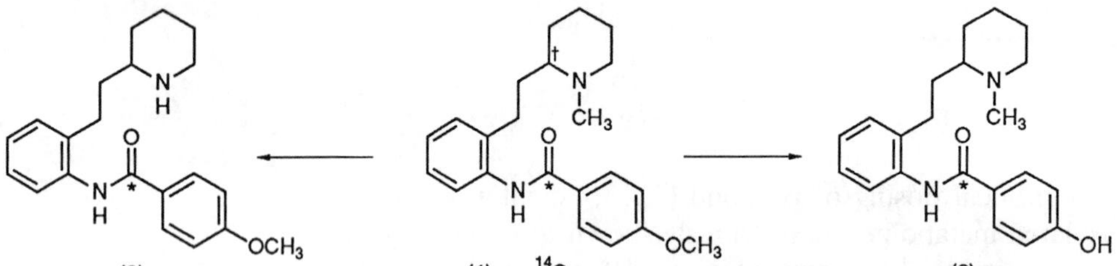

(3)

(1) * = ^{14}C
† = chiral carbon

(2)

Encainide (1) possesses a single chiral centre adjacent to the piperidine nitrogen and is normally used clinically as the racemate. Chiral derivatization with (−)-menthyl chloroformate followed by HPLC separation of the diastereoisomers was used to study the enantioselective metabolism of (1). It has been shown previously that encainide is metabolized by rat liver microsomes to give (2) and a lesser amount of (3). Analysis of (2) following incubation of 50 μM [^{14}C]encainide (1) showed that the (−)-enantiomer was always in excess of the (+)-enantiomer and the relative amounts (67:33) did not vary with time. With saturation concentrations of (1) the proportion of encainide metabolized was too small to detect any enantiomeric excess in unchanged (1). There was no evidence to suggest enantioselective formation of (3).

Reference

H. K. Jajoo, C. Prakash, R. F. Mayol, and I. A. Blair, Enantioselective metabolism of encainide by rat liver microsomes, *Biochem. Pharmacol.*, 1990, **40**, 893.

Sulofenur

Use/occurrence: Anticancer drug

Key functional groups: Chlorphenyl, indane, sulfonylurea

Test system: Mouse (oral, 100 mg kg^{-1}), rat (oral, 100 mg kg^{-1}), rhesus monkey (oral, 100 mg kg^{-1}), human (oral, 31 mg kg^{-1})

Structure and biotransformation pathway

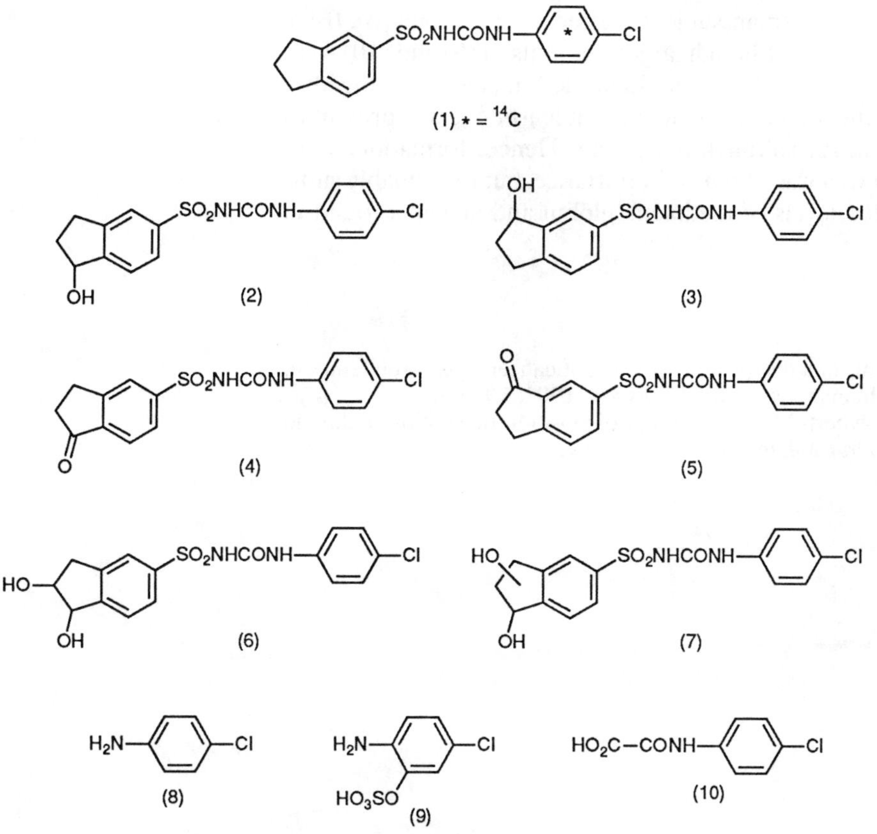

The metabolism of [^{14}C]sulofenur (1) was investigated in several species of laboratory animals and in a single human patient from a Phase I clinical trial, with particular emphasis being placed on the fate of the *p*-chloroaniline moiety, since this compound (8) is a known potent inducer of methaemoglobin formation and may, in theory, be formed from (1) *in vivo*. In treated mice, rats, and monkeys, mean totals of 56%, 62%, and 30% of the dose

167

respectively were excreted in urine during 5 days or more, and in rats with cannulated bile ducts 6% dose was excreted in bile during 24 hours. Complete recoveries of the administered radioactivity were obtained for both rodent species, but in monkeys less than half of the dose was excreted during 5 days. Radioactive components in urine and in rat bile were separated and quantified by HPLC, and identified by UV, NMR, MS, and/or chromatographic comparison with reference compounds. In all species examined, three of the four major urinary metabolites [(2)–(4)] corresponded to known biotransformation products of (1), and the fourth was identified as (5). Benzylic hydroxylation of the indane ring and subsequent oxidation to the corresponding ketoindanes were therefore the predominant metabolic pathways of (1). Two minor metabolites of the drug were identified as the dihydroxylated compounds (6) and (7). Previous studies with p-chloroaniline (8) in mice, rats, and monkeys have shown that nearly all of the administered radioactivity is excreted in the urine ($<5\%$ of which is in the form of parent compound), and that the major metabolites therein are (9) and (10). In mice, rats, monkeys, and the single human subject treated with (1), only $ca.$ 0.1% dose was present in urine as (8), although larger amounts of (9) and (10) were present. In humans, however, where the identified metabolites together accounted for $>95\%$ of the urinary radioactivity, at least 85% was present as compounds bearing an intact sulfonylurea group. Hence, formation of p-chloroaniline from (1) in $vivo$ was of minor importance, but presumably sufficient to cause the elevated levels of methaemoglobin and anaemia found in clinical trials.

References

W. J. Ehlhardt, Metabolism and disposition of the anticancer agent sulofenur in mouse, rat, monkey and human, *Drug Metab. Dispos.*, 1991, **19**, 370.

W. J. Ehlhardt and J. J. Howbert, Metabolism and disposition of p-chloroaniline in rat, mouse and monkey, *Drug Metab. Dispos.*, 1991, **19**, 366.

Furosemide

Use/occurrence:	Diuretic, antihypertensive drug
Key functional groups:	Alkyl aryl amine
Test system:	Rat (intraperitoneal, 1 mg kg^{-1}; intravenous, 1–200 mg kg^{-1})

Structure and biotransformation pathway

(1) * = ^{14}C   (2)

Furosemide (1) is excreted via the kidneys (50%) and by non-renal elimination (50%). The sulfonamide benzoic acid (1) was highly bound to human, bovine, rabbit, and rat plasma or albumin (97.4–98.4%), and also to rat tissues one hour after intraperitoneal injection (bound to adrenals, lungs, kidneys, and spleen). High doses (intravenous, 50 mg kg^{-1}) were choleretic with a saturation of hepatic drug metabolism. Rat bile contained (1) (8.6%), the anthranilic acid metabolite (2) (39%) from N-dealkylation of (1), and at least two unknown metabolites in which the furan ring remained intact. Analysis was by TLC and liquid scintillation counting. Urine and bile were incubated with glusulase, ketodase, and sulfatase as well as heated under reflux with aqueous sulfuric acid; however, there were no differences between samples and controls. The formation of a glucuronide conjugate of (1) in rats cannot be ruled out completely, as glucuronidation is a more dominant pathway in rats than in man, and the glucuronide of (1) has been reported in man.

Reference

J. Prandota and A. W. Pruitt, Pharmacokinetic, biliary excretion, and metabolic studies of ^{14}C-furosemide in the rat, *Xenobiotica*, 1991, **21**, 725.

Sulfometuron methyl

Use/occurrence: Herbicide

Key functional groups: Sulfonylurea, sulfonamide, methylpyrimidine

Test system: Goat (oral, 46–50 mg day^{-1})

Structure and biotransformation pathway

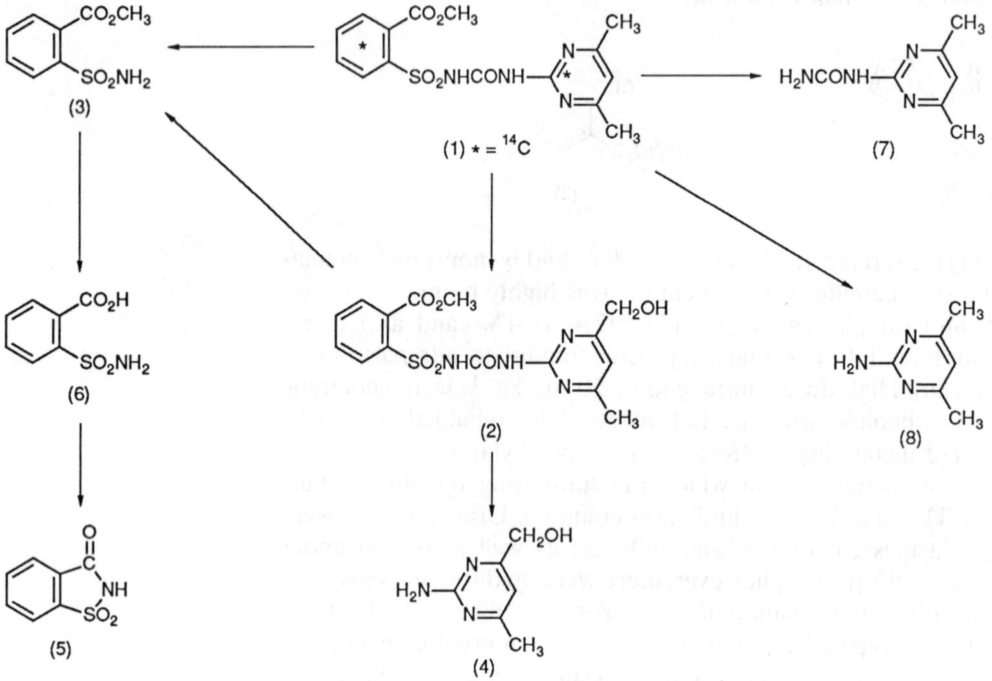

Two female goats were dosed with either the phenyl or pyrimidinyl labelled sulfometron methyl (1) for 7 days. Urine, faeces, and milk were collected throughout. The metabolites in urine and faeces were identified by reverse-phase HPLC against standard compounds. Doses of both radio-labelled forms were excreted mainly in the urine with the major metabolites being (hydroxymethyl)pyrimidine sulfometuron methyl (2) and unchanged (1). The sulfonamide (3), (hydroxymethyl)pyrimidinamine (4), and saccharin (5) were minor urinary metabolites showing that cleavage of the sulfonylurea bridge had occurred. Metabolite (6) was found only in trace amounts in the faeces. Pyrimidinylurea (7) and pyrimidinamine (8) were detected in extracts of tissues following treatment with protease.

Reference

M. K. Koeppe and C. F. Mucha, Metabolism of sulfometuron methyl in lactating goats, *J. Agric. Food Chem.*, 1991, **39**, 2304.

Recipavrin

Use/occurrence: Anticholinergic drug

Key functional groups: Benzhydryl, dimethylaminoalkyl

Test system: Rat (subcutaneous, 10 mg)

Structure and biotransformation pathway

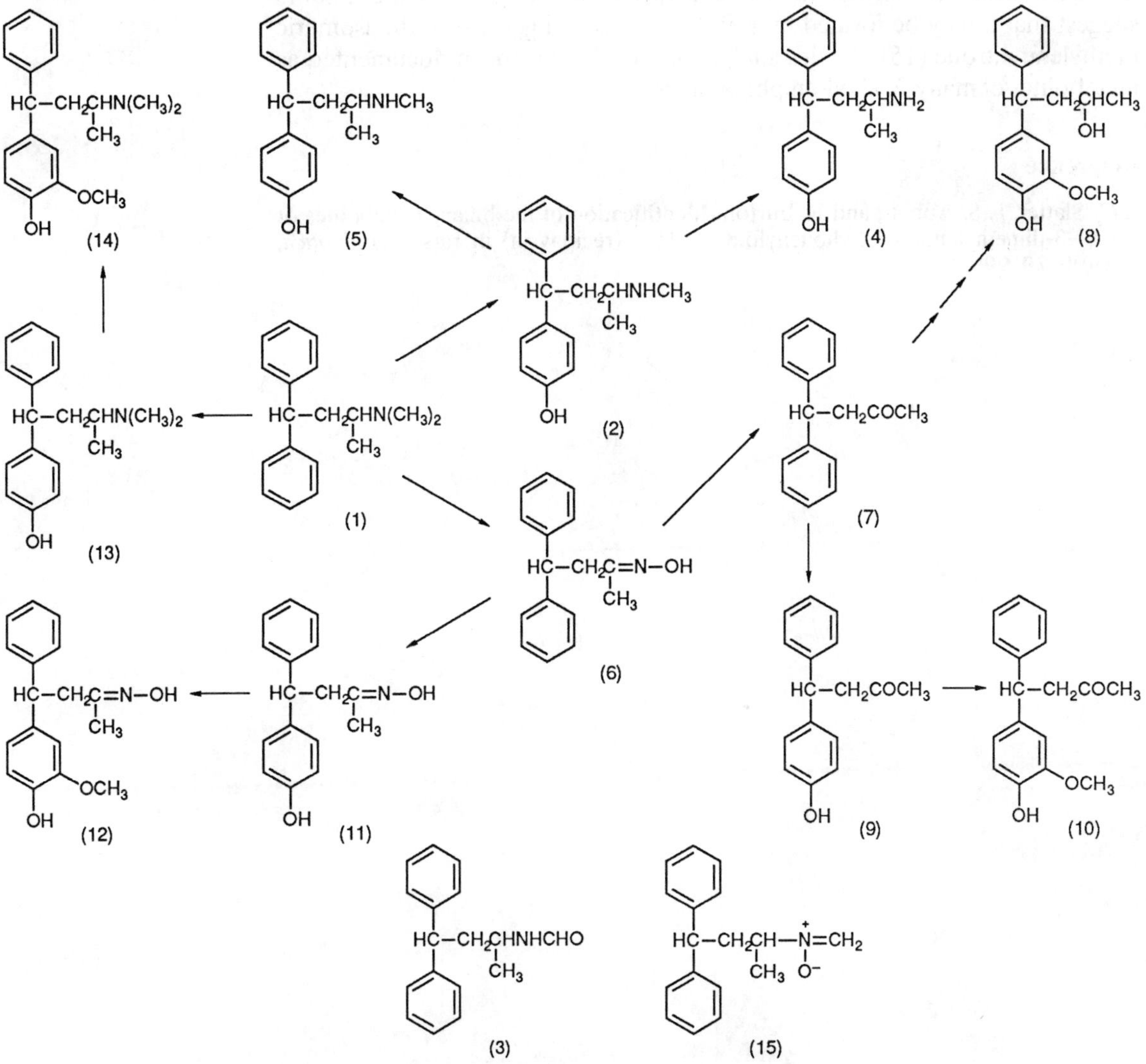

In a quantitative study of metabolic pathways, the biliary metabolites of recipavrin (1) produced by male rats were isolated and identified. Metabolites were isolated from bile by solvent extraction at pH 10, before and after treatment with deconjugating enzymes, and converted into their trimethylsilyl

171

or methylated derivatives before EI-MS; synthetic reference compounds were prepared for structure confirmation. Four general metabolic pathways for (1) were apparent (see Scheme), all of which are common to structurally related compounds in the amphetamine series. In addition to unchanged (1), 13 metabolites were identified and there was one unidentified component. Most of the metabolites were present in both free and conjugated forms but (3)–(5), (8), and (14) were present only as conjugates. The presence of conjugated forms of (2) and (4) would be surprising, and the authors suggest that there may have been incomplete extraction of non-conjugated metabolites or chemical breakdown of *N*-oxidized precursors. Although no quantitative data are presented, the oxime (6) is claimed to be the most abundant metabolite. The secondary formamide (3) is a novel compound, and the authors suggest that it may be formed by a Beckman rearrangement of the isomeric methylene nitrone (15); species analogous to (15) have been documented as metabolites of many *N*-alkyl-amphetamines.

Reference

J. G. Slatter, F. S. Abbott, and R. Burton, Identification of the biliary metabolites of (±)-3-dimethylamino-1,1-diphenylbutane HCl (recipavrin) in rats, *Xenobiotica*, 1990, **20**, 999.

Chlorpheniramine

Use/occurrence:	Antihistamine
Key functional groups:	Chlorophenyl, dimethylaminoalkyl, pyridine
Test system:	Rat (intraperitoneal, 60 mg kg^{-1} × 3), human (12 mg × 3)

Structure and biotransformation pathway

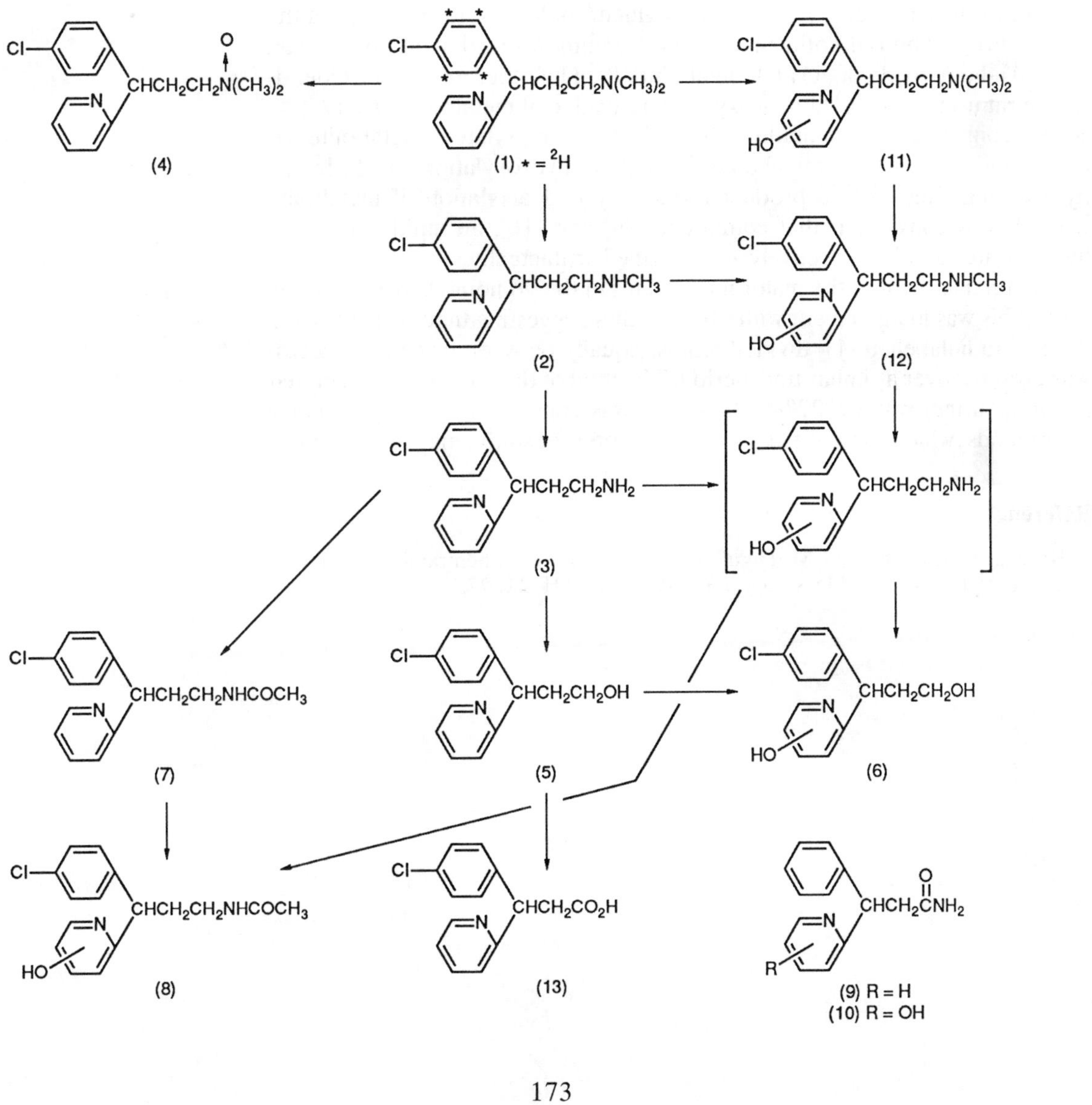

173

Chlorpheniramine [3-(p-chlorophenyl)-3-(2-pyridyl)-N,N-dimethylpropyl-amine] (1) is a potent antihistamine drug widely used in treatment of the common cold and allergic reactions. The urinary metabolites of (1) were studied following administration to rats and a female human volunteer. Study of the metabolites was facilitated by the use of a stable isotope label [$^2H_4$]-(1). Metabolites were separated by TLC, GC, and HPLC and identified by GC–MS, following enzyme deconjugation where appropriate.

Unchanged (1) was detected in both the rat urine (5.3% of dose) and human urine (11.3%). Other major metabolites found in both rat and human urine were the mono- (2) and di- (3) N-demethylated compounds. Compound (2) accounted for 18.9% and 6% of the dose in rat and human respectively, and compound (3) for 10.7% (rat) and 1.3% (human) of the dose. The alcohol (5) and the acid (13) formed by oxidative deamination of (1) were identified in both rat and human urine. The acid (13) was 3% (rat) and 3.4% (human) of the administered dose with a further 1.7% of the rat dose being present as the glucuronide of (13). This glucuronide was not detected in the human urine. The only other metabolites common to both rat and human were (4) the N-oxide of (1) and metabolite (8). The other metabolites identified in rat urine were (6) [a hydroxylated derivative of metabolite (7)], (7) [the acetyl conjugate of metabolite (3)], (8) [a hydroxylated metabolite of compound (7)], (11) [a metabolite formed by hydroxylation of (1) in the pyridyl ring], and (12) [a product formed by N-dealkylation of metabolite (11)]. It was considered that compounds (9) and (10), an amide and ring-hydroxylated amide respectively, were formed artifactually.

The total amount of the major metabolites in the 96 hour rat urine was *ca.* 40%. This was in agreement with other studies suggesting that over 90% of a dose of radiolabelled (1) (divided almost equally between urine and faeces) was excreted over a similar time period. The total of the human dose excreted (82 hour urine) was *ca.* 22%, although it was considered that highly polar compounds (which were not extracted in the present study) may be present.

Reference

F. Kasuya, K. Igarashi, and M. Fukui, Metabolism of chlorpheniramine in rat and human by the use of stable isotopes, *Xenobiotica*, 1991, **21**, 97.

Aliphatic tertiary amines, *e.g.* diphenhydramine hydrochloride

Use/occurrence:	Antihistamines
Key functional groups:	Dimethylaminoalkyl
Test system:	Human volunteers (oral, 20–120 mg)

Structure and biotransformation pathway

(1)   (2)

H$_1$ antihistamines invariably contain at least one aliphatic tertiary amino functional group. In this study of alkylamines, ethanolamines, ethylenediamines, phenothiazines, and piperazines, aliphatic tertiary amines were administered orally to two healthy male volunteers, urine was collected over 36 hours, and the metabolites were identified by HPLC, FAB-MS, and comparison with authentic synthetic compounds. The mean percentages of the doses excreted as the N^+-glucuronide for the antihistamines were: cyclizine lactate (14.3), chlorpheniramine maleate (1.0), doxylamine succinate (0.9), pheniramine maleate (1.2), promethazine hydrochloride (0.2), pyrilamine maleate (0.6), tripelennamine hydrochloride (6.6), and diphenhydramine hydrochloride (4). The structure of the latter compound is shown (1) together with its N^+-glucuronide (2). This study is significant as previously only six other N^+-glucuronide metabolites have been described; this number has therefore been doubled. One implication is that the formation of metabolites such as (2), in humans, may be far more frequent than has previously been indicated. The quaternary ammonium-linked glucuronide metabolites are generated especially when the compound contains a dimethylaminoalkyl functional group.

Reference

H. Luo, E. M. Hawes, G. McKay, E. D. Korchinski, and K. K. Midha, N^+-Glucuronidation of aliphatic tertiary amines, a general phenomenon in the metabolism of H$_1$-antihistamines in humans, *Xenobiotica*, 1991, **21**, 1281.

Crisnatol

Use/occurrence: Anticancer agent

Key functional groups: Alkyl alcohol, tert-alkyl amine, polycyclic aromatic hydrocarbon

Test system: Rat (oral, intravenous; 5 mg kg^{-1})

Structure and biotransformation pathway

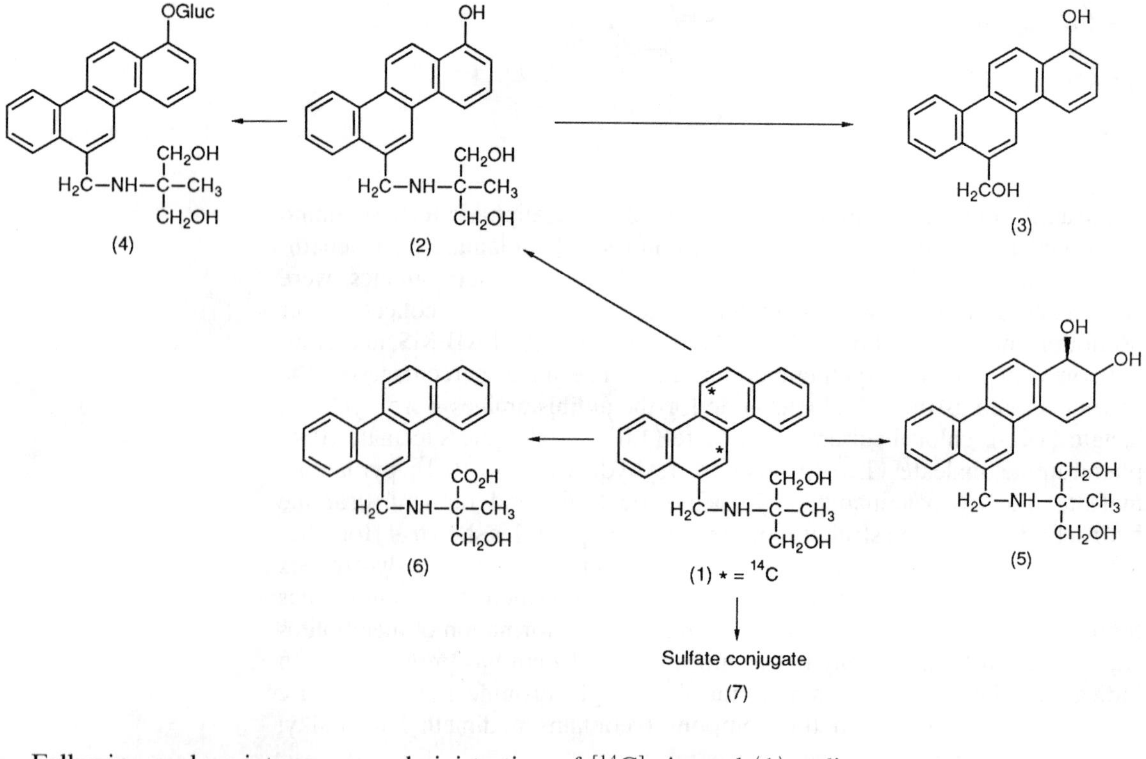

(4) (2) (3)

(6) (1) * = ^{14}C (5)

Sulfate conjugate

(7)

Following oral or intravenous administration of [^{14}C]crisnatol (1) radio-activity was mainly (81–92% dose) excreted via the faeces within 48 hours. Urine contained 6–12% dose. Unchanged (1) represented 17% of an oral dose in faeces, but quantitative data for other components were not so clearly presented. Three apparently major metabolites, (2), (3), and (5), and one lesser metabolite (6) were isolated from faeces and identified by GC–MS and ^{1}H NMR. The glucuronide conjugate (4) was isolated from urine and identified by FAB–MS and ^{1}H NMR. Enzymatic deconjugation experiments suggested that (4) was also excreted in faeces, together with a sulfate conjugate (7) of the parent compound. Metabolite (5) was the major radio-active component in urine. Phase I biotransformation of (1) involved oxidation of the aromatic ring system, leading to phenolic and dihydrodiol

176

metabolites; oxidation of the aliphatic alcohol group and deamination. Some lesser metabolites in urine and faeces were not identified.

Reference

D. K. Patel, J. L. Woolley, Jr., J. P. Shockcor, R. L. Johnson, L. C. Taylor, and C. W. Sigel, Disposition, metabolism and excretion of the anticancer agent crisnatol in the rat, *Drug Metab. Dispos.*, 1991, **19**, 491.

($\pm$)-Fenfluramine hydrochloride

Use/occurrence:	Appetite suppressant
Key functional groups:	Arylalkyl, dialkyl amine
Test system:	Human (oral, $1\ mg\ kg^{-1}$)

Structure and biotransformation pathway

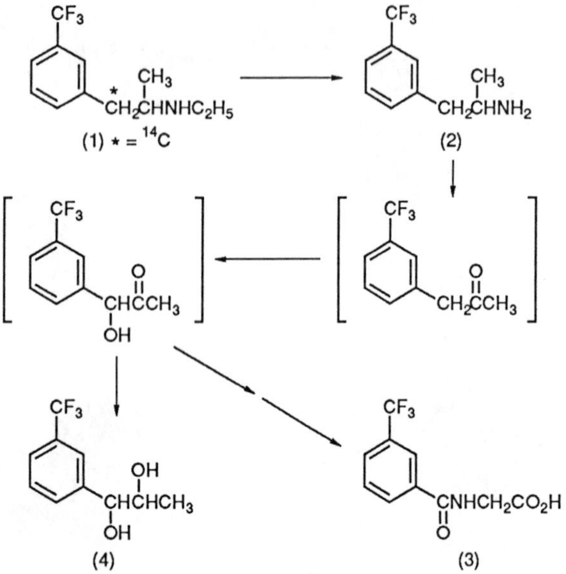

(1) $* = {}^{14}C$

(2)

(4)

(3)

More than 90% of an oral dose of fenfluramine (1) was excreted in urine of human volunteers (45% during 0–24 hours). Analysis of urine extracts by HPLC showed the presence of four metabolites. Identification of metabolites was performed by GC–MS after derivatization and by comparison with authentic standards. Three unconjugated components were identified namely fenfluramine, norfenfluramine (2), and the hippuric acid (3). The diol (4) was identified after treatment with β-glucuronidase, indicating its excretion as a conjugate. No quantitative data on the metabolites were presented.

Reference

J. Brownsill, D. Wallace, A. Taylor, and B. Campbell, Study of human urinary metabolism of fenfluramine using gas chromatography–mass spectrometry, *J. Chromatogr.*, 1991, **562**, 267.

178

3,4-(Methylenedioxy)methamphetamine

Use/occurrence:	Neurotoxin, drug of abuse
Key functional groups:	Acetal, secondary alkyl amine, arylalkyl
Test system:	Rat (subcutaneous or oral, 10 or 20 mg kg^{-1}), rat liver 9000g supernatant (0.1 mM substrate)

Structure and biotransformation pathway

All three regioisomers (2-, 5-, and 6-OH) of the monohydroxylated (1) and (2) were detected following the *in vitro* metabolism of (1). However, only the 6-OH metabolites (3) and (4) were detected *in vivo*. In further studies of the *in vitro* metabolism of (1) the trihydroxy metabolites (5) and (6) were detected; these metabolites were also detected in the urine of rats chronically dosed with (1). A further metabolite (7) was believed to be derived from (5). Incubations of (3) and (4) with rat liver 9000g supernatant were both found to yield (6), supporting the proposed metabolic scheme. Metabolites were determined, as either their acyl or silyl derivatives, by GC–MS using tandem mass spectrometry techniques. The identification of these pathways may explain the observed neurotoxicity of (1), as metabolites (5) and (6) were found to be neurotoxic to rats following their intracerebroventicular administration. (See also Vol. 2 p. 214, Vol. 3 p. 201.)

References

H. K. Lim and R. L. Foltz, *In vivo* formation of aromatic hydroxylated metabolites of 3,4-(methylenedioxy)methamphetamine in the rat: Identification by ion trap tandem mass spectrometric (MS/MS and MS/MS/MS) techniques, *Biol. Mass Spectrom.*, 1991, **20**, 677.

H. K. Lim and R. L. Foltz, Ion trap tandem mass spectrometric evidence for the metabolism of 3,4-(methylenedioxy)methamphetamine to the potent neurotoxins 2,4,5-trihydroxymethamphetamine and 2,4,5-trihydroxyamphetamine, *Chem. Res. Toxicol.*, 1991, **4**, 626.

3,4-(Methylenedioxy)methamphetamine (MDMA)

Use/occurrence:	Neurotoxin
Key functional groups:	Acetal
Test system:	Rat liver microsomes

Structure and biotransformation pathway

(1)   (2)   (3)   (4)

The selective neurotoxicity of 3,4-(methylenedioxy)methamphetamine (MDMA) (1), to serotonergic neurones is believed to be mediated by metabolites. The oxidative demethylation *in vitro* of (+)-MDMA and (−)-MDMA (20 μM) to the corresponding catecholamine, 3,4-dihydroxymethamphetamine (2), by rat liver microsomes was examined by HPLC using electrochemical detection. The reaction appeared to be dependent on cytochrome P-450 (sensitive to SKF 525A and carbon monoxide) and was significantly more rapid with the (+)-enantiomer. The product catecholamine was not stable and was further oxidized by an NADPH-dependent mechanism that required microsomal protein and which was inhibited by carbon monoxide but not by SKF 525A. Catecholamine oxidation was inhibited by ascorbic acid, by EDTA, and by superoxide dismutase but was unaffected by catalase, triethylenediamine, or benzoic acid. This supports a mechanism involving superoxide but not hydrogen peroxide, singlet oxygen, or hydroxyl radicals.

The product of the second oxidation [proposed as the quinone (3)] was reactive towards thiols forming discrete electrochemically active compounds. The product of reaction with glutathione co-chromatographed with a synthetic adduct formed between glutathione and (3) in a reaction catalysed by tyrosinase. The tyrosinase product (4) was identified from FAB and thermospray (LC–MS) mass spectra.

The study indicates that MDMA (1) is oxidized by cytochrome P-450 to a catecholamine which is further oxidized by superoxide to a quinone that may react with glutathione or other thiol compounds. Adduct formation between this quinone and thiols may contribute to some of the irreversible actions of this compound on serotonergic neurones.

Reference

M. Hiramatsu, Y. Kumagai, S. E. Unger, and A. K. Cho, Metabolism of methylenedioxymethamphetamine: Formation of dihydroxymethamphetamine and a quinone identified as its glutathione adduct, *J. Pharmacol. Exp. Ther.*, 1990, **254**, 521.

TA 870

Use/occurrence:	Dopamine pro-drug
Key functional groups:	Alkyl carbonate, alkyl amide
Test system:	Dog (intraduodenal, $mg\,kg^{-1}$)

Structure and biotransformation pathway

TA 870 (1) is a potential orally active pro-drug for dopamine (3). To investigate the biotransformation of (1) to the active entity (3), blood and plasma concentrations of unchanged (1) and its metabolites were measured (by HPLC) in the gastroduodenal vein, hepatic vein, and abdominal aorta after intraduodenal administration. ^{14}C-Labelled (1) was used but the position of labelling was not stated. The order of maximal concentrations (0–30 minutes) in gastroduodenal blood or plasma was (1) > desethoxycarbonylated TA 870 (2) > conjugated dopamine > dopamine (3) > homovanillic acid (5) > 3,4-dihydroxyphenylacetic acid (4). This was taken to indicate that catechol ester hydrolysis was the main pathway in the small intestine, and that amide hydrolysis, oxidative deamination, and catechol O-methylation were minor pathways. The order of maximal concentrations in hepatic venous plasma was conjugated (3) > (2) > (5) > (3) > (4) > conjugated (4) > (1). These results were interpreted as showing that dopamine (3) was rapidly produced and conjugated in the liver.

Some *in vitro* experiments with homogenates of liver, small intestine wall, and blood were performed but were not described in detail. The main metabolites formed were reported to be (2), (3), and 3- or 4-mono-desethoxycarbonylated (1) (no details of characterization given) and an unknown component.

Reference

M. Yoshikawa, S. Nishiyama, H. Endo, Y. Togo, and O. Takaiti, First pass metabolism and *in vitro* metabolism of a new orally active dopamine prodrug, *N*-(*N*-acetyl-L-methionyl)-*O,O*-bis(ethoxycarbonyl)dopamine in dogs, *Drug Metab. Dispos.*, 1991, **19**, 960.

N-(N-Acetyl-L-methionyl)-O,O-bis(ethoxycarbonyl)dopamine (TA 870)

Use/occurrence:	Dopamine pro-drug
Key functional groups:	Alkyl carbonate, alkyl amide
Test system:	Rat (oral, intravenous, 30 mg kg^{-1}), dog (oral, 33.5 mg kg^{-1})

Structure and biotransformation pathway

TA 870 (1) has been developed as an orally active pro-drug of dopamine (2). Acidic metabolites of TA 870 were isolated from urine and bile of rats by solvent extraction. Separated metabolites were isolated by preparative TLC and investigated by GC–MS after derivatization. Dog urine metabolites were determined by HPLC by comparison with authentic standards. The main urinary metabolites in rat and dog were dopamine (3), 3,4-dihydroxyphenylacetic acid (4), and homovanillic acid (5) and their conjugates. A small amount (2.3% dose) of the desethoxycarbonyl compound (2) was detected in dog urine after an oral dose. The total amounts of metabolites excreted by rat (r) and dog (d) in urine, free and conjugated, were dopamine (33%r, 32%d), (4) (5%r, 2%d), and (5) (12%r, 8%d). In rats and dogs the amounts of the phenylacetic acid (4) excreted, after oral doses of dopamine, were 2–18 times higher than those after equivalent doses of TA 870. This result indicated that oxidative deamination of dopamine in the liver and intestine was reduced by the acetylmethionyl protecting group on the amine moiety. A novel dopamine metabolite (6) was identified in rat urine. This dehydroxylated metabolite was probably formed by the intestinal flora.

Reference

M. Yoshikawa, H. Endo, K. Komatsu, Y. Sugawara, and O. Takaiti, Metabolism of a new orally active dopamine prodrug, N-(N-acetyl-L-methionyl)-O,O-bis(ethoxy-carbonyl(dopamine (TA 870) and dopamine after oral administration to rats and dogs, *J. Pharmacobio.-Dyn.*, 1990, **13**, 246.

2-(*N*-Propyl-*N*-2-thienylethylamino)-5-hydroxytetralin

Use/occurrence:	Dopamine agonist
Key functional groups:	Alkyl tert-amine, phenol, thiophene
Test system:	Rat (intravenous, *ca.* 20 μmol kg^{-1}; oral, 40 μmol kg^{-1}), hepatic microsomes (substrate concentration 57 and 60 μM), hepatic cytosol (114 μM), liver perfusion (100 μmol)

Structure and biotransformation pathway

(6)   (1) * = ^{3}H   (2)

(5)   (3)   (4)

The dopamine agonist, 2-(*N*-propyl-*N*-2-thienylethylamino)-5-hydroxy-tetralin (1) was administered to rats by the intravenous or oral route (using anaesthetized animals for some experiments with intravenous dosing). Clearance of (1) was rapid and independent of the dosing route or anaesthetic (urethane). The major route of excretion was biliary, accounting for 88% of the dose (6–8 hour bile), compared with 9% of the dose in the 6–8 hour urine. Less than 0.5% of the total radioactivity in the excreta was present as parent compound, indicating almost complete metabolism of (1).

The major urinary metabolite following intravenous administration of (1) was tentatively identified as a sulfate conjugate [but not metabolite (6)] from enzymic hydrolysis experiments. After oral dosing, however, the major urinary (and biliary) component was metabolite (2), the glucuronide of (1). These differences may be attributable to first-pass gastrointestinal metabolism.

184

Structure elucidation of the major metabolites was by ^{1}H NMR and FAB-MS following separation by HPLC, in combination with information from enzymic hydrolysis (β-glucuronidase/aryl sulfatase). All identified metabolites were conjugates and the major biliary metabolite was the glucuronide of (1), (2), which accounted for 50% (intravenous) or 65% (oral) of the dose. Hydroxylation of (1) at the 6-position yielded metabolite (3), a catechol intermediate which was not detected but excreted (almost exclusively in the bile) as the corresponding 5- and 6-glucuronides (4) and (5). Metabolite (3) [as the glucuronides (4) and (5)] accounted for ca. 13% (intravenous) or 9% (oral) of the dose, with (4) and (5) being detected in approximately equal amounts. Further investigation of (5) showed that the HPLC peak could be separated into two components in similar amounts, and these were identified as the diastereomeric 6-O-glucuronides of intermediate (metabolite) (3). A minor biliary metabolite was tentatively identified as the 5-sulfoconjugate (6).

In vitro experiments using perfused livers (perfusate and soluble fraction prepared from these livers), hepatic microsomes, or cytosol in the presence of the appropriate cofactors (UDPGA or PAPS respectively) demonstrated the formation of the conjugates seen in vivo. However, the major metabolites seen under oxidizing in vitro conditions using microsomes (in the presence of NADPH) were two very polar components. It was considered that these may be very hydrophilic compounds resulting from the loss of ^{3}H from the propyl side-chain and that dealkylation may be a route of metabolism of (1).

References

T. K. Gerding, B. F. H. Drenth, H. J. Roosenstein, R. A. de Zeeuw, P. G. Tepper, and A. S. Horn, The metabolic fate of the dopamine agonist 2-(N-propyl-N-2-thienyl-ethylamino)-5-hydroxytetralin in rats after intravenous and oral administration. I. Disposition and metabolic profiling, Xenobiotica, 1990, 20, 515.

T. K. Gerding, B. F. H. Drenth, R. A. de Zeeuw, P. G. Tepper, and A. S. Horn, The metabolic fate of the dopamine agonist 2-(N-propyl-N-2-thienylethylamino)-5-hydroxytetralin in rats after intravenous and oral administration. II. Isolation and identification of metabolites, Xenobiotica, 1990, 20, 525.

N-[1-Methyl-3-(3-phenoxyphenyl)prop-2-enyl]acetohydroxamic acid

Use/occurrence: Inhibitor of leukotriene synthesis

Key functional groups: Hydroxamic acid

Test system: Rabbit (oral, 50 mg kg^{-1})

Structure and biotransformation pathway

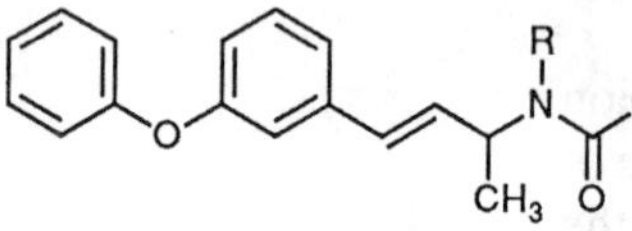

(1) R = OH
(2) R = OGluc
(3) R = H

HPLC analysis of plasma sample extracts from rabbits dosed orally with the hydroxamic acid (1) revealed the presence of two metabolites. One of these metabolites, which was more efficiently extracted at pH 3, was also detected in urine samples. Metabolites were investigated by LC–MS and GC–MS, the latter also after formation of TMS derivatives. One of the metabolites was assigned as a glucuronide (2) mainly based on its hydrolysis with β-glucuronidase to the parent compound. The other metabolite was confirmed as the acetamide (3).

Reference

P. M. Woollard, J. A. Salmon, and A. D. Padfield, Use of high-performance liquid chromatography–thermospray mass spectrometry and gas chromatography–electron-impact mass spectrometry in the identification of the metabolites of α-methylacetohydroxamic acids, potential anti-asthmatic agents, *J. Chromatogr.*, 1991, **562**, 249.

Verapamil

Use/occurrence:	Calcium channel antagonist
Key functional groups:	Alkyl tert-amine, benzyl cyanide
Test system:	Rat caecal contents (substrate concentration, 2 mM)

Structure and biotransformation pathway

The anaerobic metabolism of verapamil was studied to determine the role the intestinal flora may have in the disposition of verapamil (1). Verapamil was found to be metabolized by rat caecal contents under anaerobic conditions to nor-verapamil (2), thiocyanate anion (3), and a number of UV-absorbing products. Up to 80% of (1) was metabolized over a 6 hour incubation period, 40% of which was in the form of metabolite (2). The compound (2) disappeared with increased incubation time. No cyanide ion was detected.

In contrast, incubation of rat caecal contents under anaerobic conditions with other cyanide-containing compounds, amygdalin (4), butyronitrile (5), and acetonitrile (6), yielded both thiocyanate (3) and cyanide.

These results suggest that the cyano groups of (1) and (4)–(6) were all cleaved to give cyanide and thicyanate or thiocyanate alone in the case of (1). With verapamil, the cleavage of the cyano group may form new chemical entities which could be pharmacologically active and may lead to adverse reactions. N-Demethylation of (1) giving metabolite (2) is a reaction known to be catalysed by intestinal flora.

Reference

R. L. Koch and P. Palicharla, The anaerobic metabolism of verapamil in rat caecal contents forms nor-verapamil and thiocyanate, *J. Pharmacol. Exp. Therap.*, 1990, **254**, 612.

Labetolol

Use/occurrence:	Antihypertensive drug
Key functional groups:	Benzyl alcohol, phenol
Test system:	Bovine liver microsomes

Structure and biotransformation pathway

(1) $R^1 = R^2 = H$
(2) $R^1 = Gluc, R^2 = H$
(3) $R^1 = H, R^2 = Gluc$

Following incubation of labetolol (1) with bovine liver microsomes in the presence of [^{14}C]-UDP-glucuronic acid, four radioactive metabolites were separated by ion-pair HPLC, two of which could only be partially resolved. After isolation and further purification, each component was identified as a glucuronide conjugate of (1) by ^{1}H NMR, MS, and UV spectroscopy. One of the fully resolved conjugates was identified as the phenolic glucuronide (2), and the other as one of the two possible diastereoisomers of the alcoholic glucuronide (3). The partially resolved conjugates were both identified as (3) also, and were hence postulated to be enantiomers of the other diastereoisomer. Presumably, either only one diastereoisomer of (2) was produced or both were formed but were not separated under the conditions used. The total conversion of (1) into (2) and (3) was *ca.* 25% after the two hour incubation period, and no evidence was found for the formation of a diglucuronide conjugate of (1).

Reference

N. R. Niemeijer, T. K. Gerding, and R. A. de Zeeuw, Glucuronidation of labetolol at the two hydroxy positions by bovine liver microsomes, *Drug Metab. Dispos.*, 1991, **19**, 20.

1-[Bis(4-fluorophenyl)methyl]-4-(2,3,4-trimethoxybenzyl)piperazine dihydrochloride (KB-2796)

Use/occurrence:	Calcium antagonist
Key functional groups:	*N*-Alkylpiperazine, methoxyphenyl
Test system:	Rat (oral, 2 mg kg^{-1})

Structure and biotransformation pathway

The metabolites of KB-2796 (1), a synthetic calcium antagonist, were investigated in bile, urine, and faeces of male Wistar rats (170–299 g) after oral administration. The structures of the metabolites followed from inspection of TLC, MS, and ^{1}H NMR data, and comparison with authentic synthetic compounds. The main pathway of biotransformation of KB-2796 (1) is *O*-demethylation, which can occur at any of the three anisole positions in (1) yielding the three phenolic metabolites (2) and its two regioisomers. *C*-Hydroxylation occurs at position-5 of the trimethoxybenzyl moiety affording (3). *N*-Dealkylation of the piperazine ring affords (4) and (5) after alkyl cleavage at positions N-1 and N-4 respectively. Details of the 2-D TLC of the metabolites, the *m/z* fragments, and the assigned ^{1}H NMR were given for all the metabolites. Each of the metabolites was less potent than KB-2796 (1) as a cerebral vasodilator.

Reference

T. Kawashima, O. Satomi, and N. Awata, Isolation and identification of the new metabolites of 1-[bis(4-fluorophenyl)methyl]-4-(2,3,4-trimethoxybenzyl)piperazine dihydrochloride (KB-2796) from rat bile, urine and feces, *J. Pharmacobio.-Dyn.*, 1991, **14**, 449.

Enciprazine dihydrochloride

Use/occurrence:	Anxiolytic drug
Key functional groups:	Alkyl aryl ether, *N*-alkylpiperazine, methoxyphenyl
Test system:	Rat (oral, 20 mg kg^{-1}; intravenous, 20 mg kg^{-1}), dog (oral, 10 mg kg^{-1}; intravenous, 10 mg kg^{-1}), human volunteers (oral, 50 mg)

Structure and biotransformation pathway

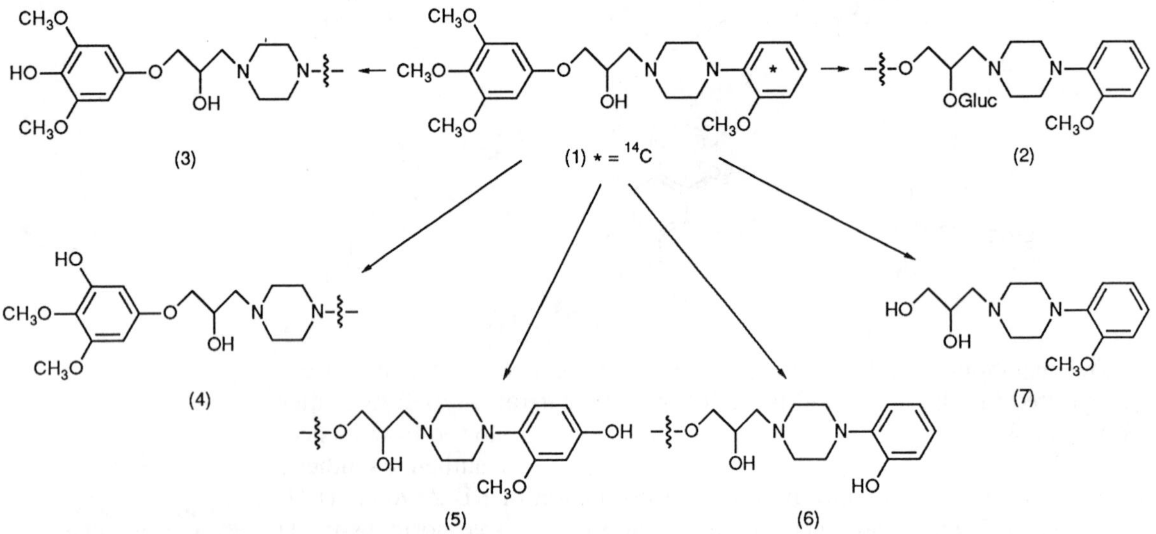

Enciprazine (1) is an anxiolytic drug with similar potency to diazepam, but it is not a benzodiazepine. The major metabolites of this substituted *o*-methoxyphenylpiperazine (OMPP) do not include OMPP, a 5-HT$_1$ serotonin agonist, although other drugs containing this moiety produce OMPP. The glucuronide (2), derived from enciprazine (1), was the major metabolite in human urine (11% of the dose in human urine); however, in dogs this was not a significant pathway. The glucuronide (2) was efficiently secreted into rat bile, but was absent in rat urine. The two regioisomeric desmethyl metabolites (3) (0.4%) and (4) (3.3%) and the hydroxylated metabolite (5) (4.6%) were formed by all species, with interspecies variation in the relative proportions. The desmethylphenylpiperazine metabolite (6) (1.9%) was found only in human urine. A product of *O*-dealkylation, the glycol (7) (2.8%), from elimination of the trimethoxyphenyl group, was found in rat bile and human urine.

The Phase I oxidative metabolites (3)–(7) were all present as their glucuronides. Detection and identification of the metabolites was by HPLC and MS. OMPP does not contribute to the anxiolytic activity of enciprazine. How the metabolites contribute, if at all, to the activity of (1) remains to be established.

Reference

J. A. Scatina, S. R. Lockhead, M. N. Cayen, and S. F. Sisenwine, Metabolic disposition of enciprazine, a nonbenzodiazepine anxiolytic drug, in rat, dog and man, *Xenobiotica*, 1991, **21**, 1591.

Mexiletine

Use/occurrence:	Antiarrhythmic drug
Key functional groups:	sec-Alkyl amine, alkyl aryl ether
Test system:	Rat (intraperitoneal, 10 mg kg^{-1}), mouse, rat, guinea-pig, hamster, and rabbit liver microsomes, human volunteers (oral, 200 mg)

Structure and biotransformation pathway

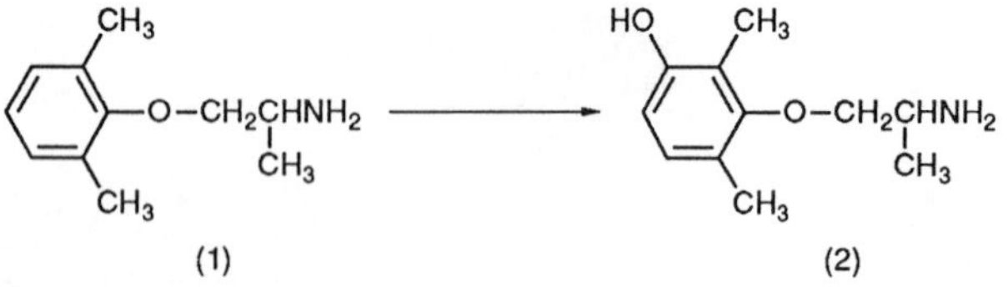

(1)   (2)

The *meta*-hydroxy metabolite (2) was isolated from the urine (acid- or enzyme-hydrolysed) of rats which had received intraperitoneal doses of mexiletine (1). The metabolite was identified using MS, ^{1}H NMR, and IR. Metabolite (2) was also produced in varying amounts by hepatic microsomes from various animal species, with the highest quantities being formed by the hamster, followed by rat and rabbit and very much lower quantities by mouse and guinea-pig. Metabolite (2) was also present as an acid-labile conjugate in human urine where it was estimated to represent 1.1–1.7% dose during 48 hours after oral administration. No quantitative data were presented for excretion of (2) in the rat.

Reference

O. Grech-Bélanger, J. Turgeon, M. Lalande, and P. M. Bélanger, *Meta*-Hydroxy-mexiletine, a new metabolite of mexiletine, *Drug Metab. Dispos.*, 1991, **19**, 458.

Bucromarone

Use/occurrence:	Antiarrhythmic drug
Key functional groups:	Alkoxyphenyl, arylmethyl, dialkylaminoalkyl
Test system:	Rat (oral and intravenous, 4.4 mmol kg^{-1})

Structure and biotransformation pathway

(2)

(1) $* = {}^{14}C$

(4)

(3)

After oral or intravenous administration of [^{14}C]bucromarone (1) hydrochloride or succinate to rats, at least 90% of the dose was excreted in the faeces during 72 hours and less than 7% dose in the urine. Biliary excretion accounted for *ca.* 75% of an intravenous dose of (1) and slightly less after an oral dose. HPLC analysis of pooled 0–4 hour bile after enzymatic hydrolysis with β-glucuronidase/sulfatase indicated the presence of four major radioactive components, and samples of each were isolated for MS and chromatographic comparison with reference compounds. Three of these components were unequivocally identified as unchanged (1) and metabolites (2) and (3), which represented 15%, 20%, and 46% of the biliary radioactivity respectively, whereas the fourth component, which accounted for only 4% of the biliary radioactivity, was tentatively identified as (4) from its chromatographic mobility alone. The same four compounds were also determined to be the main radioactive components in plasma at 1 hour after oral and intravenous drug administration. Preliminary studies have shown that the mono- and di-desbutyl metabolites, (2) and (3), possess simiar antiarrhythmic properties to (1) itself.

Reference

J. C. Maurizis, C. Nicolas, M. Verny, M. Ollier, M. Faurie, M. Payard, and A. Veyre, Biodistribution and metabolism in rats and mice of bucromarone, *Drug Metab. Dispos.*, 1991, **19**, 94.

Carvedilol

Use/occurrence:	Antihypertensive drug
Key functional groups:	Alkyl aryl ether, sec-alkyl alcohol, dialkyl amine, indole
Test system:	Rat (oral, 10 mg kg^{-1}; intravenous, 1 mg kg^{-1})

Structure and biotransformation pathway

(1) X = OH, Y = H; * = ^{14}C
(2) X = H, Y = OH; * = ^{14}C

(3) X = OH, Y = H
(4) X = H, Y = OH

(5) X = OH, Y = H
(6) X = H, Y = OH

The objectives of this study were to determine the biliary excretion of the R(+)- and S(−)-enantiomers of carvedilol [(1) and (2) respectively] and the stereochemical composition of the principal metabolites excreted in the bile of rats after dosing with reconstituted enantiomerically radiolabelled pseudo-racemates of the drug. Thus, [^{14}C]-(1)/non-labelled (2) and [^{14}C]-(2)/non-labelled (1) were separately administered orally and intravenously to groups of bile duct-cannulated rats, and the bile obtained was analysed by non-chiral reverse-phase HPLC. The total biliary excretion of radioactivity from the radiolabelled R(+)- and S(−)-enantiomers was 41.4% and 41.5% respectively of the oral dose, and 43.7% and 40.0% respectively of the intravenous dose. After oral dosing of the pseudoracemates, no enantiomeric difference was found in the rates of biliary excretion of radioactivity, whereas after intravenous dosing this rate was significantly greater for the R(+)-enantiomer than for its optical antipode, possibly as a result of the larger volume of distribution of the latter isomer. After administration by both routes, the major radioactive components detected in bile were the known metabolites, 1-hydroxycarvedilol O-glucuronide [(3), (4)] and 8-hydroxycarvedilol O-glucuronide [(5), (6)], which together accounted for more than 60% of the dose. The ratios of these metabolites (as measured by HPLC with UV detection) were similar after oral or intravenous administration of the two enantio-merically radiolabelled pseudoracemates, but there were considerable differences in the *S/R* enantiomer ratio of each metabolite, *viz.* 0.59 and 0.43 for the 1-hydroxy metabolite after oral and intravenous dosing respect-ively, and 3.29 and 2.63 respectively for the 8-hydroxy metabolite. These results suggest that formation of the former metabolite is selective for the *R*(+)-enantiomer, while that of the latter is selective for the *S*(−)-enantio-mer, and that stereoselective metabolism in the carbazole ring hydroxylation pathways is responsible for the stereochemical composition of the glucuro-nides.

Reference

M. Fujimaki, S. Shintani, and H. Hakusui, Stereoselective metabolism of carvedilol in the rat, *Drug Metab. Dispos.*, 1991, **19**, 749.

Carvedilol

Use/occurrence:	Antihypertensive drug
Key functional groups:	Indole
Test system:	Rat (oral, 10 mg kg^{-1})

Structure and biotransformation pathway

(1) $* = {}^{14}C$, † = chiral carbon

(3)

The biliary metabolites of [^{14}C]-R,S-carvedilol (1) were investigated in cannulated male rats; similar studies were conducted with non-radiolabelled forms of the individual R- and S-enantiomers in order to investigate the stereoselectivity of the metabolic pathways. Metabolites were isolated from desalted bile by Sephadex chromatography and HPLC, and structures were assigned by a combination of FAB- and EI-MS and by ^{1}H NMR. 85% of the administered radioactivity was excreted in bile, predominantly as two phenolic glucuronide conjugates (2) and (3) which resulted from initial hydroxylation of the carbazole ring, and these accounted for 39% and 22% of the dose respectively. Unchanged (1) accounted for less than 1% of the dose. Both (2) and (3) were hydrolysed by β-glucuronidase. Administration of R-(1) led to highly selective excretion of (2), whereas S-(1) resulted predominantly in excretion of (3) rather than (2). This finding suggests that the majority of the 8-hydroxy metabolite produced by R,S-(1) arises from its S-(1) component. The preferential 8-hydroxylation of S-(1) may explain why its oral bioavailability in rats is only half that of R-(1).

Reference

M. Fujimaki and H. Hakusui, Identification of two major biliary metabolites of carvedilol in rats, *Xenobiotica*, 1990, **20**, 1025.

Metoprolol

Use/occurrence:	β-Adrenergic blocker
Key functional groups:	sec-Alkyl alcohol, isoalkylamino, alkyl aryl ether, dialkyl ether
Test system:	Dog (oral, 1.37 mg kg^{-1}; intravenous, 0.51 mg kg^{-1})

Structure and biotransformation pathway

(1) R = ^{1}H
(2) R = ^{2}H

† = chiral carbon

(3)

(4)

(5)

In view of the known inhibitory effects of some calcium channel blockers on the hepatic clearance of a number of other drugs, the objective of the present study was to establish whether any such pharmacokinetic interaction exists between verapamil and metoprolol (1). To this end, plasma concentration–time profiles of the enantiomers of (1) and the urinary excretion of its known principal metabolites (3)–(5) were determined by quantitative GC using a mass-selective detector following single oral and intravenous doses to dogs of deuterium-labelled pseudo-racemic metoprolol (2), *viz.* an equimolar mixture of (2S)-[^{2}H$_6$]-(1) and (2R)-[^{2}H$_0$]-(1), with and without simultaneous oral administration of racemic verapamil (3 mg kg^{-1}). Co-administration of verapamil resulted in a 50–70% inhibition in both the oral and systemic clearance of (2) and in complete abolition of its first-pass effect. The hepatic clearance of S-(2) tended to be higher than that of R-(2), and its inhibition by verapamil was selective towards the former enantiomer, resulting in no over-all enantioselectivity in hepatic clearance when verapamil was co-administered. After intravenous administration of (2), metabolites (3), (4), and (5) in urine accounted for 43%, 14%, and 6% of the dose respectively, and a further 9% dose was recovered as unchanged drug. Similar proportions of these four compounds were found in urine after oral dosing, so together they accounted for *ca.* 70% of the dose after each route of administration. When verapamil was administered simultaneously, a modest (14–34%) decrease in the recovery of the three metabolites was found, relative to the control values, whereas the recovery of parent drug increased three- to four-fold. *N*-Dealkylation was highly enantioselective towards *S*-(1) (*R/S* ratio *ca.* 0.6), whereas *O*-demethylation and α-hydroxylation showed a small degree of selectivity towards the *R*-enantiomer (*R/S* ratio *ca.* 1.1), resulting in the observed slight *S*-enantioselectivity in the overall hepatic clearance. The formation clearances of all three metabolites were inhibited by 60–80% following co-administration of verapamil, and this inhibitory effect was found to be greatest for the *O*-demethylation pathway, which (like the *N*-dealkylation pathway) was apparently selective towards the *S*-enantiomer. These observations suggest that the greater verapamil-induced inhibition of hepatic clearance found for *S*-metoprolol compared with its optical antipode was due to the *S*-enantioselective inhibition in *O*-demethylation.

Reference

S. S. Murthy, W. L. Nelson, D. D. Shen, J. M. Power, C. M. Cahill, and A. J. McLean, Pharmacokinetic interaction between verapamil and metoprolol in the dog, *Drug Metab. Dispos.*, 1991, **19**, 1093.

Naftopidil

Use/occurrence:	Antihypertensive drug
Key functional groups:	Alkyl aryl ether, N-alkylpiperazine methoxyphenyl, naphthalene
Test system:	Mouse (oral, 5 mg kg^{-1}), rat (oral, 5 mg kg^{-1}; intravenous, 1 mg kg^{-1}), dog (oral 10 mg kg^{-1}; intravenous, 2.5 mg kg^{-1}), man (oral, 50 mg; intravenous, 5 mg)

Structure and biotransformation pathway

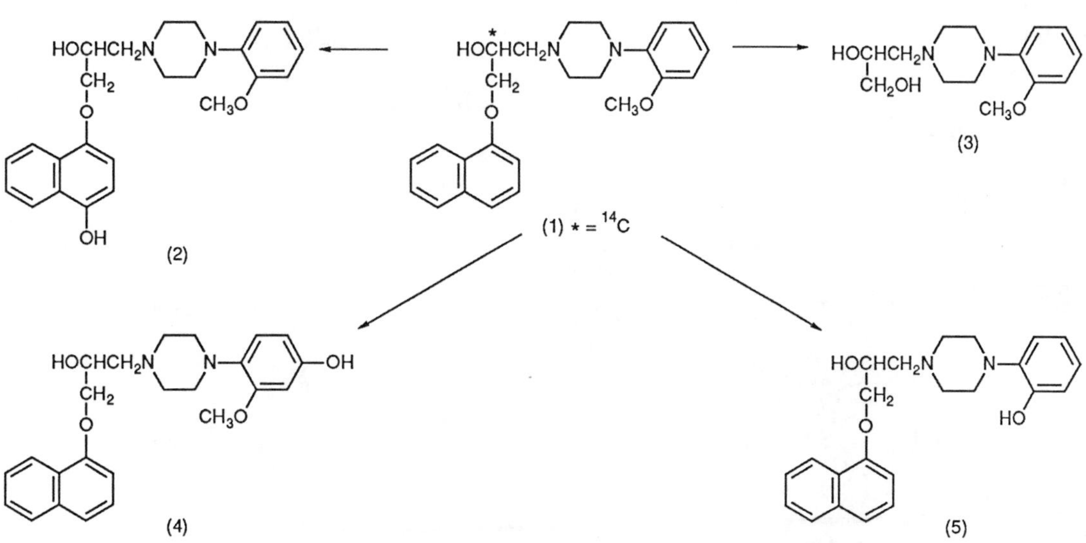

Oral and intravenous doses of [^{14}C]naftopidil (1) were mainly excreted in the faeces by rats and dogs. In the mouse and man excretion of radioactivity was more evenly divided between urine and faeces. Urine apparently contained mainly conjugates. The Phase I metabolites were isolated after enzymatic hydrolyis and were identified by MS, although no details were given. Faeces contained unchanged (1) and Phase I metabolites. The biotransformation of (1) was qualitatively similar in all four species. The identified metabolites (2)–(5) were formed by a variety of oxidative pathways, including hydroxylation of the naphthalene and phenyl ring systems, O-demethylation, and cleavage of the naphthyl ether linkage. A considerable proportion of 'unknown' material was also reported. Aromatic hydroxylation was generally

199

the most important route while demethylation only occurred to a minor extent.

Reference

G. Niebch, M. Locher, G. Peter, and H. O. Borbe, Metabolic fate of the novel anti-hypertensive drug Naftopidil, *Arzneim.-Forsch.*, 1991, **41(II)**, 1027.

Amitriptyline

Use/occurrence:	Antidepressant drug
Key functional groups:	Dibenzocycloheptane, dimethylaminoalkyl
Test system:	Human female patients (oral, 100–300 mg), human volunteers (oral, 10 and 50 mg)

Structure and biotransformation pathway

(5) Gluc = glucuronyl

Previous studies in humans and animals have shown the principal hydroxylated metabolites of amitriptyline (1) to be the isomeric alcohols *E*- and *Z*-10-hydroxyamitriptyline (2) and -nortriptyline. These metabolites occur in human urine predominantly as conjugates hydrolysable with β-glucuronidase/arylsulfatase. The *N*-glucuronide of amitriptyline (3) has also been described. This study confirms the occurrence of amitriptyline-*N*-glucuronide and describes the *N*-glucuronide conjugates of *E*- and *Z*-10-hydroxyamitriptyline (4) and 10,11-*trans*-dihydroxyamitriptyline (5) as urinary metabolites of amitriptyline. Conjugated metabolites were isolated from urine using a combination of solid-phase extraction, TLC, and HPLC and purified glucuronides were characterized by ^{1}H NMR and FAB-MS. In addition, conjugates were hydrolysed with β-glucuronidase and the liberated aglycones compared with reference compounds by TLC. The *N*-glucuronide of amitriptyline (3) represented 3–21% of the dose during 24 hours after dosing. Conjugates of 10-hydroxyamitriptyline totalled 4–12% dose, about half of which was present as the *N*-glucuronide.

In supplementary experiments, the same volunteer received an oral dose of *E*-10-hydroxyamitriptyline (20 mg) and an intravenous infusion of amitriptyline *N*-glucuronide (17.5 mg, 0.72 mg min^{-1}) which was interrupted when flushing and tachycardia developed. 10-Hydroxyamitriptyline-*N*-glucuronide was recovered as a minor metabolite from urine (oral dose, 2.8% in 16 hours; intravenous dose, 1.4% in 12 hours). From rate calculations, it was proposed that *N*-glucuronidation of (2) and 10-hydroxylation of (3) each contributed to the formation of (4).

Reference

U. Breyer-Pfaff, B. Becher, E. Nusser, K. Nill, B. Baier-Weber, D. Zaunbrecher, H. Wachsmuth, and A. Prox, Quaternary *N*-glucuronides of 10-hydroxylated amitriptyline metabolites in human urine, *Xenobiotica*, 1990, **20**, 727.

Doxepin

Use/occurrence:	Antidepressant drug
Key functional groups:	Dimethylaminoalkyl
Test system:	Human patients (oral, 100 or 300 mg day^{-1})

Structure and biotransformation pathway

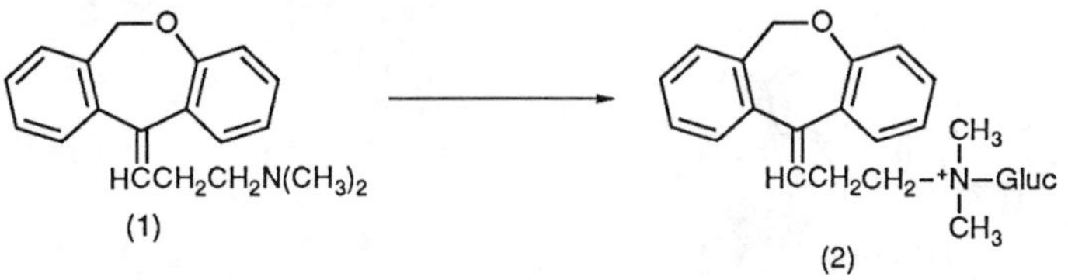

(1)   (2)

This short communication describes the isolation and characterization (by FAB-MS and ^{1}H NMR) of the quaternary ammonium-linked glucuronide conjugate (2) from the urine of human patients undergoing therapy with doxepin (1). A reference standard of the glucuronide was also synthesized chemically from doxepin. HPLC analysis of urine indicated that the 24 hour excretion of (2) accounted for *ca.* 20% of the daily dose of (1).

Reference

H. Luo, E. M. Hawes, G. McKay, E. D. Korchinski, and K. D. Midha, The quaternary ammonium-linked glucuronide of doxepin: a major metabolite in depressed patients treated with doxepin, *Drug Metab. Dispos.*, 1991, **19**, 722.

Doxepin

Use/occurrence:	Antidepressant drug
Key functional groups:	Arylalkene, dibenzoxepin, dimethylaminoalkyl
Test system:	Human (oral, 50 mg)

Structure and biotransformation pathway

(1) R = CH$_3$
(3) R = H

(2) R = CH$_3$
(4) R = H

After oral administration of equal amounts of Z-[^{2}H$_0$]doxepin (1) and E-[^{2}H$_4$]doxepin (2) hydrochlorides to eight healthy male volunteers, serum and urine levels of each parent drug and its N-desmethyl metabolite, (3) and (4) respectively, were monitored by quantitative GC–MS. The results showed that the geometrical isomers of doxepin did not undergo interconversion *in vivo*, but E-doxepin was metabolized to Z-N-desmethyldoxepin in significant quantities, the urinary Z/E ratio ranging from 0.08 to 3.06. By contrast, only small amounts of E-N-desmethyldoxepin were formed from Z-doxepin in most, but not all, subjects. It is proposed that isomerization occurred during the N-demethylation process, and that as mass spectroscopic evidence has recently been found for the presence of a hydrated metabolite of doxepin in urine, such an intermediate may be involved in the dealkylation pathway. Subsequent dehydration regenerating the exocyclic double bond could then theoretically lead to the formation of both E- and Z-desmethyl metabolites.

Reference

H. Ghabrial, C. Prakash, U. G. Tacke, I. A. Blair, and G. R. Wilkinson, Geometric isomerization of doxepin during its N-demethylation in humans, *Drug Metab. Dispos.*, 1991, **19**, 596.

Miscellaneous Alicyclics, Aromatics, and Macrocyclics

Illudin S

Use/occurrence:	Fungal toxin
Key functional groups:	Cyclohexenone, cyclopropane
Test system:	Rat liver 9000 g supernatant

Structure and biotransformation pathway

The toxin Illudin S (1) recovered from the mushroom *Lampteromyces japonicus* is characterized by a cyclopropane ring. The compound is reported to have antibacterial and antitumour properties. This study describes the isolation and identification of metabolites of illudin S formed by rat liver 9000 g supernatant.

Illudin S (1 mM) was incubated for 30 minutes at 37 °C with the rat liver preparation (250 ml, equivalent to 33 g liver) at a pH of 7.4. Metabolites (2) and (3) were extracted with ethyl acetate, purified by TLC and recrystallization, and characterized by melting point, UV, IR, NMR (^{1}H, ^{13}C, and 2D-NMR), and high-resolution mass spectrscopy. Appropriate controls and a time-course experiment indicated that formation of (2) and (3) were enzyme-dependent and complete within the incubation period.

Metabolites (2) and (3) were identified as cyclopropane ring-cleavage compounds. Compound (3) is novel in having a chlorine inserted into the molecule. A mechanism for the metabolism of (1) is proposed which involves an initial reduction at the α,β-unsaturated carbonyl system to give a reactive intermediate (4) from which elimination of the 2-hydroxy group and nucleophilic cleavage of the cyclopropane ring might lead to (2) and (3).

Other metabolites of (1) are likely since (2) and (3) accounted for only *ca.* 50–60% of the substrate.

Reference

K. Tanaka, T. Inoue, S. Kadota, and T. Kikuchi, Metabolism of illudin S, a toxic principle of *Lampteromyces japonicus*, by rat liver. I. Isolation and identification of cyclopropane ring-cleavage metabolites, *Xenobiotica*, 1990, **20**, 671.

MC-969

Use/occurrence:	Analogue of the calcemic drug 1α-hydroxyvitamin D_3
Key functional groups:	Cyclopropylalkyl
Test system:	Cultured hepatocytes (Hep 3B cell line, substrate concentrations 5 and 10 μg ml^{-1})

Structure and biotransformation pathway

(2) X = OH, Y = H
(3) X = H, Y = OH

(5) (1) (4)

Two analogues of hydroxylated forms of vitamin D, (1) and (5), were synthesized in which the carbon atoms at 25, 26, and 27 positions were incorporated into a cyclopropane ring. Both compounds were found to have potent calcemic effects *in vivo* in rats. It was expected that (1) would be hydroxylated in the liver at the 25 position to produce (5) as the active form, in a manner similar to that seen in the activation of 1α-hydroxyvitamin D_3. The liver vitamin D_3-25-hydroxylase has been shown to possess quite a broad substrate specificity and activity is found in both mitochondrial and microsomal cell fractions. The microsomal vitamin D_3-25-hydroxylase has been purified and is thought to be P-450h. In this study there was no evidence of any 25-hydroxylated metabolites. Three metabolites of (1) found *in vitro* in the human hepatocyte model were all products of 24-hydroxylation. The identity of the metabolites (2), (3), and (4) was established by HPLC co-chromatography and by mass spectrometry of native and chemically modified metabolites. Preliminary competition studies with vitamin D_3 suggested that the vitamin D_3-25-hydroxylase may have been involved in 24-hydroxylation of (1).

Reference

S. Strugnell, M. J. Calverley, and G. Jones, Metabolism of a cyclopropane-ring-containing analog of 1α-hydroxyvitamin D_3 in a hepatocyte cell model, *Biochem. Pharmacol.*, 1990, **40**, 333.

Pravastatin sodium

Use/occurrence: Cholesterol-lowering drug

Key functional groups: Cycloalkene

Test system: Rat hepatocytes

Structure and biotransformation pathway

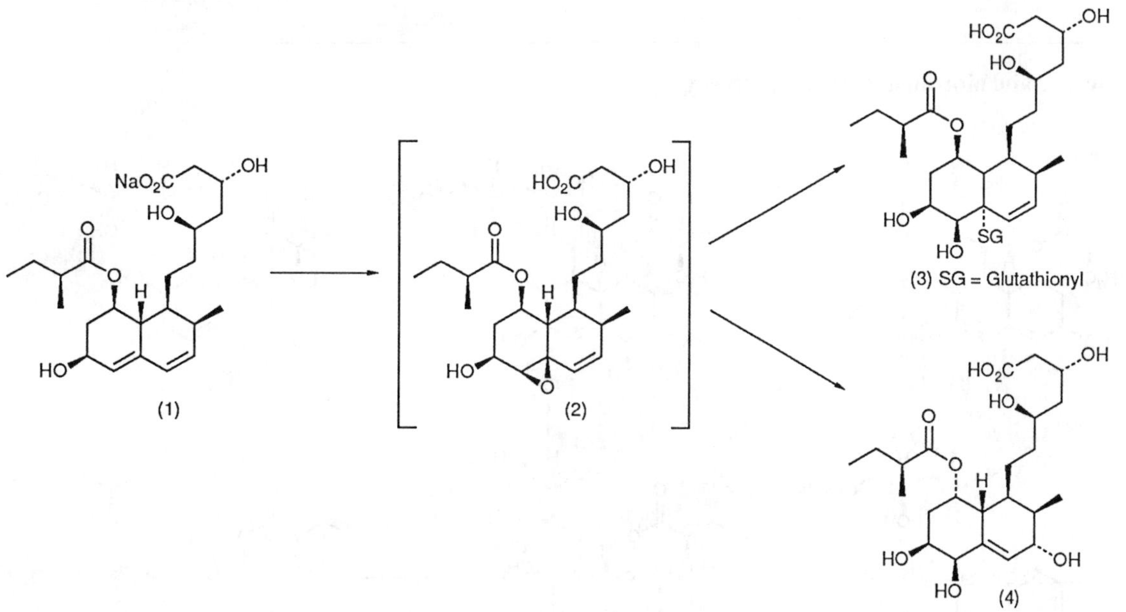

Pravastatin sodium (1) is a potent inhibitor of 3-hydroxy-3-methylglutaryl-coenzyme A reductase and has been developed as a cholesterol-lowering agent in humans. Two major metabolites have been identified by ^{1}H and ^{13}C NMR following incubation of pravastatin sodium with isolated rat hepatocytes. These were the 4′aα-glutathione conjugate (3) and the 3′,5′-dihydrodiol (4). It is proposed that a common metabolic intermediate between (1) and the two metabolites (3) and (4) is the 4′aβ,5β-epoxide (2) which was synthesized chemically. The present result is unusual in that it appears to be the first example of a glutathione conjugate of a drug shown to have an angular glutathione-S-yl group.

Reference

T. Nakamura, K. Yoda, H. Kuwano, K. Miyaguchi, S. Muramatsu, H. Takahagi, and T. Kinoshita, Metabolism of pravastatin sodium in isolated rat hepatocytes. II. Structure elucidation of the metabolites by n.m.r. spectroscopy, *Xenobiotica*, 1991, **21**, 227.

Pravastatin sodium

Use/occurrence:	Cholesterol-lowering drug
Key functional groups:	Alkyl carboxylic acid, allylic alcohol, cycloalkene, hexahydronaphthalene, isoalkyl ester
Test system:	Human (oral, 19.2 mg; intravenous, 9.9 mg)

Structure and biotransformation pathway

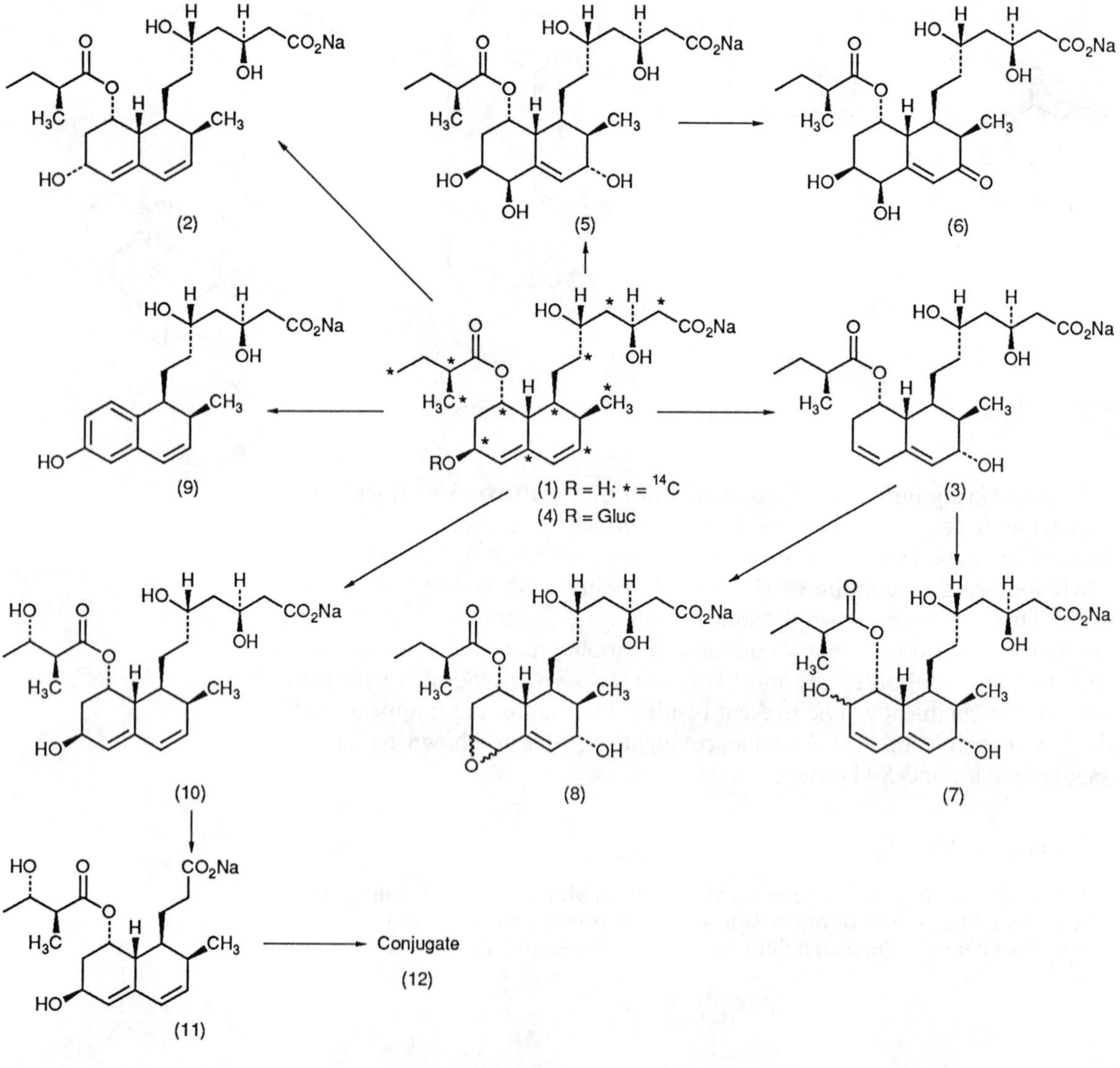

210

Following single oral or intravenous doses of [^{14}C]pravastatin sodium (1) to healthy male human subjects, mean totals of 20% and 59% of the dose respectively were excreted in urine and 71% and 34% dose respectively were excreted in the faeces. Patterns of radioactive components in pooled samples of plasma, urine, and faeces were determined by HPLC, and the results showed that the principal drug-related component in each of these body fluids corresponded to parent compound. In the case of urine, unchanged (1) accounted for 21% of the urinary radioactivity after oral dosing and 69% after intravenous dosing. Two metabolites also identified in this way were the isomers (2) and (3), but both of these are known non-enzymatic acid-catalysed degradation products of (1) and so may have been formed before absorption. In addition to these three compounds and their lactones (which were only present in trace quantities), 15 other metabolites were resolved, none of which accounted for more than 6% of the total urinary radioactivity. Samples of the more important metabolites were isolated after an aliquot of urine from the radiotracer study was combined with urine obtained from other healthy human subjects who had received single oral doses of non-labelled (1) (40 mg). The pooled urine was concentrated on XAD-2 resin, extracted with ethyl acetate, and subjected to extensive purification by HPLC. In this way, unchanged (1), its isomers (2) and (3), and eight other metabolites [(4)–(11)] were identified by a combination of HPLC, UV, NMR, and MS (including comparison with the synthetic reference compounds in some cases). The principal biotransformation pathways (excluding isomerization) of (1) in man therefore involve: ($\omega - 1$) oxidation of the ester side-chain; β-oxidation of the heptanecarboxylic acid side-chain; ring oxidation and subsequent aromatization; oxidation of a ring hydroxy group to the ketone; and conjugation (including glucuronidation). In contrast to the structurally related compound lovastatin, none of the major metabolites of (1) showed any significant activity when tested as inhibitors of 3-hydroxy-3-methylglutaryl-coenzyme A reductase, the rate-limiting enzyme in cholesterol biosynthesis.

Reference

D. W. Everett, T. J. Chando, G. C. Didonato, S. M. Singhvi, H. Y. Pan, and S. H. Weinstein, Biotransformation of pravastatin sodium in humans, *Drug Metab. Dispos.*, 1991, **19**, 740.

(+)-(1*R*,4a*S*,10a*R*)-1,2,3,4,4a,9,10,10a-
Octahydro-1,4a-dimethyl-7-(1-methylethyl)-
6-sulfo-1-phenanthrenecarboxylic acid
6-sodium salt pentahydrate (TA-2711)

Use/occurrence:	Anti-ulcer drug
Key functional groups:	Alkyl carboxylic acid
Test system:	Rat (oral, 100 mg kg^{-1}), dog (oral, 100 mg kg^{-1})

Structure and biotransformation pathway

TA-2711 (1) is a new anti-ulcer drug and the absorption of an oral dose to rats was 3.4–7.0%. After oral dosing, the plasma radioactivity peaked at 6 hours in rats and at 2 hours in male beagle dogs. There was little transfer of TA-2711 (1) into tissue from plasma (after either an oral or an intravenous dose); whole body autoradiograms of rats showed that most of the oral dose was localized in the gastrointestinal tract. The urinary excretion of radioactivity was low, the faecal excretion extremely high. The sole metabolite (0.25% of the dose in dogs) was a β-glucuronide, tentatively assigned as the acyl glucuronide (2). This metabolite was also present in rat urine (0.03% of the dose). No biliary or faecal metabolites were detected in rats after an oral dose and the urinary metabolite (2) was not detected after intravenous administration of TA-2711 (1) to rats. TLC and HPLC analytical methods were used. The low oral absorption of (1) supports its local action as an anti-ulcer drug.

References

Y. Ito, Y. Sugawara, O. Takaiti, and S. Nakamura, Metabolic fate of a new anti-ulcer drug (+)-(1*R*,4a*S*,10a*R*)-1,2,3,4,4a,9,10,10a-octahydro-1,4a-dimethyl-7-(1-methyl-ethyl)-6-sulfo-1-phenanthrenecarboxylic acid 6-sodium salt pentahydrate (TA-2711). 2. Distribution in the rat stomach, *J. Pharmacobio.-Dyn.*, 1991, **14**, 547.

Y. Ito, T. Fukushima, Y. Sugawara, O. Takaiti, and S. Nakamura, Metabolic fate of a new anti-ulcer drug (+)-(1*R*,4a*S*,10a*R*)-1,2,3,4,4a,9,10,10a-octahydro-1,4a-dimethyl-7-(1-methylethyl)-6-sulfo-1-phenanthrenecarboxylic acid 6-sodium salt pentahydrate (TA-2711). 1. Disposition, metabolism and protein-binding in rats and dogs, *J. Pharmacobio.-Dyn.*, 1991, **14**, 533.

Mitoxantrone

Use/occurrence:	Anticancer agent
Key functional groups:	Anthraquinone, alkyl alcohol, dialkyl amine
Test system:	Pig (intravenous, 4–6.5 mg kg^{-1})

Structure and biotransformation pathway

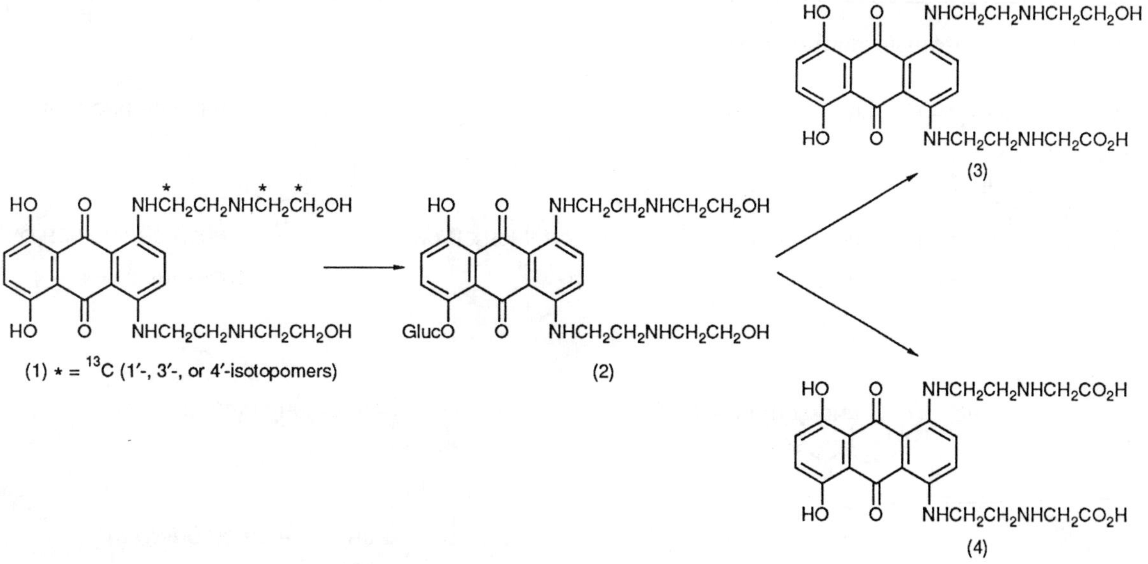

(1) $* = {}^{13}$C (1′-, 3′-, or 4′-isotopomers)

(2)

(3)

(4)

Mitoxantrone (1) was administered to anaesthetized mini-pigs by intravenous bolus, and urine and bile were collected over a 5 hour period. Urine was dialysed and the metabolites were separated using a step gradient with a PRP-1 resin column, before being further isolated by HPLC and identified by ^{13}C NMR and MS.

Apart from unchanged (1), the main urinary metabolite was the glucuronide conjugate (2) together with the mono- and di-acids (3) and (4). In bile unchanged (1) was the major component with small amounts of (3) and (4).

Reference

J. Blanz, K. Mewes, G. Ehninger, B. Proksch, B. Greger, D. Waidelich, and K.-P. Zeller, Isolation and structure elucidation of urinary metabolites of mitoxantrone, *Cancer Res.*, 1991, **51**, 3427.

213

Mitoxantrone

Use/occurrence:	Anticancer drug
Key functional groups:	Alkyl alcohol, anthraquinone, dialkyl amine, aryl amine, phenol
Test system:	Rat (intravenous, 9.7 mg kg^{-1}), pig (intravenous, 5.2 and 6.5 mg kg^{-1}), human (12 mg m^{-2})

Structure and biotransformation pathway

(1) $R^1 = R^2 = CH_2OH$
(2) $R^1 = CO_2H$, $R^2 = CH_2OH$
(3) $R^1 = R^2 = CO_2H$

(4)

(5) GS = glutathionyl

(6)

(7)

Following administration of [^{14}C]mitoxantrone (1) to rats, 2.3% of the dose was excreted in urine during 5 days, and HPLC analysis with UV and radiochemical detection showed that neither of the known human urinary metabolites (2) and (3) were present in this urine. However, several other more-polar metabolites and one less-polar metabolite were detected. The latter metabolite was also found in the urine of human cancer patients undergoing chronic treatment with (1), and in the urine and bile of treated pigs. A purified sample of this metabolite isolated from human urine was identified as (4) by MS comparison with the synthetic reference compound, using positive and negative CI-MS and CAD daughter ion spectra. Further evidence for this structure was obtained by HPLC and UV–visible spectroscopic comparison with authentic (4), which was prepared by incubating (1) with H_2O_2 and horseradish peroxidase, and its structure unequivocally characterized by ^{13}C NMR and MS–MS. When the incubation was performed in the presence of glutathione, two glutathione conjugates of (1) were formed, which were similarly identified as (5) and (6). A mechanism was proposed for the *in vitro* enzyme-catalysed oxidation of (1) to the observed products (4)–(6), which involves formation of a highly reactive quinone-di-imine intermediate (7) via a two-electron oxidation of the phenylenediamine moiety. As previous studies have shown that the known cytochrome P-450 inhibitor metyrapone prevents conjugation of (1) and its cytotoxic effects, it is likely that microsomal P-450 is the enzyme system responsible for the oxidative biotransformation of (1) *in vivo*. Since it is also known that the electrophilic quinone-di-imine intermediate (7) can bind covalently to nucleic acids, alkylation of DNA and/or cellular proteins may be an important mode of action of (1).

Reference

J. Blanz, K. Mewes, G. Ehninger, B. Proksch, D. Waidelich, B. Greger, and K.-P. Zeller, Evidence for oxidative activation of mitoxantrone in human, pig and rat, *Drug Metab. Dispos.*, 1991, **19**, 871.

Idarubicin hydrochloride

Use/occurrence:	Antileukaemic, antitumour agent
Key functional groups:	Alkyl ketone
Test system:	Rat (intravenous, 2 mg kg^{-1}), mouse (intravenous, 3 mg kg^{-1}), rabbit (intravenous, 0.75 mg kg^{-1}), dog (intravenous, 0.24 mg kg^{-1}), human cancer patients (intravenous, 15 mg m^{-2}; oral, 44 mg m^{-2})

Structure and biotransformation pathway

(1) $R^1 = R^2 = H$
(2) $R^1 = OCH_3$, $R^2 = H$
(3) $R^1 = OCH_3$, $R^2 = OH$

(4)

Idarubicin (1), 4-demethoxydaunorubicin, is an antitumour anthracyclin. Idarubicinol, the 13-dehydro secondary alcohol (4) is the major urinary metabolite of (1) in humans. The alcohol (4) has similar antitumour activity to (1) in animal models. The biotransformation of (1) into (4) is stereoselective affording the 13 S-enantiomer shown. The 13 R/13 S epimers were isolated from rat bile and determined by HPLC. They were quantified in urine from man (5% of the dose), rats (5%), mice (0.5%), rabbits (15%), and dogs (10%) after intravenous administration and in man after administration of oral doses. In man, the 13 R-epimer of (4) was 4.1% of the total secondary alcohols after intravenous administration and 3.8–5.0% after oral doses. The reduction is believed to be catalysed by cytoplasmic aldo-keto reductases and microsomal cytochrome P-450 reductase by analogy with the reduction of daunorubicin (daunomycin) (2) and doxorubicin (adriamycin) (3).

Reference

M. Strolin Benedetti, E. Pianezzola, D. Fraier, M. G. Castelli, and P. Dostert, Stereoselectivity of idarubicin reduction in various animal species and humans, *Xenobiotica*, 1991, **21**, 473.

Irinotecan

Use/occurrence: Antitumour agent

Key functional groups: Carbamate

Test system: Rat (intravenous, 10 mg kg^{-1})

Structure and biotransformation pathway

(1) * = ^{14}C

(2)

(3)

7-Ethyl-10-[4-(1-piperidino)-1-piperidino]carbonyloxycamptothecin (irinotecan) (1) is a new antitumour agent synthesized from the plant alkaloid camptothecin. The present study was designed to characterize the biliary metabolites of (1).

Following intravenous administration of [^{14}C]-(1) to bile duct-cannulated rats, 62.2% of the dose was excreted in the 0–48 hour bile. Over the same period, 33.3% of the dose was recovered in the urine and 9% in the faeces. Biliary metabolites were separated by TLC and HPLC. The known metabolite 7-ethyl-10-hydroxycamptothecin (2) was observed in rat bile together with a new metabolite (3) identified as the glucuronide conjugate of (2) by β-glucuronidase hydrolysis, ^{1}H NMR, and FAB-MS. Approximately 55% of the biliary ^{14}C excreted in 24 hours was unchanged (1) and metabolites (2) and (3) accounted for 9% and 22% respectively. Compound (1) probably undergoes metabolism to (2) by the action of carboxylesterase and the resultant hydroxy group is then conjugated with glucuronic acid [metabolite (3)]. Of these metabolites (2) has the strongest cytotoxic activity being some 100–1000 times more potent than (1) and about 100 times more potent than (3). Thus metabolism of (1) represents an activation step for the drug.

Reference

R. Atsumi, W. Suzuki, and H. Hakusui, Identification of the metabolites of irinotecan, a new derivative of camptothecin, in rat bile and its biliary excretion, *Xenobiotica*, 1991, **21**, 1159.

Territrem A, B, and C

Use/occurrence: Mycotoxins

Key functional groups: Dioxolane, methoxyphenyl, methylcyclohexenone

Test system: S9 fraction of rat liver microsomes

Structure and biotransformation pathway

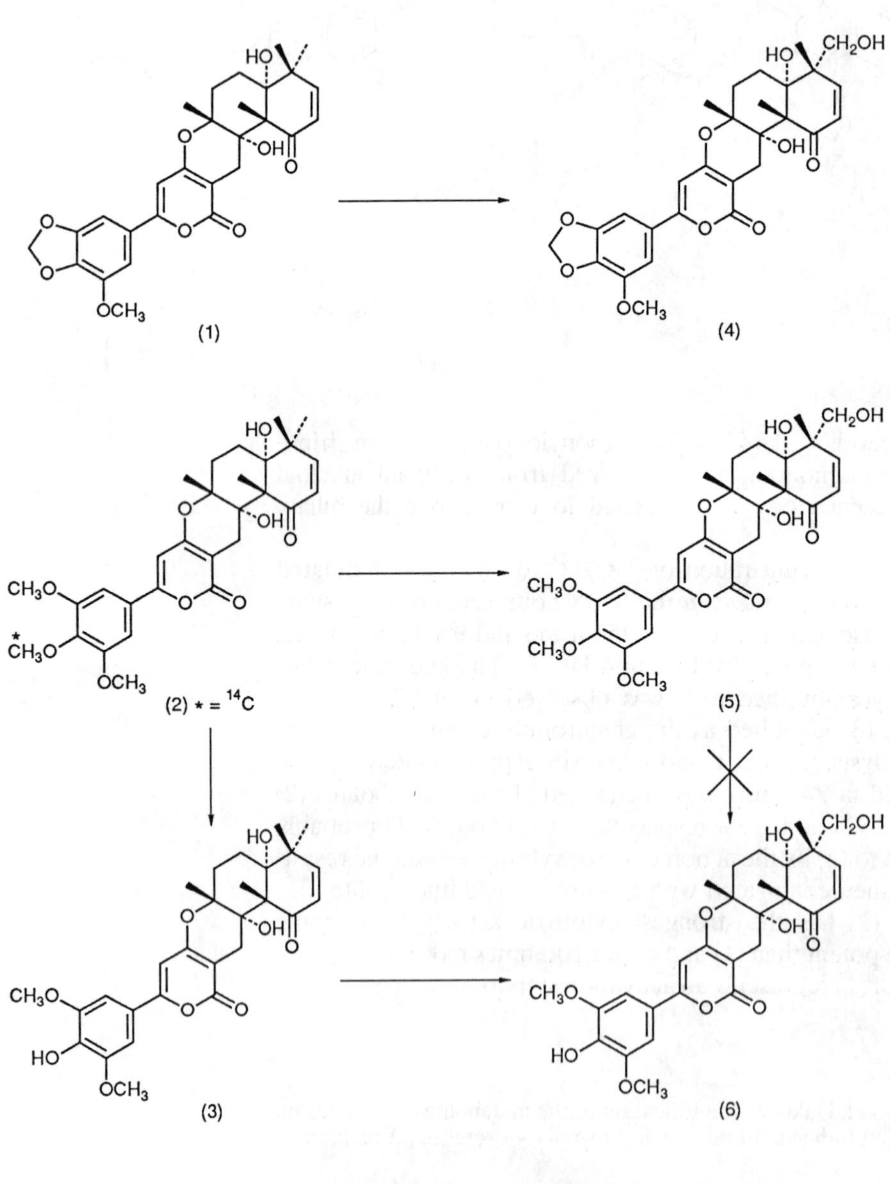

(1)

(4)

(2) ∗ = ^{14}C

(5)

(3)

(6)

The biotransformation of the tremorgenic mycotoxins territrem A, B, and C, (1)–(3) respectively, was investigated *in vitro* by incubating each with the S9 fraction from the livers of untreated rats and of those pretreated with either phenobarbital, 3-methylcholanthrene, or polychlorinated biphenyl (Arochlor 1254). Solvent extracts of the incubation media were subjected to TLC analysis, and the separated fluorescent components isolated and purified by HPLC for structural characterization by UV, NMR, and MS. In this way it was shown that (1) was converted into two metabolites, the most important of which was identified as (4); the minor component was not identified. Similarly, (2) was converted into four metabolites, three of which were identified as (3), (5), and (6), and (3) was converted into a single component (6). Formation of the products was dependent on the presence of the S9 fraction, NADP and glucose-6-phosphate, as well as the territrem, in the incubation mixture, and the reactions were enhanced by pretreating the rats with phenobarbital. In view of these results, it is suggested that cytochrome P-450 mono-oxygenase enzymes are responsible for the hydroxylation and *O*-demethylation metabolic pathways.

Reference

K. H. Ling, C. M. Chiou, and Y. L. Tseng, Biotransformation of territrems by S9 fraction from rat liver, *Drug Metab. Dispos.*, 1991, **19**, 587.

18β-Glycyrrhetic acid

Use/occurrence:	Hydrolysis product of main constituent of liquorice extract
Key functional groups:	Methylcyclohexane
Test system:	Rat liver post-mitochondrial supernatant

Structure and biotransformation pathway

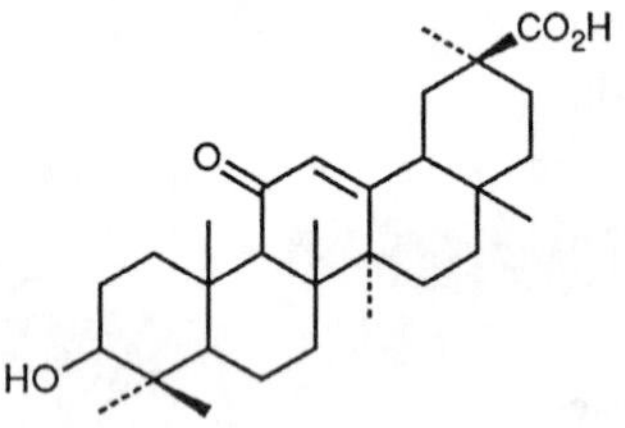

(1) 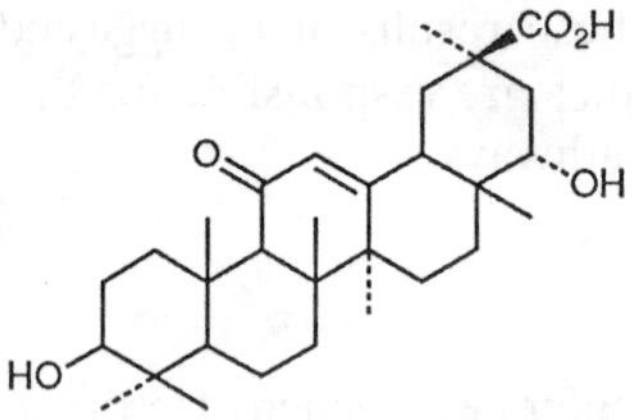　　　　　　　　　(2) 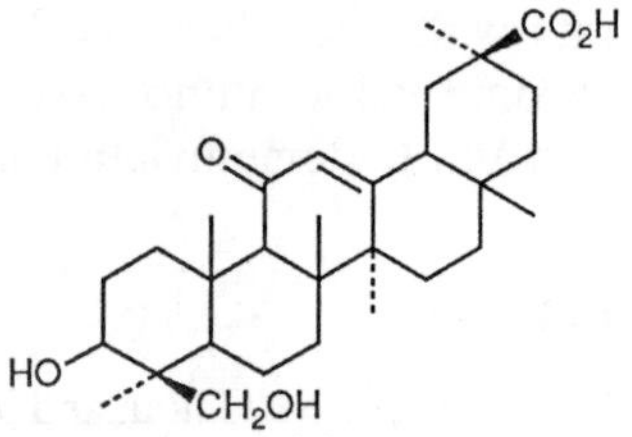　　　　　　　　　(3)

Humans ingest glycyrrihizin as a main constituent of liquorice extract. However, glycyrrhizin is not detected in the sera of human subjects after oral administration of glycyrrhizin but the aglycone (1) is found. In this study two metabolites of 18β-glycyrrhetic acid (1) were identified as the monohydroxylated products (2) and (3). The metabolites were identified following both MS and NMR analysis.

Reference

T. Akao, M. Aoyama, T. Akao, M. Hattori, Y. Imai, T. Namba, Y. Tezuka, T. Kikuchi, and K. Kobashi, Metabolism of glycyrrhetic acid by rat liver microsomes-II: 22 α- and 24-hydroxylation, *Biochem. Pharmacol.*, 1990, **40**, 291.

Trospectomycin

Use/occurrence:	Antibacterial drug
Key functional groups:	Cycloalkyl amine, cyclohexanol
Test system:	Dog (intravenous, 5 and 50 mg kg^{-1}; intramuscular, 5, 25, 50, and 100 mg kg^{-1}), rabbit (intravenous, 5 mg kg^{-1})

Structure and biotransformation pathway

(1) T = ^{3}H

Following intravenous or intramuscular administration of [^{3}H]trospecto-mycin sulfate (1) (5 mg kg^{-1}) to dogs, *ca.* 80% of the dose was excreted in urine during 120 hours and only 4–7% dose was excreted in the faeces in both cases. In rabbits, urinary excretion represented *ca.* 60% of the dose and faecal excretion *ca.* 4% dose. As the dose level to dogs was increased, urinary excretion declined, until at 100 mg kg^{-1} it only accounted for 60% of the dose. TLC analysis of dog urine indicated the presence of only two radio-active components, one of which corresponded to unchanged drug. The other component was probably a decomposition product rather than a true biotransformation product, as no metabolites were detected in dog plasma and it has been shown previously that (1) degrades on storage in rat urine, even at −20 °C. It was considered likely that only the parent drug was filtered at the glomerulus in the kidney, and that the decomposition product was formed in the bladder and during storage. By contrast, no unchanged (1) was observed in rabbit urine. Since no metabolites were detected in rabbit plasma, degradation in this case was probably due to the known chemical instability of the drug in the alkaline conditions encountered in rabbit urine (dog and rat urine, on the other hand, are usually acidic). Because of this apparent lack of metabolism, the total radioactivity data were used in the calculation of pharmacokinetic parameters. Thus, the bioavailability of an intramuscular dose of (1) in dogs was 100%, and the kinetics were linear over the dose range studied. Renal clearance constituted *ca.* 90% of the total clearance, and, since the drug is renally eliminated in rats, rabbits, dogs, and humans after intravenous administration, plasma clearance could, predictably, be well described by an allometric relationship.

Reference

D. J. Nichols, M. Burrows, A. Bye, D. A. Constable, L. G. Dring, and P. Jeffrey, Pharmacokinetics and fate of [³H]trospectomycin sulfate, a novel aminocyclitol antiobiotic, in male and female dogs and rabbits, *Drug Metab. Dispos.*, 1991, **19**, 781.

Arteether

Use/occurrence:	Antimalarial drug
Key functional groups:	Acetal, alkyl peroxide, methylcyclohexane
Test system:	Rat (intravenous, 11.6 mg kg^{-1})

Structure and biotransformation pathway

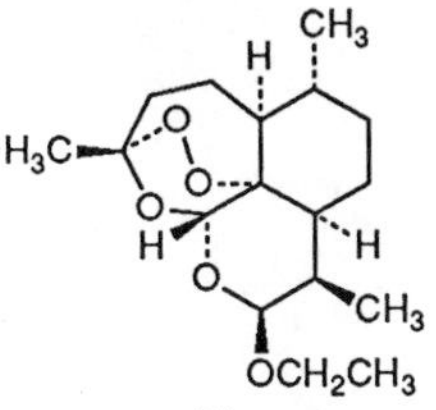

(1) 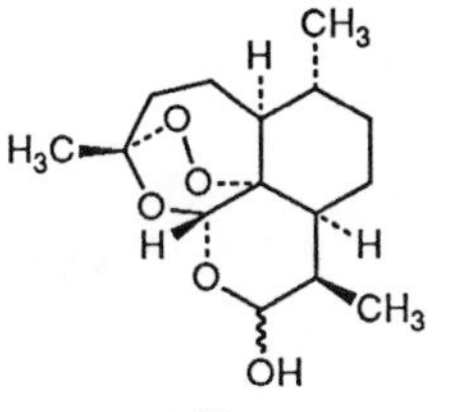   (2) 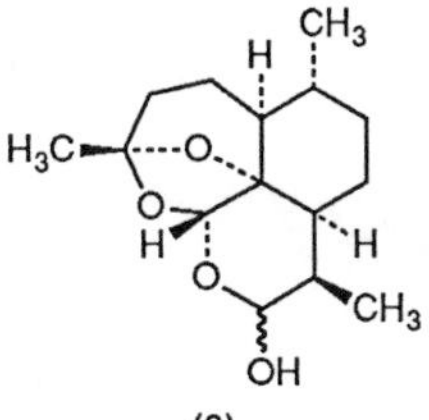   (3) 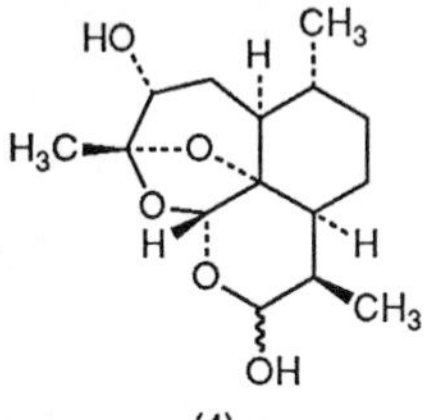   (4)

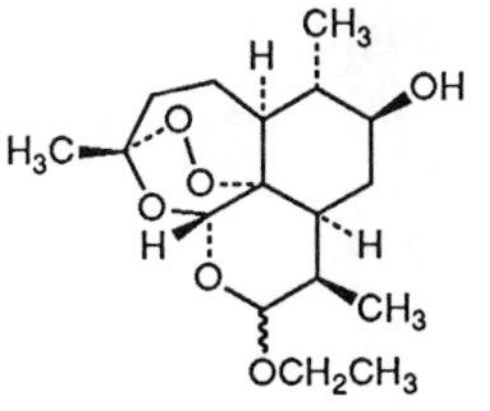

(5) 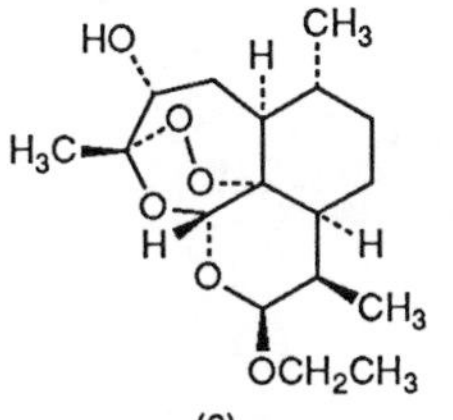   (6) 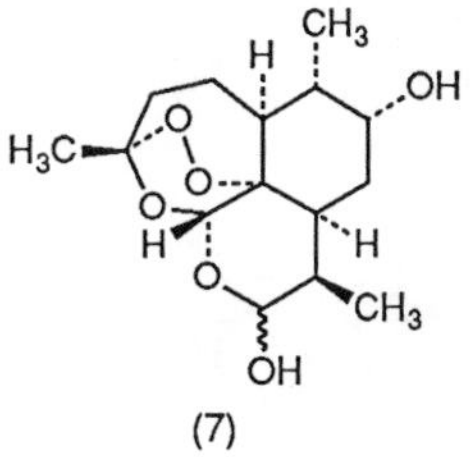   (7) 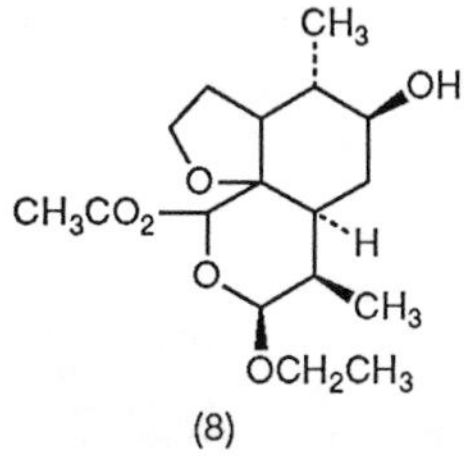   (8)

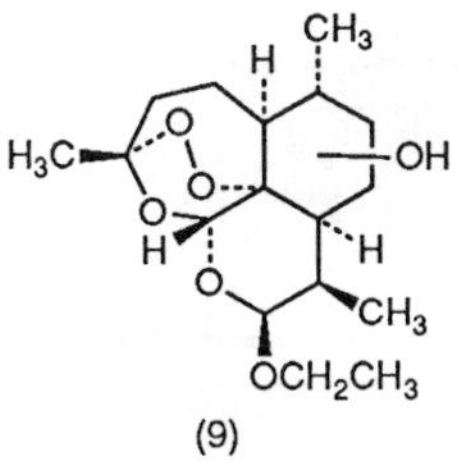

(9) 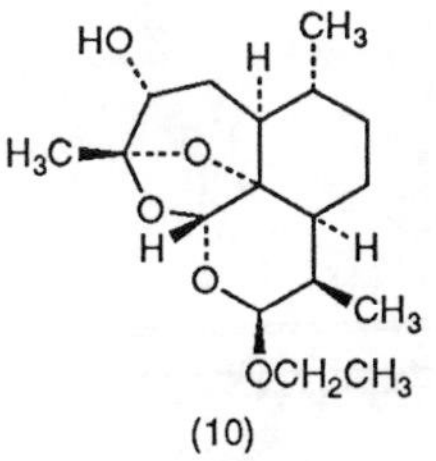   (10) 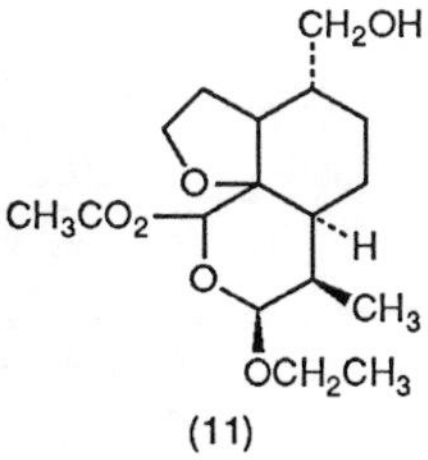   (11) 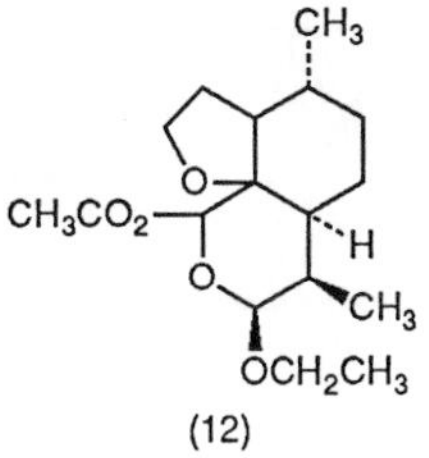   (12) 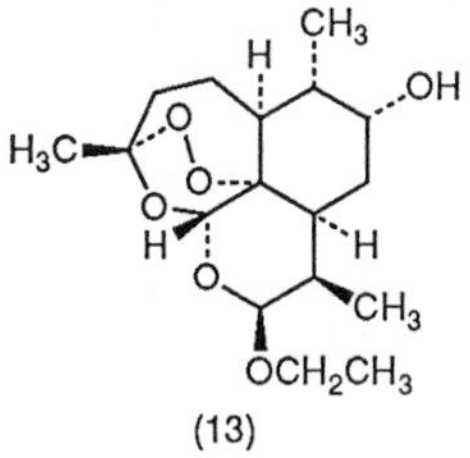   (13)

Plasma metabolites of arteether (1) were analysed by HPLC–MS, and identified by comparison with synthetic standards (nine metabolites) and/or on the basis of their mass spectra and chromatographic properties alone. Fifteen minutes after intravenous administration of (1) plasma was found to contain 12 metabolites [(2)–(13)] in the 10–1000 ng ml^{-1} range. Within 60 minutes two of these metabolites, (6) and (7), attained a concentration higher than parent (1), while several others attained similar concentrations to that of parent. The pseudo-first-order half-life of (1) was 10 minutes, whereas the half-lives of most of its metabolites were in the range 15–30 minutes. The *in vitro* antimalarial activity of several of the metabolite standards was tested, and all of these tested were active in the low ng ml^{-1} range.

Reference

H. T. Chi, K. Ramu, J. K. Baker, C. D. Hufford, I. Lee, Z. Yan-Lin, and J. D. McChesney, Identification of the *in vivo* metabolites of the antimalarial arteether by thermospray high performance liquid chromatography/mass spectrometry, *Biol. Mass Spectrom.*, 1991, **20**, 609.

Hexahydrocannabinol

Use/occurrence:	Model cannabinoid
Key functional groups:	Arylalkyl, methylcyclohexane, n-pentyl
Test system:	Liver microsomes from mouse, rat, guinea-pig, rabbit, and hamster

Structure and biotransformation pathway

(1)

Many cannabinoid compounds contain alkene functions and for them a major pathway of metabolism is hydroxylation at an allylic carbon atom. The objective of this study was to determine the pattern of hydroxylation in a cannabinoid without an alkene function. The equatorial C-11 methyl isomer of hexahydrocannabinol (1) was chosen, as sterically it most closely resembles the shape of tetrahydrocannabinoids. GC–MS was used for characterization of the metabolites after trimethylsilylation. Quantitatively, only the relative proportions of monohydroxylated metabolites are presented. Hydroxylation of the C-11 methyl group was a major process in all the five species, but least important in the hamster and guinea-pig. In the guinea-pig hydroxylation of the pentyl side-chain was important with all five positions being hydroxylated to varying degrees. In the hamster hydroxylation at the 8-position was most important with both 8α-and 8β-isomers being produced. The 8α-hydroxy compound was also a major metabolite in mouse and rat. A novel feature of the metabolism of (1) in comparison with other cannabinoids was the appearance at low levels of a metabolite hydroxylated at the 4-position of the aromatic ring. This occurred mainly in the guinea-pig and rabbit, but also in rat and hamster. The authors concluded that the stereochemical orientation of the cannabinoid molecule with the cytochrome P-450 active site was the main determinant of the position of hydroxylation, rather than the predicted greater reactivity of allylic carbon atoms.

Reference

D. J. Harvey and N. K. Brown, *In vitro* Metabolism of the Equatorial C_{11}-Methyl Isomer of Hexahydrocannabinol in Several Mammalian Species, *Drug Metab. Dispos.*, 1991, **19**, 714.

Ethyl-Δ^8-tetrahydrocannabinol, Ethyl-Δ^9-tetrahydrocannabinol

Use/occurrence:	Model compounds
Key functional groups:	Arylalkyl, methylcyclohexene, methylpyran
Test system:	Mouse (intraperitoneal, 100 mg kg^{-1})

Structure and biotransformation pathway

(1)   (2)

(3) $R^1 = CO_2H$, $R^2 = H$, $R^3 = H$
(4) $R^1 = CH_2OH$, $R^2 = H$, $R^3 = H$
(5) $R^1 = CH_2OH$, $R^2 = H$, $R^3 = OH$
(6) $R^1 = CO_2H$, $R^2 = H$, $R^3 = OH$

(7) $R^4 = CO_2OH$, $R^5 = H$, $R^6 = H$
(8) $R^4 = CH_2OH$, $R^5 = H$, $R^6 = H$
(9) $R^4 = CH_2OH$, $R^5 = OH$, $R^6 = H$
(10) $R^4 = CO_2H$ (further hydroxylation position unknown)
(11) $R^4 = CO_2H$ [isomer of (10)]

Ethyl-Δ^8-tetrahydrocannabinol (1) and ethyl-Δ^9-tetrahydrocannabinol (2) were separately administered to mice; one hour later the mice were killed and the livers removed. The livers were homogenized in saline before solvent extraction. Following clean-up the extracted metabolites were converted into either trimethylsilyl (TMS), [^{2}H$_9$]TMS, methyl ester/TMS, or dihydro/TMS derivatives. Metabolites of (1), were (3)–(6), which are similar with respect to positions substituted on the terpene ring as to those produced from higher homologues (see Vol. 2, pp. 275–283). The major metabolite identified, accounting for *ca.* 95% of the metabolite fraction, was the carboxylic acid. Metabolites hydroxylated in the ethyl group were not detected. Metabolites of (2) were (7)–(11), which were also similar with respect to positions substituted on the terpene ring to those produced from higher homologues (see Vol. 2, pp. 275–283), with the exception that less metabolism occurred at C-8 and a higher percentage of the total metabolic fraction was accounted for by the carboxylic acid (7). This metabolite accounted for >95% of the metabolite fraction.

Reference

N. K. Brown and D. J. Harvey, *In vivo* metabolism of the ethyl analogues of Δ^8-tetrahydrocannabinol and Δ^9-tetrahydrocannabinol in mouse, *Biol. Mass Spectrom.*, 1991, **20**, 324.

Cannabidiol monomethyl ether (CBDM), Cannabidiol dimethyl ether (CBDD)

Use/occurrence:	Model compounds
Key functional groups:	Arylalkyl, methylcyclohexene
Test system:	Guinea-pig (intraperitoneal, 100 mg kg^{-1}) (CBDD), guinea-pig (hepatic microsomes, substrate concentration 55.5 μg ml^{-1}) (CBDM and CBDD)

Structure and biotransformation pathway

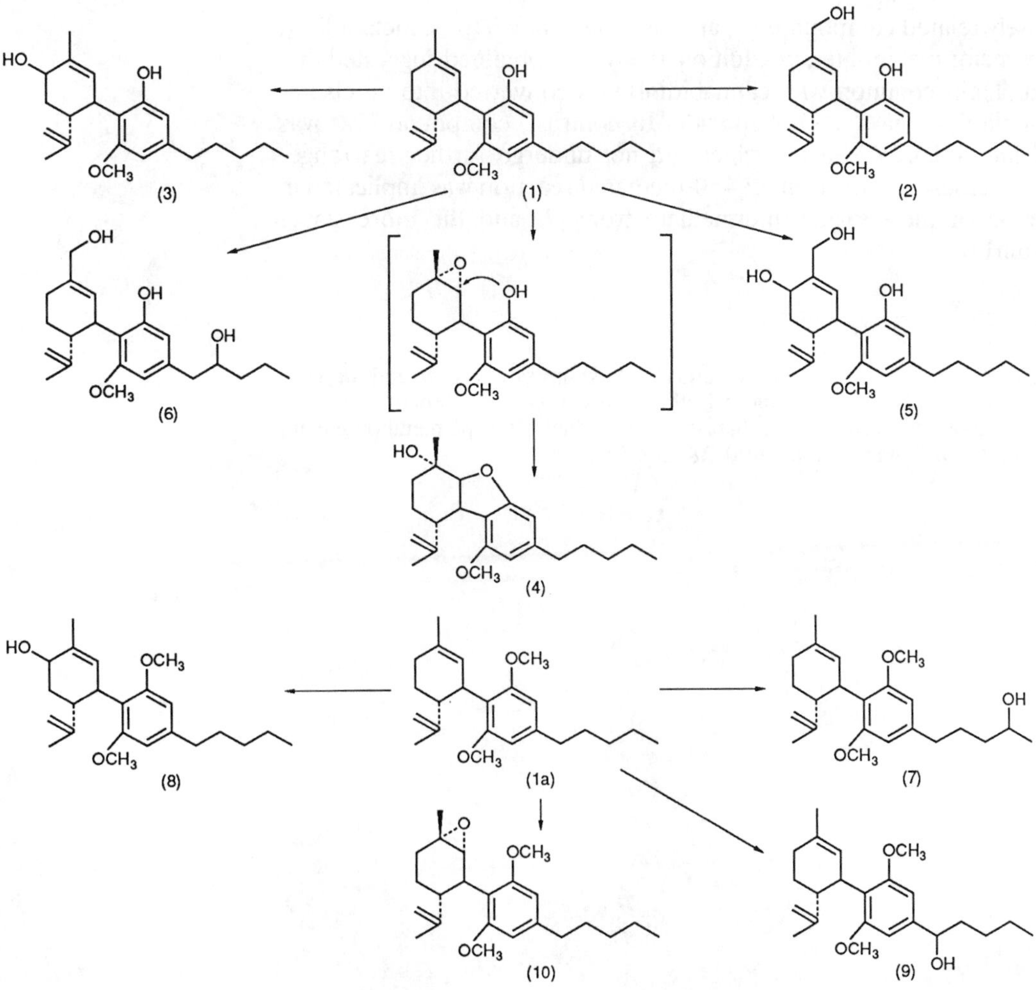

227

Cannabidiol, one of the major components of marihuana, is a potential anticonvulsant agent. In the present study the *in vitro* metabolism of the monomethyl (1) and dimethyl (1a) ethers of cannabidiol was investigated using guinea-pig microsomes in the presence of an NADPH regenerating system. Metabolites were identified by comparison with synthetic standards and GC–MS of the trimethylsilyl (TMS) derivatives. The most abundant metabolites of (1) were the 7-OH- (2) and 6-OH- (mixture of 6α- and 6β-isomers) (3) derivatives which were formed in approximately equal quantities. Cannabielsoin monomethyl ether (4) was formed in much smaller amounts [*ca.* 7% of (2)] and 6,7-dihydroxy-(1) (5) and 7,2″-dihydroxy-(1) (6) were also identified [each *ca.* 2% of the peak area of (2)].

In vitro metabolism of the dimethyl ether (1a) by guinea-pig liver microsomes showed that the major metabolite was 4″-hydroxy-(1a) (7). Other metabolites formed [quantified as percentage peak area of metabolite (7)] were 6-hydroxy-(1a) (30%, present as a mixture of 6α- and 6β-isomers), metabolite (8); 1″-hydroxy-(1a) (4%), metabolite (9); and the $1S,2R$-epoxy-(1a) (1%), metabolite (10). The liver from a guinea-pig treated *in vivo* with (1a) also contained metabolite (10). The animal was killed 1 hour after a single dose of 100 mg kg^{-1}.

The closely related compounds (1) and (1a) were shown to be metabolized to different major metabolites. In addition, the results obtained suggested that compound (1) (in common with cannabidiol) was converted into an elsoin-type metabolite (4) via a $1S,2R$-epoxide. In contrast, compound (1a) was converted into a $1S,2R$-epoxide which did not undergo further rearrangement. In both cases, a cytochrome P-450-mediated reaction was implicated in the formation of the epoxide intermediate from (1) and the more stable epoxide from (1a).

Reference

H. Gohda, S. Narimatsu, I. Yamamoto, and H. Yoshimura, *In vivo* and *in vitro* metabolism of cannabidiol monomethyl ether and cannabidiol dimethyl ether in the guinea pig: on the formation mechanism of cannabielsoin-type metabolite from cannabidiol, *Chem. Pharm. Bull.*, 1990, **38**, 1697.

11-Oxo-Δ^8-tetrahydrocannabinol

Use/occurrence:	Metabolite of Δ^8-THC
Key functional groups:	Alkyl aldehyde
Test system:	Mouse liver microsomes

Structure and biotransformation pathway

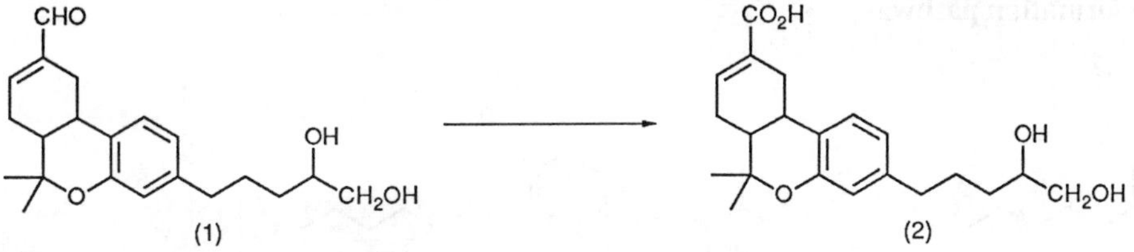

Mouse liver microsomes catalysed the oxidation of the aldehyde (1) to the acid (2). GC–MS was used to identify the product. The reaction required NADPH and molecular oxygen and was inhibited by SKF 525-A, α-naphthoflavone, and metapyrone. When (1) was incubated with microsomes under $^{18}O_2$, one ^{18}O atom was incorporated into the acid product (2). Phenobarbital pretreatment increased the oxidation activity on the basis of microsomal protein, but did not affect it on the basis of cytochrome P-450 content. 3-Methylcholanthrene pretreatment did not significantly affect the oxidation activity. The authors suggest that the reaction might be catalysed by a cytochrome P-450 enzyme, rather than enzymes usually associated with aldehyde oxidation.

Reference

K. Watanabe, N. Hirashi, S. Narimatsu, I. Yamamoto, and H. Yoshimura, Mouse hepatic microsomal enzyme that catalyses oxidation of 11-oxo-Δ^8-tetrahydrocannabinol to Δ^8-tetrahydrocannabinol-11-oic acid, *Drug Metab. Dispos.*, 1991, **19**, 218.

Cannabichromene

Use/occurrence:	Natural product
Key functional groups:	Alkene, benzopyran, n-pentyl
Test system:	Rabbit and mouse liver microsomes

Structure and biotransformation pathway

(1)

(2)

(3)

	R^1	R^2	R^3	R^4	R^5	R^6	R^7	R^8	R^9
(4)	OH	H	H	H	H	H	H	H	H
(5)	OH	H	H	H	H	H	H	H	H
(6)	H	OH	H	H	H	H	H	H	H
(7)	H	OH	H	H	H	H	H	H	H
(8)	H	H	OH	H	H	H	H	H	H
(9)	H	H	H	OH	H	H	H	H	H
(10)	H	H	H	H	OH	H	H	H	H
(11)	H	H	H	H	H	OH	H	H	H
(12)	H	H	H	H	H	H	OH	H	H
(13)	H	H	H	H	H	H	H	OH	H
(14)	H	H	H	H	H	H	H	H	OH
(15)	H	H	OH	H	OH	H	H	H	H
(16)	H	H	H	OH	OH	H	H	H	H
(17)	H	H	OH	H	H	H	OH	H	H
(18)	H	H	H	OH	H	H	OH	H	H
(19)	H	H	OH	H	H	H	H	OH	H
(20)	H	H	H	OH	H	H	H	OH	H
(21)	H	H	OH	H	H	H	H	H	OH
(22)	H	H	H	OH	H	H	H	H	OH

Metabolites of cannabichromene (1) were analysed by GC–MS as their trimethylsilyl (TMS) and [^{2}H$_9$]TMS derivatives. Most of the metabolites were hydroxylated compounds whose mass spectra gave little information on metabolite structure as fragmentation was dominated by formation of a sub-

230

stituted chromenyl ion. However, in this paper 21 metabolites (2)–(22) were successfully identified using both deuterium-exchange reactions and hydrogenation of the metabolites to their correspondig tetrahydroderivatives.

Reference

N. K. Brown and D. J. Harvey, Identification of cannabichromene metabolites by mass spectrometry: Identification of eight new dihydroxy metabolites in the rabbit, *Biol. Mass Spectrom.*, 1991, **20**, 275.

Paxilline

Use/occurrence:	Mycotoxin
Key functional groups:	Indole
Test system:	Sheep bile (50 μg; 24 hours)

Structure and biotransformation pathway

(1)   (2)

Paxilline (1), a tremogenic mycotoxin that occurs in endophyte-infected ryegrass, is believed to be a biosynthetic precursor of indole-terpenoid alkaloids which cause neurological disorders ('staggers') in sheep and cattle. When [^{14}C]-(1) was incubated in bile from pasture-fed sheep, a metabolite was produced, which after isolation by chloroform extraction and purification by TLC and HPLC was shown to be (2), a di-oxygenated derivative in which the indole 2,3 double bond had opened to give an eight-membered ring. Structural assignment was by EI-MS, in comparison with a synthetic standard of (2), which was shown to be much less tremogenic than (1). Bile of sheep fed on hay and concentrates before slaughter did not form (2). Furthermore, thermal inactivation of bile enzymes did not impair transformation of (1) by the bile of pasture-fed sheep. The authors suggest that the reaction is mediated by dissolved molecular oxygen, and that it may occur *in vivo*. If the latter is true, then this biotransformation may be an important detoxification pathway for (1) and other indole-terpenoid tremogens.

Reference

P. G. Mantle, S. J. Burt, K. M. MacGeorge, J. N. Bilton, and R. N. Sheppard, Oxidative transformation of paxilline in sheep bile, *Xenobiotica*, 1990, **20**, 809.

Codeine

Use/occurrence:	Analgesic/antitussive drug
Key functional groups:	Methoxyphenyl, *N*-methyl alkyl amine
Test system:	Mouse, rat, guinea-pig, rabbit (subcutaneous, 10 mg kg^{-1})

Structure and biotransformation pathway

Although codeine (1) is widely used as an analgesic and antitussive, the excretion of conjugated metabolites of this drug has not been examined by modern analytical techniques. Metabolites were separated by HPLC and compared with reference standards.

Codeine and its metabolites were analysed in the 24 hour urines of mice, rats, guinea-pigs, and rabbits following a subcutaneous dose. Codeine-2-glucuronide (2) was a minor metabolite (0.2% of the dose) in the rat and unchanged (1) accounted for 1.6% of the dose. The major rat urinary metabolites were morphine-3-glucuronide (3) (23.9% of the dose) and morphine (4) (4.3%). A similar metabolic pattern of (1) was observed in mouse urine, in which the dominant conjugate was (3) (7.6% of the dose). Unchanged (1) (6.8%), (2) (1.6%), and (4) (0.8% of the dose) were also detected. However, the major metabolite in mouse was norcodeine (6) (9% of the dose), which was not detected in rats, guinea-pigs, or rabbits.

The major metabolite of (1) in both guinea-pigs and rabbits was (2), which accounted for 39.8% and 24.5% of the dose respectively. Morphine-6-glucuronide (5) was also detected in guinea-pigs and rabbits (0.7% and 1.9% in the respective species). Compounds (1) (1.6%, guinea-pigs and 2.2%, rabbits) (3) (1.6%, guinea-pigs and 17.9%, rabbits), and (4) (0.2%, guinea-pigs and 1.3%, rabbits) were all detected.

Codeine is metabolized in all four species by glucuronidation and by oxidative N- and O-demethylation, but the quantitative excretion of metabolites was different in the four species investigated. Compound (2) is considered to be pharmacologically less active than (1), but (5) is a more potent analgesic than (1). Metabolite (5) has a strong affinity for the opiate receptor and may account for the majority of the analgesic effect in man.

Reference

K. Oguri, N. Hanioka, and H. Yoshimura, Species differences in metabolism of codeine: urinary excretion of codeine glucuronide, morphine-3-glucuronide and morphine-6-glucuronide in mice, rats, guinea-pigs and rabbits, *Xenobiotica*, 1990, **20**, 683.

Codeine

Use/occurrence:	Analgesic/antitussive drug
Key functional groups:	Allylic alcohol, methoxyphenyl, morphinan
Test system:	Guinea-pig (subcutaneous, 20 mg kg^{-1})

Structure and biotransformation pathway

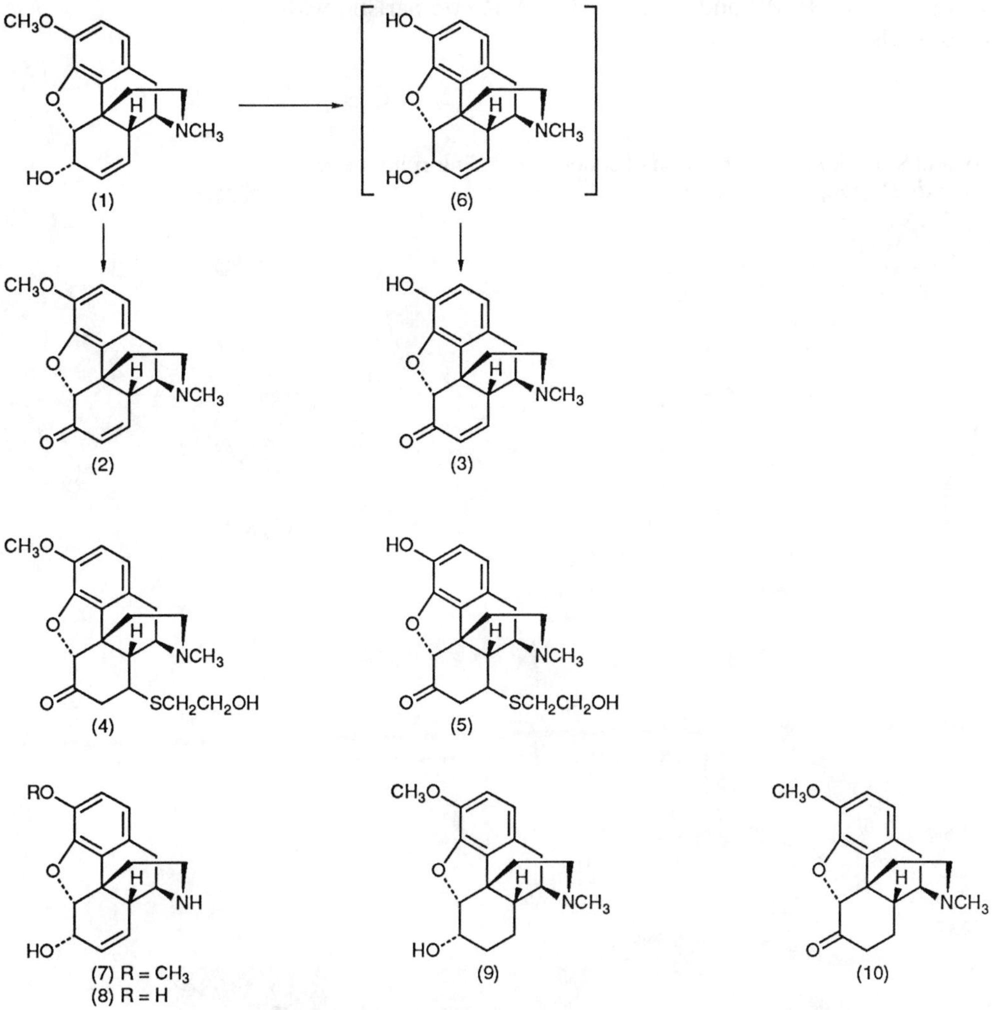

Previous investigations into the biotransformation of codeine (1) have been unable to determine codeinone (2) and morphinone (3) in biological fluids because of their chemical reactivity. Since these metabolites are toxic and able to bind covalently with the sulfhydryl group in opiate receptors, the objective of this study was to confirm unequivocally their formation from (1) *in vivo*. Thus, subcutaneous doses of (1) were administered to guinea-pigs with cannulated bile ducts, and the collected bile was subsequently treated with 2-mercaptoethanol to convert any (2) and (3) present into their stable adducts (4) and (5) respectively. Quantitative HPLC analysis of the treated bile showed that during 6 hours post-administration means of 10.5% and 2.7% of the dose were excreted in bile as (2) and (3) respectively. These experiments also confirmed the presence in bile of unchanged (1) and the known metabolites (6)–(10), each of which accounted for 0.2–4.0% of the dose. Metabolites present in the bile collected from guinea-pigs that had received multiple subcutaneous doses of (1) were separated by reverse-phase column chromatography, and fractions containing metabolites (2) and (3) (as their mercaptoethanol adducts) were further purified by HPLC for confirmation of their identities by FAB-MS and ^{1}H and ^{13}C NMR comparison with the synthetic compounds.

Reference

T. Ishida, M. Yano, and S. Toki, *In vivo* formation of codeinone and morphinone from codeine, *Drug Metab. Dispos.*, 1991, **19**, 895.

Quinine sulfate

Use/occurrence:	Cinchona alkaloid/malaria chemotherapy
Key functional groups:	Piperidine
Test system:	Human volunteers (oral, 600 mg)

Structure and biotransformation pathway

(1) $R^1 = R^2 = H$
(2) $R^1 = OH, R^2 = H$
(3) $R^1 = H, R^2 = OH$

Quinine (1) is able to eliminate resistant strains of the malaria parasite and the drug is regaining prominence. The major metabolite of (1) in human urine is (2). This is the only metabolite in human plasma and saliva following oral dosing of (1). Previously, (2) was proposed as the metabolite of (1) (assigned only by analogy with quinidine), rather than (3), which was first suggested in 1951 by Brodie and co-workers as the quinine metabolite. Preparative TLC, CI (ammonia) MS, UV, and ^{1}H NMR spectroscopy have allowed the unequivocal assignment of the oxidative metabolite of (1) to the structure (2) and not (3). However, the stereochemistry of the metabolite, *i.e.* of the tertiary alcohol at C-3, is yet to be established. As many biological hydroxylations proceed with retention of configuration, and as no change in configuration was observed in the metabolic hydroxylation of the analogous quinidine (a dextrorotatory stereoisomer of quinine which differs only by epimerization at the secondary alcohol), the stereochemistry is expected to follow that of quinine (1).

Reference

O. O. Bolaji, C. P. Babalola, and P. A. F. Dixon, Characterization of the principal metabolite of quinine in human urine by ^{1}H-n.m.r. spectroscopy, *Xenobiotica*, 1991, **21**, 447.

Strychnine

Use/occurrence: Alkaloid

Key functional groups: Cycloalkene,
Cycloalkylamine, indole

Test system: Dog, guinea-pig, mouse,
rabbit, and rat liver
microsomes

Structure and biotransformation pathway

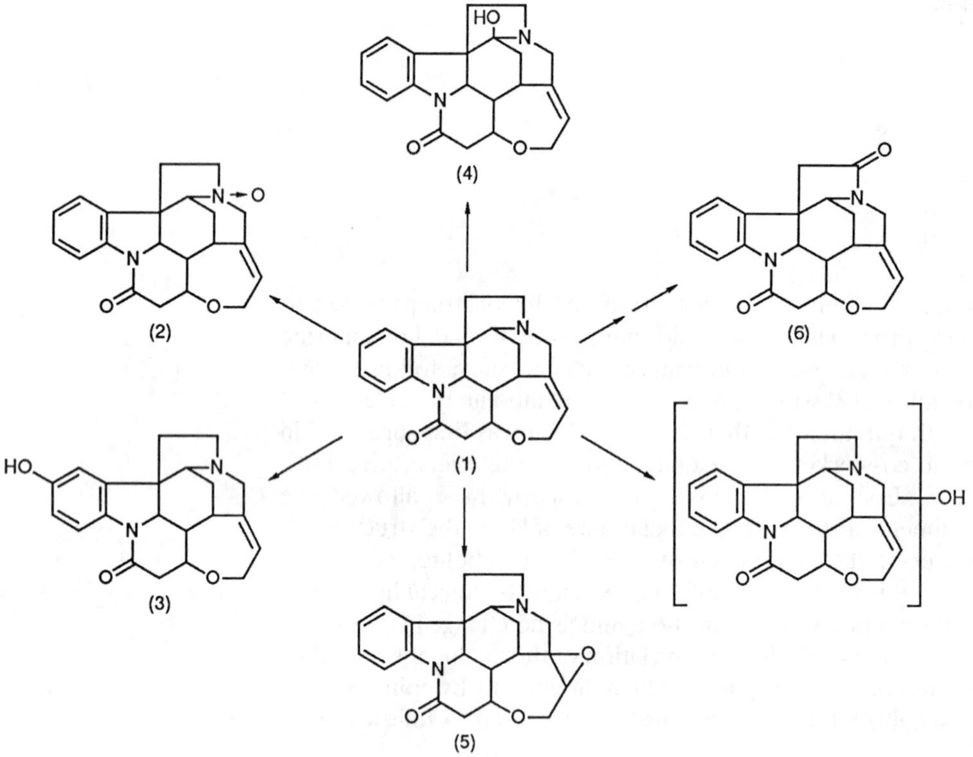

After incubation of strychnine (1) with rabbit liver microsomes metabolites were extracted with chloroform/isopropyl alcohol and isolated following chromatographic separation using TLC. Five metabolites were identified by GC–MS and comparison with reference compounds. The three major metabolites were the *N*-oxide (2) and the ring-hydroxylated components (3) and (4). Two minor metabolites were the epoxide (5) and the keto compound (6). Comparison of metabolite profiles formed by other species showed that (4) was the major metabolite from rat and mouse whereas in guinea-pig and rabbit it was the phenol (3) and in dog the *N*-oxide (1). The metabolic activity was much higher in guinea-pig microsomes than in those from other species.

Reference

Y. Tanimoto, T. Ohkuma, K. Oguri, and H. Yoshimura, Species diference in metabolism of strychnine with liver microsomes of mice, rats, guinea pigs, rabbits and dogs, *J. Pharmacobio.-Dyn.*, 1990, **13**, 136.

Strychnine

Use/occurrence:	Rodenticide
Key functional groups:	Cycloalkene, cycloalkylamine, indole
Test system:	Rat, mouse, guinea-pig, rabbit, and dog liver microsomes

Structure and biotransformation pathway

(2) (1) (6)

(3) (4) (5)

The metabolism of the alkaloid strychnine (1) has been studied using phenobarbital (PB) or 3-methylcholanthrene (MC) pretreated rat liver microsomes. The tertiary alcohol 16-hydroxystrychnine (2), the enol 22-hydroxystrychnine (3), and the oxirane strychnine 21,22-epoxide (4) were induced two-fold on PB pretreatment. MC pretreatment resulted in only 1.4-fold induction of each metabolic oxidation activity. Up to 10.5-fold induction of 2-hydroxylation activity affording the phenol 2-hydroxystrychnine (5) was observed using liver microsomes from mice, guinea-pigs, rabbits, and dogs. N-Oxidation activity affording strychnine N-oxide (6) was also significantly increased. The P-450 isozymes P-450 I (P-450 IIB1) and P-450 II (P-450 IIB2), purified from PB-pretreated rat livers, displayed selective activities towards hydroxylation and N-oxidation with pH optima between 8.4 and 8.6. In this reconstituted system, only ethylmorphine N-demethylation displayed a similar pH profile. Therefore, the optimum conditions for a P-450 isozyme to interact with its substrates need not necessarily be the same for each substrate.

Reference

Y. Tanimoto, H. Kaneko, T. Ohkuma, K. Oguri, and H. Yoshimura, Site-selective oxidation of strychnine by phenobarbital inducible cytochrome-P-450, *J. Pharmacobio.-Dyn.*, 1991, **14**, 161.

Strychnine

Use/occurrence:	Alkaloid
Key functional groups:	Cycloalkene, cycloalkylamine, indole
Test system:	Guinea-pig liver microsomes

Structure and biotransformation pathway

The metabolism of strychnine (1) by guinea-pig liver microsomes was investigated. Metabolites were separated by TLC and HPLC. A novel metabolite was isolated by repeated preparative TLC and identified by IR spectroscopy, ^{1}H NMR, and EI-MS.

Following incubation of guinea-pig liver microsomes with (1) in the presence of a NADPH regenerating system, five metabolites were detected. 2-Hydroxystrychnine (2), 16-hydroxystrychnine (3), strychnine 21,22-epoxide (4), and strychnine N-oxide (5) were identified; these metabolites had been characterized in a previous study with rabbit liver microsomes as the enzyme source.

In addition, guinea-pig liver microsomes catalysed the formation of 22-hydroxystrychnine (6). This metabolite had previously been observed in a study with rabbit liver microsomes but remained unidentified at that time. The results indicate that metabolic oxidation at the double bond of strychnine can occur via two routes: epoxidation yielding (4) and direct insertion of oxygen giving metabolite (6). Although two oxidation routes are well known for aromatic compounds (producing the corresponding epoxide and phenol metabolite), this was thought to be a unique example of such oxidations at a double bond in an alicyclic compound.

Reference

Y. Tanimoto, T. Ohkuma, K. Oguri, and H. Yoshimura, A novel metabolite of strychnine, 22-hydroxystrychnine, *Xenobiotica*, 1991, **21**, 395.

Zeranol

Use/occurrence:	Non-steroidal anabolic agent
Key functional groups:	Cycloalkyl alcohol, lactone, phenol
Test system:	Rat and pig liver microsomes and cytosols

Structure and biotransformation pathway

(1) R = H / OH

(2) R = OH / H

(3) R = O

Non-radiolabelled zeranol (1) and its known metabolites taleranol (2) and zearalanone (3) were separately incubated with rat and pig liver microsomes and cytosols in the presence of UDP-[^{14}C]glucuronic acid or [^{35}S]Na$_2$SO$_4$ and the radioactive products formed were analysed by HPLC. [^{3}H]-(1) was similarly incubated in the presence of non-labelled UDPGA or sulfate. The results showed that with pig and rat microsomes (1)–(3) were each converted into a mono-glucuronide conjugate that corresponded chromatographically to a conjugate isolated from the plasma of treated pigs that was hydrolysed by β-glucuronidase. Furthermore, the rate of conversion was similar for all three substrates, suggesting that conjugation occurred at one of the phenolic hydroxy groups. Similarly, incubation of (1), (2), or (3) with rat hepatic cytosol and [^{35}S]Na$_2$SO$_4$ yielded mono-sulfate conjugates, and the same compounds were also formed in pig hepatic cytosol but at lower rates than in rat cytosol.

Reference

G. F. Bories, E. F. Perdu-Durand, J. F. Sutra, and J. E. Tulliez, Evidence for glucuronidation and sulfation of zeranol and metabolites (taleranol and zearalanone) by rat and pig hepatic subfractions, *Drug Metab. Dispos.*, 1991, **19**, 140.

242

Navelbine

Use/occurrence:	Antitumour drug
Key functional groups:	Methyl carboxylate
Test system:	Man (intravenous infusion)

Structure and biotransformation pathway

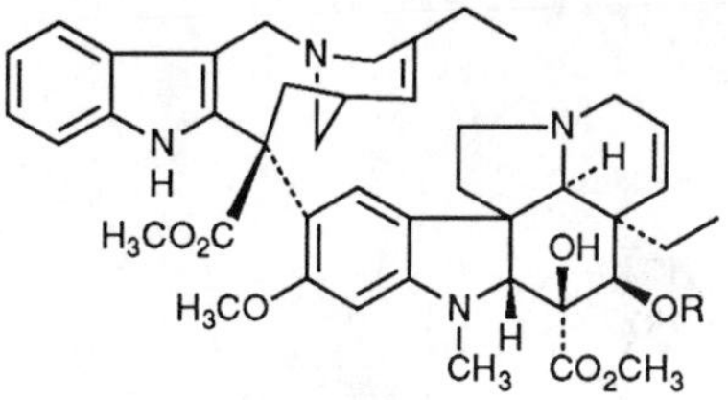

(1) R = COCH3
(2) R = H

Following a 15 minute intravenous infusion of navelbine (1) serum and urine were extracted and analysed by HPLC and comparison with reference compounds. Only (1) was detected in serum. Urine contained mainly (1) (10.9% of the dose in the first 48 hours) together with trace amounts (0.24% of the dose) of deacetyl-navelbine (2).

Reference

E. Jehl, E. Quoix, D. Leveque, G. Pauli, F. Breillout, A. Krikorian, and H. Monteil, Pharmacokinetic and preliminary metabolic fate of Navelbine in humans as determined by high pressure liquid chromatography, *Cancer Res.*, 1991, **51**, 2073.

Avilamycin

Use/occurrence:	Antibiotic used in animal feedstuff
Key functional groups:	Orthoester
Test system:	Pig (80 p.p.m. diet), rat (oral, 100 mg kg^{-1})

Structure and biotransformation pathway

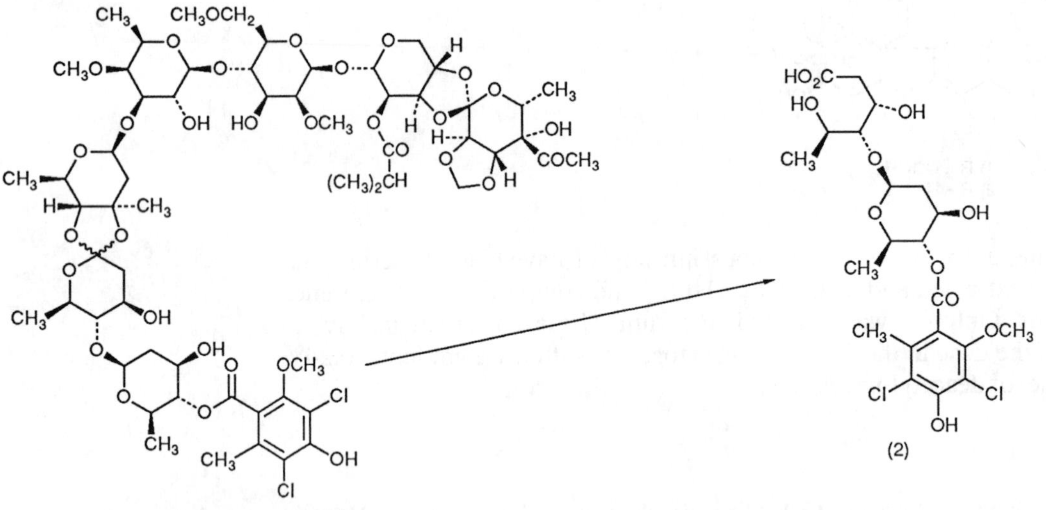

(1)

(2)

Following dietary administration of avilamycin (1) to pigs, tissues and excreta were extracted and a preliminary separation of metabolites was made using column chromatography followed by TLC. The majority of the dose (93%) was excreted in the faeces as a number of components, unchanged (1) (*ca.* 8% of dose), flambic acid the major metabolite (2), and a group of three metabolites which remained unidentified but which appeared to contain only the oligosaccharide and/or eurekanate portion of the molecule. Rats were shown to have a similar pattern of excretion and metabolite profile.

Reference

J. D. Magnussen, J. E. Dalidowicz, T. D. Thomson, and A. L. Donoho, Tissue residues and metabolism of Avilamycin in swine and rats, *J. Agric. Food Chem.*, 1991, **39**, 306.

244

Heterocycles

Nicotine

Use/occurrence:	Tobacco alkaloid
Key functional groups:	*N*-Methylpyrrolidine, pyridine
Test system:	Stumptailed macaque (intravenous, 0.3 mg kg^{-1})

Structure and biotransformation pathway

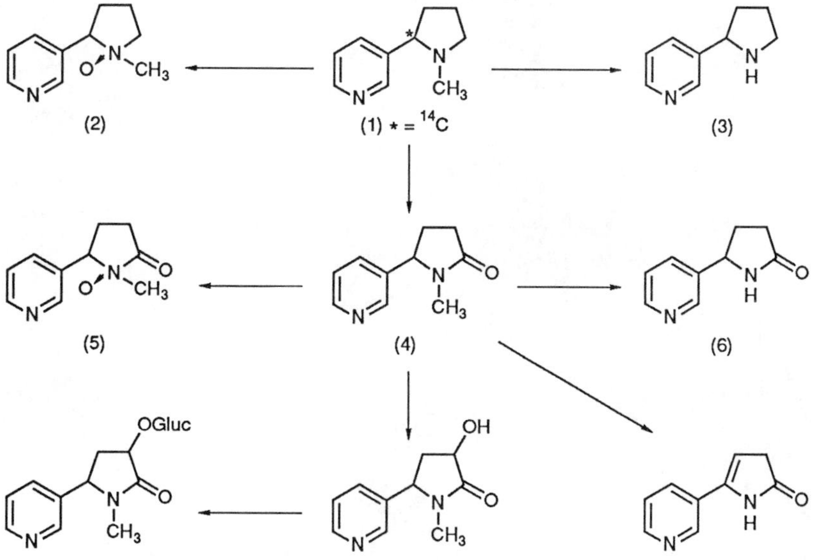

The objective of this study was to develop an animal model for nicotine metabolism in man. HPLC analysis of urine separated unchanged nicotine (1) and eight metabolites, which were assigned on the basis of comparison of retention times with standards. 3-Hydroxycotinine glucuronide (8) was quantitatively by far the most important urinary metabolite, as it represented *ca.* 40% dose in male animals and *ca.* 30% in females. Unchanged (1) accounted for 7–8% dose and the other metabolites generally represented less than 5% dose each. Phenobarbital pretreatment of the animals resulted in excretion of slightly increased amounts of cotinine *N*-oxide (5), desmethylcotinine (6), 3-hydroxycotinine (7), and desmethylcotinine-$\Delta^{2',3'}$-enamine (9). Amounts of nicotine and the other metabolites excreted in urine were slightly reduced. Cimetidine pretreatment resulted in marginally reduced excretion of some metabolites. The authors considered the production of metabolites (8) and (9) by the macaques to be of special interest as these had recently been reported to be formed in man. Overall, the authors suggested that the stumptailed macaque may be a useful model for certain aspects of nicotine metabolism in humans.

Reference

M. Seaton, G. A. Kyerematen, M. Morgan, E. V. Jeszenka, and E. S. Vessell, Nicotine metabolism in stumptailed macaques, *Macaca arctoides*, *Drug Metab. Dispos.*, 1991, **19**, 946.

Nicotine

Use/occurrence:	Tobacco alkaloid
Key functional groups:	N-Methylpyrrolidine, pyridine
Test system:	Human volunteers (smokers and non-smokers) (intravenous, $0.19\ mg\ kg^{-1}$)

Structure and biotransformation pathway

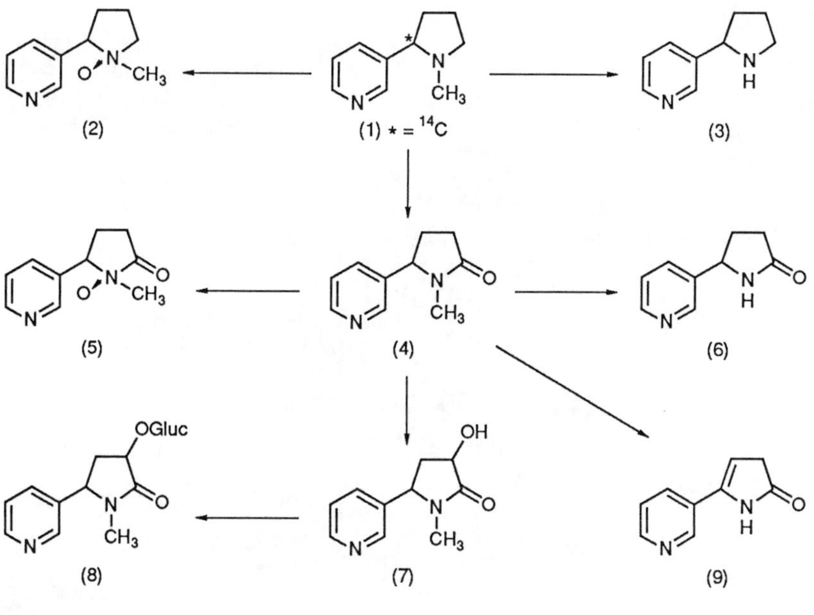

HPLC analysis of urine from both smokers and non-smokers separated unchanged nicotine (1) and eight metabolites (2)–(9). Two of these metabolites, 3-hydroxycotinine glucuronide (8) and desmethylcotinine $\Delta^{2',3'}$-enamine (9), were reported to be novel. Evidence of structure was obtained from CI-MS on all metabolites and additionally using enzymatic deconjugation (8) or LC–MS (9). Metabolites (8) and (9) were considered to be of particular interest because, in smokers, they both persisted longer than cotinine and could therefore have potential as indicators of passive exposure to cigarette smoke. Quantitatively nicotine, cotinine, and the two novel metabolites were the most important components in the urine from both smokers and non-smokers, but there were significant differences between the two groups. Non-smokers excreted more unchanged nicotine and less of the glucuronide (8) than did smokers.

Reference

G. A. Kyerematen, M. L. Morgan, B. Chattopadhyay, J. D. de Bethizy, and E. S. Vessell, Disposition of nicotine and eight metabolites in smokers and non-smokers: identification in smokers of two metabolites that are longer lived than cotinine, *Clin. Pharmacol. Ther.*, 1990, **48**, 641.

Nicotine 1′-oxide

Use/occurrence: Nicotine metabolite

Key functional groups: N-Oxide

Test system: Rabbit (oral, 1.44 mg kg^{-1}; intravenous infusion, 0.72 mg kg^{-1} over 4 hours; intraperitoneal, 1.44 mg kg^{-1})

Structure and biotransformation pathway

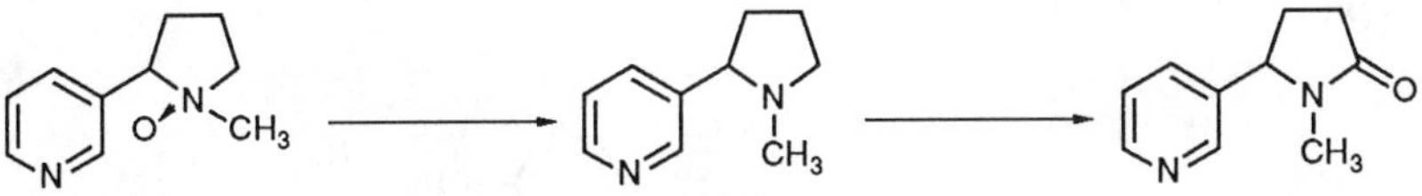

GC was used to determine concentrations of nicotine N-oxide (1), nicotine (2), and cotinine (3) in plasma and urine. Following intravenous infusion of (1) 49.9% dose was recovered unchanged in the 0–6 hour urine, while (2) and (3) represented 0.6% and 0.04% dose respectively. After oral administration of (1), 7.9% dose was excreted unchanged and 0.2% and 0.8% dose as (2) and (3) respectively. Fractional conversion values for (1) into (2) and (3), calculated from plasma kinetic data, were 2–3% each after intravenous infusion but 45% and 2% respectively after oral dosing. The oral bioavailability of (1) was 15%. From these results the authors concluded that presystemic reduction (bacterial or intestinal) was important in the metabolism of (1) after oral administration.

Reference

M. J. Duan, L. Yu, C. Savanpridi, P. Jacob III, and N. L. Benowitz, Disposition kinetics and metabolism of nicotine-1′-N-oxide in rabbits, *Drug Metab. Dispos.*, 1991, **19**, 667.

Medetomidine

Use/occurrence:	α_2-Adrenoceptor agonist/ analgesic drug
Key functional groups:	Arylmethyl, imidazole
Test system:	Rat (subcutaneous, 80 μg kg^{-1} or 5 mg kg^{-1}), rat hepatic fractions (9000 g supernatant or microsomes; substrate concentration, 15 μM)

Structure and biotransformation pathway

Rat hepatic 9000 g supernatant and microsomes converted [^{3}H]medetomidine (1) into a single major metabolite, 3-hydroxymedetomidine (2), in the presence of NADPH. The radiolabelled product was identified by co-chromatography with the authentic reference compound on TLC and HPLC as well as FAB-MS.

Following administration of a high dose (5 mg kg^{-1}) of (1) to rats, 29.2% of the dose was excreted in the urine within 24 hours. Radio-HPLC analysis of urine showed that the major urinary metaboites were the glucuronide of (2) [metabolite (3), accounting for *ca.* 35% of total radiolabel] and medetomidine carboxylic acid [metabolite (4), *ca.* 40% of the total ^{3}H]. Metabolite (3) was shown to be converted into (2) following treatment with β-glucuronidase and was further identified by ^{1}H NMR and FAB-MS.

Unchanged (1) accounted for *ca.* 2% or 11% of the dose at the low or high dose level respectively. Minor metabolites (representing $\leqslant$ 10% of the dose) were tentatively suggested to be a mercapturate of (2) and an *N*-methyl and phenolic derivative of (1).

The major metabolites of (1), *i.e.* (2) and (4), are devoid of α-receptor activity and hence conversion of (1) represents an inactivation of the drug.

The pattern of metabolism is similar to that of the close structural analogue detomidine.

Reference

J. S. Salonen and M. Eloranta, Biotransformation of medetimidine in the rat, *Xenobiotica*, 1990, **20**, 471.

Antipyrine

Use/occurrence: Hepatic probe drug, analgesic, antipyretic drug

Key functional groups: Methylpyrazolinone

Test system: Rat (intraperitoneal, 40 mg kg^{-1}), rat liver microsomes (0–16 mM substrate)

Structure and biotransformation pathway

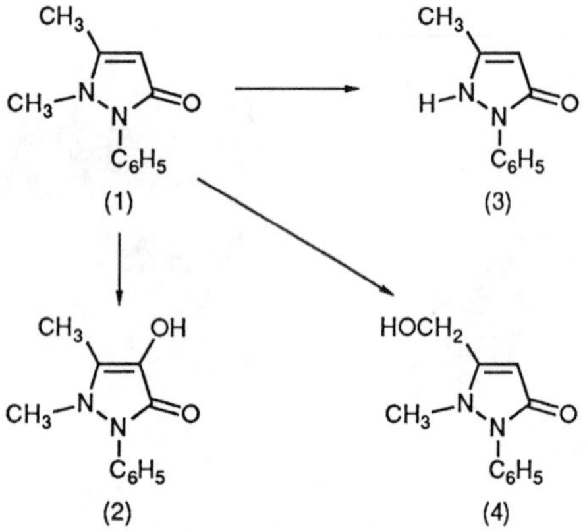

Antipyrine (1) metabolism depends on at least three isoenzymes of cytochrome P-450 forming the main metabolites (2)–(4). Sex differences in the metabolism of antipyrine (1) were investigated in two strains of rats. Antipyrine (1) and its metabolites (2)–(4) in urine and microsomes were measured by an HPLC assay. Clearance of antipyrine was 46% higher in male than in female rats; this was associated with a 40% higher urinary excretion of (4) in male rats, the other metabolites being excreted to a similar extent. The correlation between *in vitro* and *in vivo* clearance for (4) was good but unconvincing for (2) and (3).

Reference

J. T. M. Buters and J. Reichen, Sex difference in antipyrine 3-hydroxylation: An *in vivo–in vitro* correlation of antipyrine metabolism in two rat strains, *Biochem. Pharmacol.*, 1990, **40**, 771.

N-[4-(5-Nitro-2-furyl)-2-thiazolyl]-formamide

Use/occurrence:	Model carcinogen
Key functional groups:	Formamide, nitrofuran, thiazole
Test system:	Rat, guinea-pig (oral, 100 mg kg^{-1}), rat and guinea-pig liver and kidney microsomes

Structure and biotransformation pathway

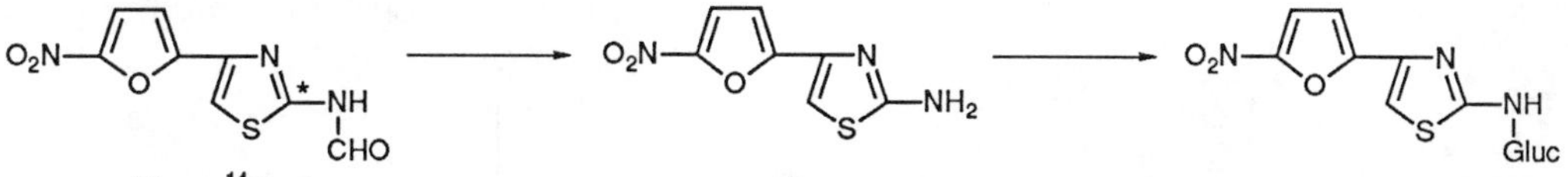

(1) * = ^{14}C   (2)   (3)

Following oral dosing of N-[4-(5-nitro-2-furyl)-2-thiazolyl]formamide (FANFT) (1) to rats or guinea-pigs several metabolites were produced. ANFT (2) was a major component in urine of both species, the amount present being reduced when (2) was the compound dosed. A unique metabolite was detected in the urine of guinea-pigs and identified by MS as the N-glucuronide (3).

The activity of the ANFT-UDP-glucuronosyltransferase was shown to be located only in the microsomal fraction of guinea-pig liver and kidney. The activity of this enzyme may explain the resistance of the guinea-pig to bladder cancer induced by (1) or (2).

Reference

R. M. Dawley, T. V. Zenser, M. B. Mattammal, V. M. Lakshmi, F. F. Hsu, and B. B. Davis, Metabolism and disposition of bladder carcinogens in rat and guinea pig: possible mechanism of guinea pig resistance to bladder cancer, *Cancer Res.*, 1991, **51**, 514.

Niridazole

Use/occurrence: Antischistosomal drug

Key functional groups: Imidazolidinone, thiazole

Test system: Mouse (intraperitoneal, 50 mg kg^{-1}), human patients (25 mg kg^{-1} day^{-1})

Structure and biotransformation pathway

This paper describes the identification of the carboxymethylurea (4) as a urinary metabolite of niridazole (1) in the mouse and human patients. In mouse urine (4) accounted for *ca.* 12–14% dose. The metabolite was actually isolated from human urine and identified by EI- and CI-MS after methylation with BF$_3$/methanol. The structure was confirmed by chemical synthesis. The metabolite was reported to be unique among known niridazole metabolites in lacking the intact imidazolidinone ring. The structure allows for keto–enol tautomerism in which the enol form is stabilized at alkaline pH by conjugation with the nitrothiazole ring. This was proposed as the explanation for the observed pH-dependent 80 nm shift in the UV–visible spectrum of the metabolite. The authors proposed the biotransformation pathway shown, after observing that formation of (4) from (2) could be catalysed by NAD$^+$-dependent aldehyde dehydrogenase.

Reference

J. W. Tracy, B. A. Catto, and L. T. Webster, Jr., Formation of *N*-(5-nitro-2-thiazolyl)-*N'*-carboxymethylurea from 5-hydroxyniridazole, *Drug Metab. Dispos.*, 1991, **19**, 508.

1-[2,4-Dichloro-5-[N-(methylsulfonyl)-amino]phenyl]-1,4-dihydro-3-methyl-4-(difluoromethyl)-5H-triazol-5-one (F6285)

Use/occurrence:	Pre-emergence herbicide
Key functional groups:	Difluoromethyl, methyl sulfonamide, methyl triazole
Test system:	Rat (oral, 10 mg kg^{-1}), goat (oral, 90 mg day^{-1}), hen (oral, 4 mg day^{-1})

Structure and biotransformation pathway

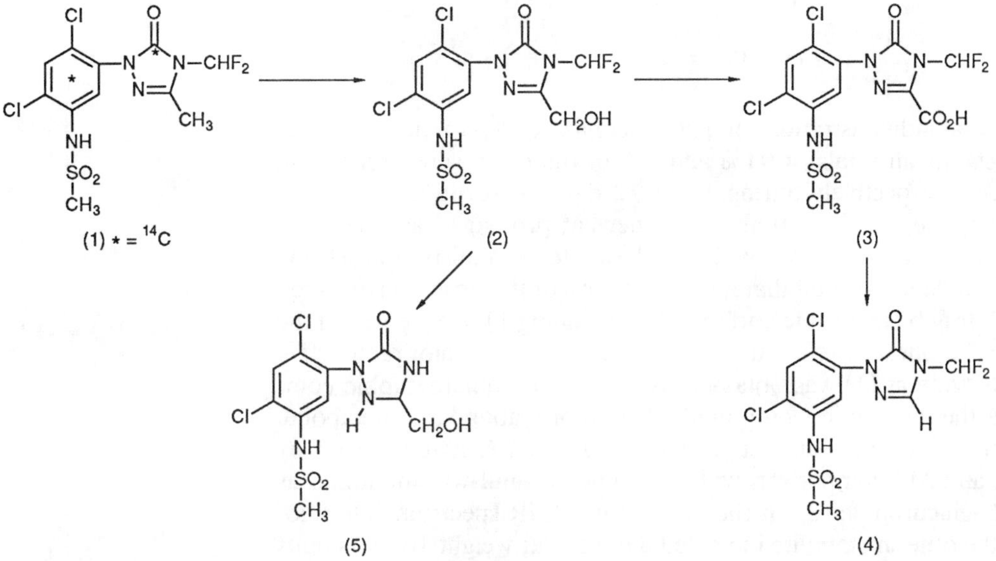

(1) * = ^{14}C   (2)   (3)   (5)   (4)

Following oral administration of (1) to rats and goats, over 90% of the dose was excreted in urine. Analysis of metabolites by TLC and HPLC followed by MS showed that the 3-hydroxymethyl metabolite (2) was the major component (82–97%) of the radioactivity in the excreta, together with the carboxylic acid (3) (0.3–5%). Goats and hens excreted trace amounts of (4) and unchanged (1). Rats did not form (4), and (1) was only detected in the faeces of females. Metabolite (5) was only found in the urine of female rats and the faeces of males. Goat milk contained mainly (2) with traces of (1) and (3).

Reference

L. Y. Leung, J. W. Lyga, and R. A. Robinson, Metabolism and distribution of the experimental triazalone herbicide F6285 [1-[2,4-dichloro-5-[N-(methylsulphonyl)-amino]phenyl]-1,4-dihydro-3-methyl-4-(difluoromethyl)-5H-triazol-5-one] in the rat, goat and hen, *J. Agric. Food Chem.*, 1991, **39**, 1509.

Fluconazole

Use/occurrence:	Antifungal drug
Key functional groups:	Benzyl alcohol, triazole
Test system:	Human (oral, 50 mg)

Structure and biotransformation pathway

Following oral administration of $[^{14}C]$fluconazole (1) to healthy male human subjects, mean totals of 91% and 2% of the dose were excreted in urine and faeces respectively during 11 or 12 days. Absorption was hence effectively complete, but excretion was somewhat prolonged as only two-thirds of the dose was recovered within 3 days after administration. HPLC analysis of urine demonstrated that a mean of 80% of the dose was excreted as unchanged drug by this route and that the remaining 11% dose was in the form of two metabolites. Confirmation of the identity of the major radioactive component in urine as (1) was obtained by MS and chromatographic comparison with the authentic compound. The more abundant metabolite liberated fluconazole on treatment with β-glucuronidase, as evidenced by TLC, HPLC, and MS comparison with authentic (1), and was identified as fluconazole O-glucuronide (2) on the basis of its NMR spectrum. Thermospray MS of the other metabolite indicated a molecular weight 16 mass units greater than that of the parent drug, and as it was converted into (1) during GC–MS or on treatment with $TiCl_3$ this metabolite was identified as a thermally labile N-oxide of fluconazole (3). No metabolic cleavage products of fluconazole were observed.

Reference

K. W. Brammer, A. J. Coakley, S. G. Jezequel, and M. H. Tarbit, The disposition and metabolism of $[^{14}C]$fluconazole in humans, *Drug Metab. Dispos.*, 1991, **19**, 764.

Etridiazole

Use/occurrence:	Agricultural fungicide
Key functional groups:	Alkyl ether, thiadiazole, trichloromethyl
Test system:	Human (oral, *ca.* 0.5 and 0.15 μmol kg^{-1}), rat (oral, *ca.* 2, 4, 12, 120, 600, and 1200 μmol kg^{-1})

Structure and biotransformation pathway

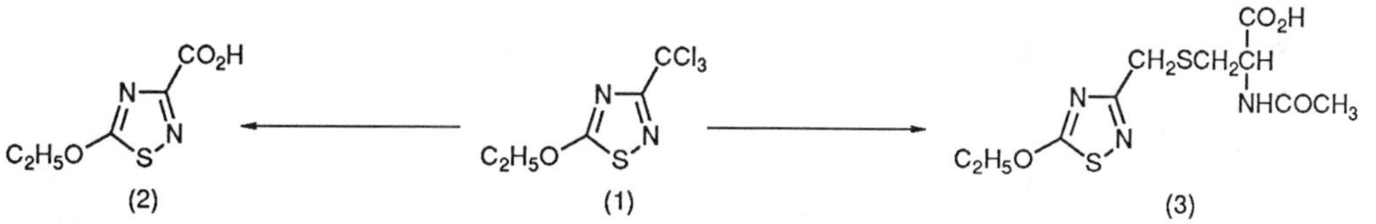

Following oral administration of (1) to both rats and humans, urine was collected and analysed for (1), (2), and (3) by a GC–MS assay. Parent compound (1) (< 1% of dose), (2) (*ca.* 22% of the dose), and (3) (a minor metabolite) were all detected and identified in rat urine. Only (2) could be detected in human urine, accounting for *ca.* 13% of the dose. Urinary excretion of (2) was linear with dose in rats over the dose range studied, and may therefore be a useful marker in monitoring human exposure.

Reference

R. T. H. Van Welie, R. Mensert, P. Van Duyn, and N. P. E. Vermeulen, Identification and quantitative determination of a carboxylic and a mercapturic acid metabolite of etridiazole in urine of rat and man. Potential tools for biological monitoring, *Arch. Toxicol.*, 1991, **65**, 625.

259

DuP 753

Use/occurrence:	Angiotensin II receptor antagonist
Key functional groups:	Tetrazole
Test system:	Monkey liver slices

Structure and biotransformation pathway

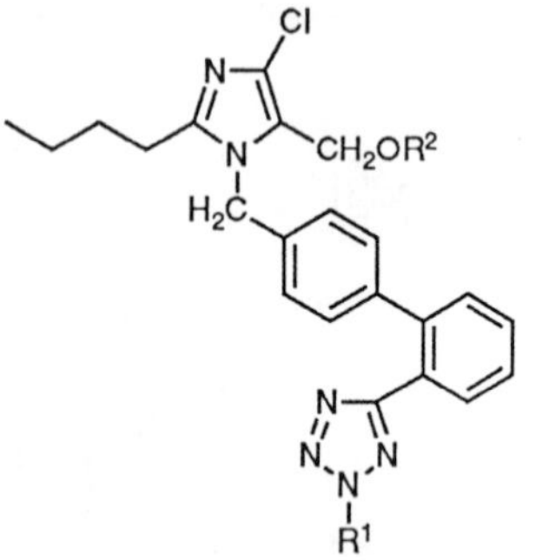

(1) $R^1 = R^2 = H$
(2) R^1 = Glucuronyl, R^2 = H
(3) R^1 = H, R^2 = Glucuronyl

This short communication describes the synthesis and structural elucidation of three glucuronic acid conjugates of DuP 753 (1), namely the ether glucuronide and the tetrazole-N^1-and N^2-β-glucuronides. HPLC and ^{1}H NMR were then used to show that the major metabolite of (1) formed during incubation of (1) with monkey liver slices was the N^2-β-glucuronide (2). The authors claimed that this was the first example of the formation of a tetrazole-N^2-β-glucuronide. The ether glucuronide (3) was also apparently formed in the incubation but this was not emphasized in the paper.

Reference

R. A. Stearns, G. A. Doss, R. R. Miller, and S.-H. L. Chiu, Synthesis and identification of a novel tetrazole metabolite of the angiotensin II receptor antagonist DuP 753, *Drug Metab. Dispos.*, 1991, **19**, 1160.

N-Benzylpiperidine

Use/occurrence:	Model compound
Key functional groups:	Benzyl amine, piperidine
Test system:	Rat liver microsomes and biomimetic chemical systems

Structure and biotransformation pathway

The purpose of this study was to investigate the mechanism of oxidation (particularly at the β-position) of the piperidine ring (which is incorporated in the structures of various drugs and natural products) in liver microsomes and a biomimetic chemical system. The main substrate studied was N-benzyl-piperidine (1), the chemical system used was the cytochrome P-450 model, meso-tetraphenylporphinatoiron(III) chloride/2,6-dimethyliodosylbenzene, and the liver microsomes were obtained from male Wistar rats pretreated with phenobarbital. Chemical oxidation of (1) yielded many products, the most abundant being piperidine (2) and the ketones (3)–(5). The yields of the latter three compounds were much greater than those of the hydroxylated derivatives (6) and (7), which were enhanced when the reaction products were treated with NaBH$_4$. After incubation of (1) with hepatic microsomes, the metabolite profile determined by GC–MS was similar to that resulting from the chemical model system, although the yield of (5) was reduced in the former case. The microsomal oxidation reactions were all inhibited when incubation was performed in the presence of SKF 525-A or when NADP was omitted, indicating that they are dependent on cytochrome P-450. Atmospheric oxygen was shown not to be involved in the formation of (4) in the chemical system by comparing the amounts formed under aerobic and anaerobic conditions. The use of deuterium-labelled analogues showed that in the chemical and microsomal systems the hydrogen abstraction step at the β-position was not the rate-limiting step in the conversion of (1) into (4). Therefore, a mechanism involving the intermediacy of the iminium ion (8) was proposed. Consistent with this proposal, relatively good yields of the adduct (9) were obtained when (1) was incubated with microsomes in the presence of NaCN (which does not inhibit microsomal cytochrome P-450 activity), and oxidation of (8) (as its perchlorate salt or enamine free base) with the chemical system yielded the keto analogues (3) and (4). Complete mechanisms are proposed for the oxidation of (1) and other, related structures, *viz*. phencyclidine and dipropylbenzylamine.

Reference

H. Masumoto, S. Ohta, and M. Hirobe, Application of chemical cytochrome P-450 model systems to studies on drug metabolism. IV. Mechanism of piperidine metabolism pathways *via* an iminium intermediate, *Drug Metab. Dispos.*, 1991, **19**, 768.

1-Methyl-4-phenyl-1,2,3,6-tetrahydropyridine (MPTP)

Use/occurrence:	Model compound
Key functional groups:	*N*-Methyl tetrahydropyridine
Test system:	Cultured mouse astrocytes

Structure and biotransformation pathway

(4)　　　(1)　　　(2)　　　(3)

The biotransformation of 1-methyl-4-phenyl-1,2,3,6-tetrahydropyridine (MPTP) (1) is essential for its neurotoxic effects. The route of metabolism which generates the neurotoxic metabolite is mediated by monoamine oxidase (MAO) B and immunochemical studies have provided evidence that the enzyme is located in both serotonergic neurones and the astrocyte cells present in brain tissue. The present study was conducted to investigate the role of mouse brain astrocytes in the metabolism of (1).

Following exposure of primary cultures of mouse astrocytes to (1), three metabolites were observed following separation by HPLC. These were the dihydropyridinium ion, MPDP$^+$ (2); the pyridinium ion, MPP$^+$ (3), the ultimate neurotoxin; and MPTP *N*-oxide (4). At all time points, the total level of (2), (3), and (4) was similar to the amount of (1) which had disappeared from the incubation, suggesting the lack of significant contribution by any other metabolites to the overall biotransformation of (1). Production of (3) and (4) occurred at constant rates throughout the 4 day incubation period whereas the level of (2) remained relatively low.

It should be noted that the levels of MAO A in astrocytes are much higher than the corresponding levels of MAO B. However, MAO A has been shown to be active in the formation of (2) and (3) although metabolism of (1) occurs at a much lower rate than in the presence of MAO B. Thus clorgyline (a specific MAO A inhibitor) blocked the formation of (2) and (3) more effectively than deprenyl (a specific MAO B inhibitor) in mouse astrocytes.

These data suggest that metabolic pathways other than the MAO B dependent formation of (3) can contribute to the biotransformation of (1) by mouse astrocytes in an *in vitro* system. The formation of metabolite (4) was shown to be mediated primarily by a flavin containing mono-oxygenase.

Reference

D. A. Di Monte, E. Y. Wu, I. Irwin, L. E. Delanney, and J. W. Langston, Biotransformation of 1-methyl-4-phenyl-1,2,3,6-tetrahydropyridine in primary cultures of mouse astrocytes, *J. Pharmacol. Exp. Ther.*, 1991, **258**, 594.

1-(1-Phenylcyclohexyl)-2,3,4,4-tetra-hydropyridinium perchlorate

Use/occurrence:	Model compound
Key functional groups:	Tetrahydropyridinium
Test system:	Rat brain and liver mitochondria (0.2 mM substrate)

Structure and biotransformation pathway

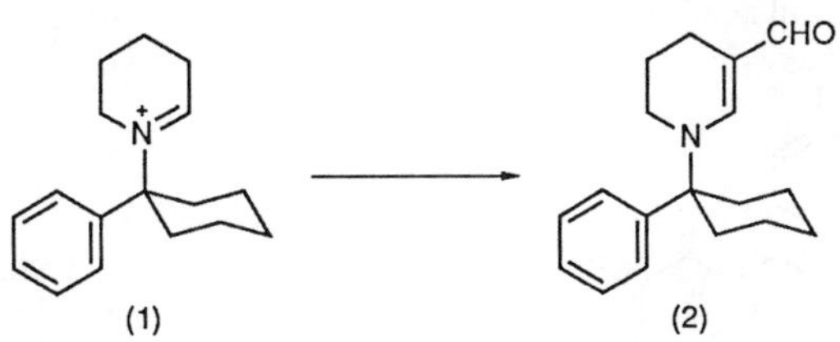

Rat brain and liver mitochondrial preparations were found to be capable of metabolizing (1) to the *C*-formylated metabolite (2). The C-5 formylation could also be achieved by reacting N^5-tetrahydrofolinic acid directly with (1), suggesting that the reaction in the mitochondrial incubations was one of trans-formylation. Identity of the metabolite was confirmed by GC–MS and by comparison with a synthetic standard.

Reference

Z. Zhao, L. Y. Leung, A. Trevor, and N. Castagnoli, Jr., *C*-Formylation in the presence of rat brain mitochondria of the 2,3,4,5-tetrahydropyridinium metabolite derived from the psychotomimetic drug phencyclidine, *Chem. Res. Toxicol.*, 1991, **4**, 426.

Haloperidol

Use/occurrence: Model compound

Key functional groups: Hydroxypiperidine

Test system: Rat liver microsomes (0.5 mM substrate)

Structure and biotransformation pathway

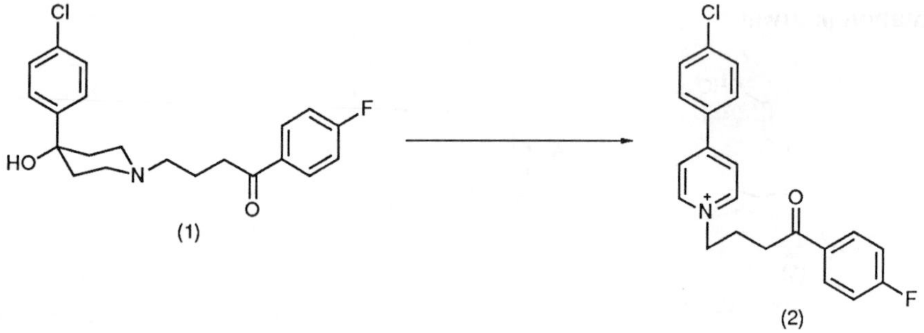

Rat liver microsomal preparations were found to be capable of metabolizing haloperidol (1) to the corresponding pyridinium metabolite (2), which is potentially neurotoxic. Interestingly neither (1) nor its 1,2,3,4-tetrahydropyridine dehydration products were substrates for purified bovine liver MAO B in the same way as the classic pro-neurotoxin MPTP (which is activated to MPP^+), which may be due to the bulky substituents. Haloperidol and its putative metabolites were analysed by HPLC with diode array UV detection (see Vol. 4, p. 274).

Reference

B. Subramanyam, T. Woolf, and N. Castagnoli, Jr., Studies on the *in vitro* conversion of haloperidol to a potentially neurotoxic pyridinium metabolite, *Chem. Res. Toxicol.*, 1991, **4**, 123.

Haloperidol

Use/occurrence: Neuroleptic

Key functional groups: Chlorophenyl, fluorophenyl, hydroxypiperidine

Test system: Mouse hepatic microsomes

Structure and biotransformation pathway

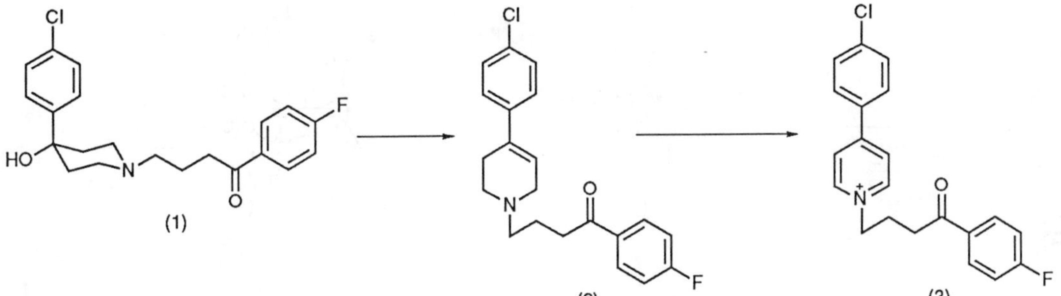

Following incubation of haloperidol (1) with mouse hepatic microsomes the metabolic products were isolated using solid-phase extraction and analysed by HPLC–MS. Two metabolites were identified, the first formed by dehydration of the alcohol producing the tetrahydropyridine (2) and the second the pyridinium ion (3). The dehydration of the alcohol to form a double bond is believed to be a new metabolic pathway.

Reference

J. Fang and J. W. Gorrod, Dehydration is the first step in the bioactivation of halo-peridol to its pyridinium metabolite, *Toxicol. Lett.*, 1991, **59**, 117.

Cocaine

Use/occurrence:	Anaesthetic/drug of abuse
Key functional groups:	Alkaloid, aryl carboxylate, methyl ester, N-methylpiperidine
Test system:	Human liver microsomes

Structure and biotransformation pathway

(1) * = ^{1}H
(2) * = ^{3}H

(3)

(4)

It is known that N-oxidation of cocaine (1) to norcocaine (3) and thence to N-hydroxynorcocaine (4) occurring in mice is closely associated with the hepatotoxic effects of (1), and that these effects increase with the level of oxidation, $i.e.$ (4) > (3) ≥ (1). The purpose of the present study was to determine whether this metabolic pathway also occurs in human liver microsomes *in vitro*. Incubations were performed in the presence of NADPH and the esterase inhibitor sodium fluoride, and the formation of metabolites (3) and (4) was monitored by quantitative HPLC. In this way it was shown that (1) was converted into both of these metabolites in every human liver microsomal preparation examined, and that the apparent K_m values for each oxidation reaction were greater than those for the corresponding processes in mouse hepatic microsomes under the same conditions, whereas V_{max} values were lower. Both reactions were inhibited in the absence of NADPH and in the presence of carbon monoxide, SKF 525-A, or n-octylamine, indicating that cytochrome P-450 was the enzyme responsible for both processes. Interestingly, whereas omission of NaF from the human liver microsomal suspensions resulted in a 50% reduction in the rate of N-demethylation of (1) and a small reduction in the rate of N-hydroxylation of (3), this omission had virtually no effect on either process during incubation with mouse liver microsomes. Studies were also performed to determine whether irreversibly protein-bound metabolites of (1) were formed in human liver tissue (as they are in the corresponding mouse preparations) using [^{3}H]cocaine (2) as substrate. The results showed that both the rate and extent of irreversible binding of radioactivity were similar in human and mouse hepatic microsomes, and that formation of the metabolite responsible was mediated by cytochrome P-450. In view of the observed similarities in the N-oxidative metabolism of

(1) in humans and mice, it was suggested that both species share the same bioactivation mechanism of cocaine-induced liver damage.

Reference

S. M. Roberts, R. D. Harrison, and R. C. James, Human microsomal *N*-oxidative metabolism of cocaine, *Drug Metab. Dispos.*, 1991, **19**, 1046.

Nifedipine

Use/occurrence:	Calcium antagonist, antihypertensive, antianginal drug
Key functional groups:	Dihydropyridine, methyl ester
Test system:	Human volunteers (oral, 20 mg)

Structure and biotransformation pathway

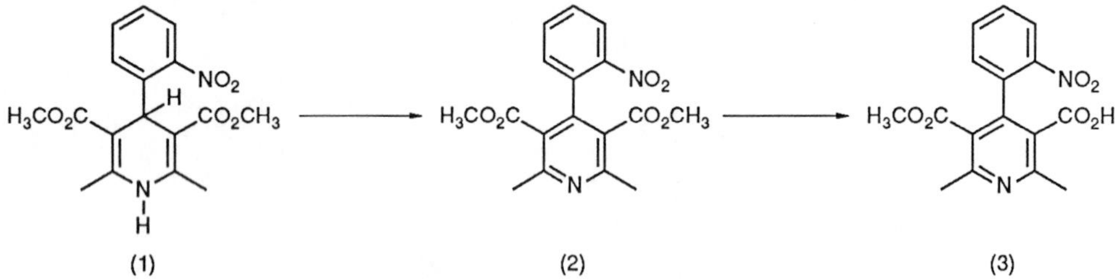

(1)　　　　　　　　　　(2)　　　　　　　　　　(3)

Nifedipine (1) disposition following oral administration exhibits wide inter-subject variability: the area under the plasma concentration–time curve of (1) varied eight-fold. The major metabolite (3) is formed by aromatization of (1) to (2) and then ester hydrolysis of (2) to (3). The non-linear first-pass metabolism of (1) in healthy human volunteers was evident in the formation of (2), but not reflected in the formation of (3). Therefore, (2) is biotransformed into (3), but (3) arises from (2) and also by a hitherto unknown route.

Reference

J. H. M. Schellens, I. M. M. van Haelst, J. B. Houston, and D. D. Breimer, Nonlinear first-pass metabolism of nifedipine in healthy subjects, *Xenobiotica*, 1991, **21**, 547.

Nimodipine

Use/occurrence:	Antihypertensive drug
Key functional groups:	Alkyl ester, dihydropyridine, methyl ether, methylpyridine, nitrophenyl
Test system:	Rat, dog, and rhesus monkey (oral, 5, 10, and 20 mg kg^{-1})

Structure and biotransformation pathway

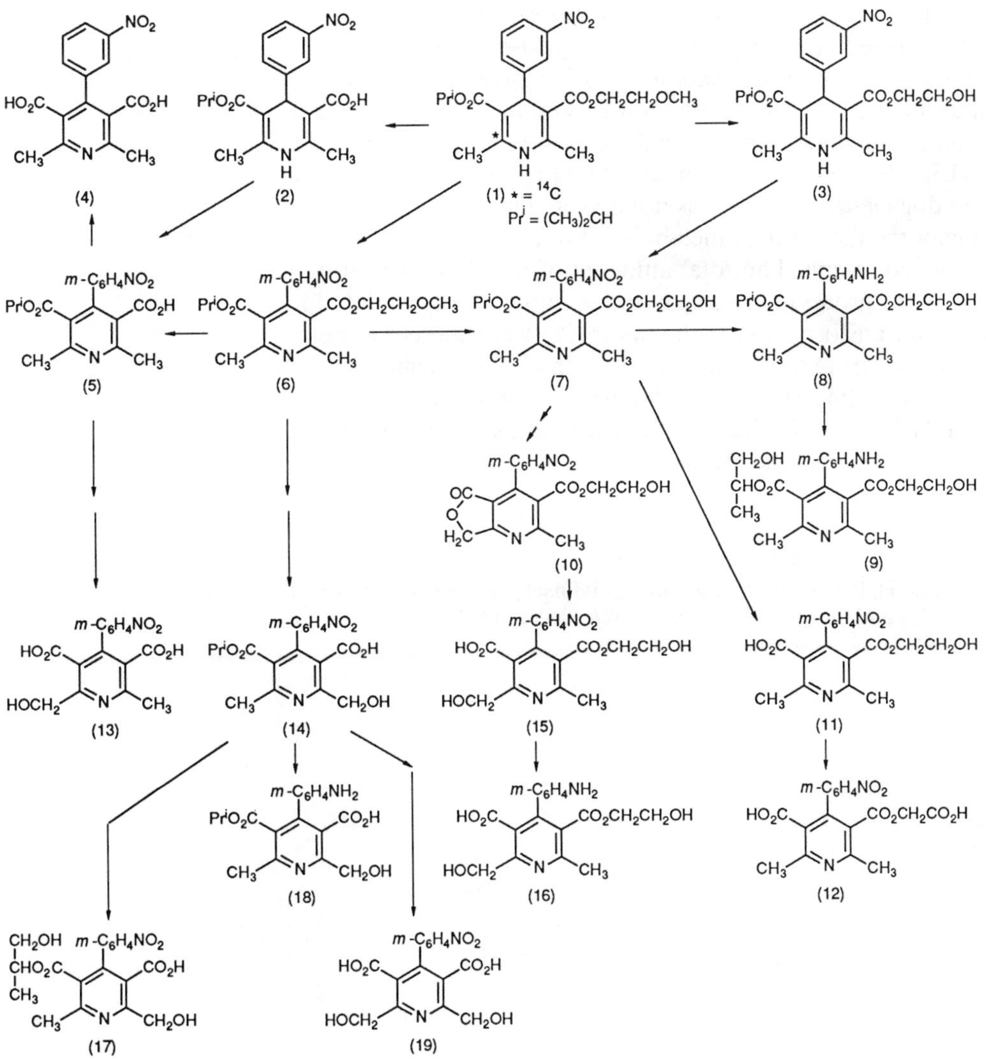

271

After oral administration of [^{14}C]nimodipine (1) to rats, dogs, and monkeys, *ca.* 21% of the dose was excreted in urine during 24 hours by the former two species and 47% of the dose was excreted in urine during 10 days by the latter species. Following intraduodenal administration of (1) to bile duct-cannulated rats, 76% of the dose was recovered in the bile during 6 hours, and 8% of an oral dose of (1) to a single dog killed at 4 hours was recovered in the bile obtained from the gall bladder. Patterns of radioactive components in the excreta and rat plasma were determined by TLC, metabolites were isolated and purified by TLC and HPLC, and their structures were characterized by MS, NMR, and chromatographic comparison with reference compounds (where available). Intact drug was detected in rat plasma, but not in any of the excreta samples, which demonstrates that nimodipine, like most dihydropyridine calcium channel blockers, undergoes extensive biotransformation in each species before excretion. A total of 18 metabolites [(2)–(19)] were identified, which together accounted for *ca.* 75% of the urinary radioactivity, 50% of the biliary radioactivity, and 80% of the plasma radioactivity. Unchanged (1) and the dihydropyridine metabolites (2) and (3) were only detected in rat plasma at 1 hour post-dose, when (14) was the major metabolite, but the only metabolite bearing an intact dihydropyridine ring found in the excreta was (3), which was present in bile as its glucuronic acid conjugate. The main urinary metabolites were as follows: rat, (11) and (15); dog, (15), (16), and (18); monkey, (5), (11), (16), and (18). The important rat and dog metabolite (15) was not detected in monkey urine, and conversely the major monkey urinary metabolite (5) was only present in small amounts in rat and dog urine. The total amounts of metabolites bearing a reduced aromatic nitro group followed the order: monkey > dog > rat. The principal Phase I biotransformation pathways of (1) were therefore: dehydrogenation of the dihydropyridine ring, ester hydrolysis, *O*-demethylation and oxidation of the resulting primary alcohol to the carboxylic acid, hydroxylation of the methyl groups in the heterocyclic ring and of the isopropyl ester, and reduction of the aromatic nitro group.

Reference

D. Scherling, K. Bühner, H. P. Krause, W. Karl, and C. Wünsche, Biotransformation of nimodipine in rat, dog and monkey, *Arzneim.-Forsch.*, 1991, **41**, 392.

Nitrendipine

Use/occurrence:	Antihypertensive drug
Key functional groups:	Alkyl ester, dihydropyridine, methyl ester, methylpyridine, nitrophenyl
Test system:	Mouse (oral, 5 mg kg^{-1}), rat (oral, 5 mg kg^{-1}), dog (oral, 5 mg kg^{-1}), rat liver microsomes, rat hepatocyte, and isolated perfused rat liver

Structure and biotransformation pathway

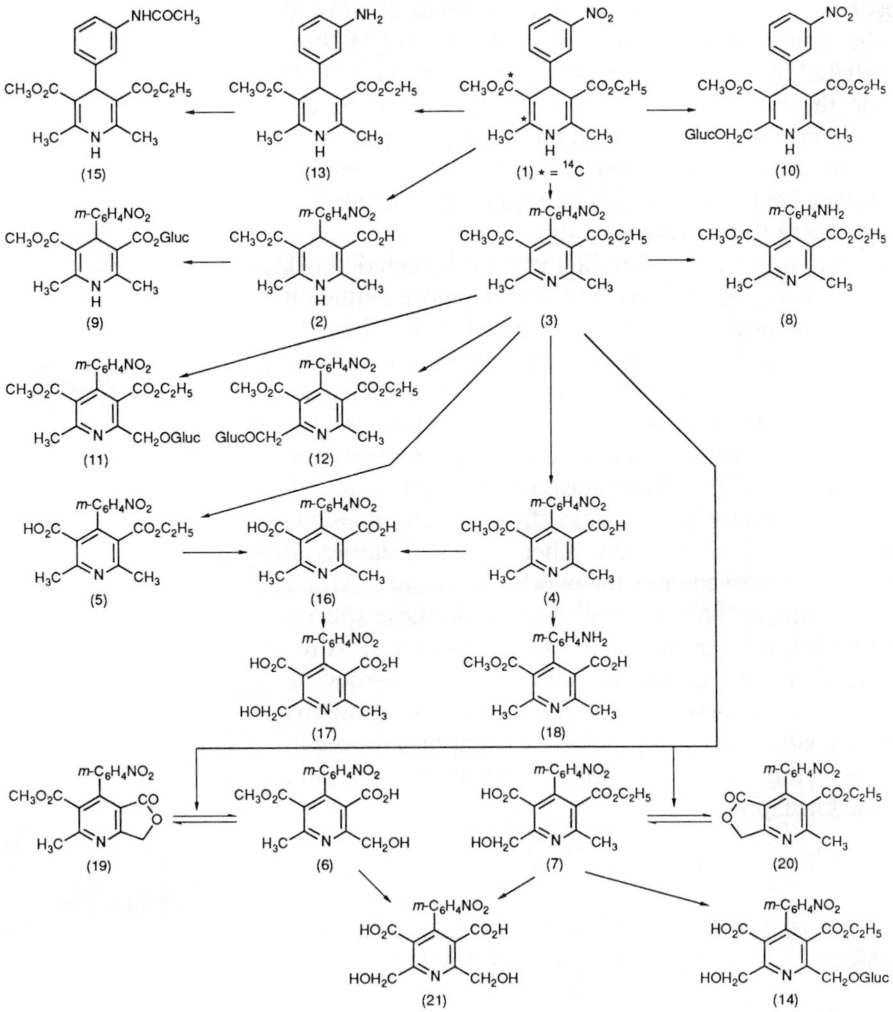

(15) (13) (1) $* = {}^{14}C$ (10)

(9) (2) (3) (8)

(11) (12)

(5) (16) (4)

(17) (18)

(19) (6) (7) (20)

(21) (14)

273

Plasma, urinary and/or biliary metabolites from mice, rats, and dogs dosed orally or intraduodenally with [^{14}C]nitrendipine (1), together with those from isolated perfused rat liver, were separated by TLC and HPLC, further purified by HPLC, and identified by chromatographic (TLC, HPLC, GC) and/or spectroscopic (EI-, CI-, and/or FAB-MS and ^{1}H NMR) comparison with synthetic reference compounds. Little or no unchanged (1) was detected in the excreta, and less than 5% of the radioactivity in rat plasma at 1 hour post-dose was in the form of parent drug. Nitrendipine was therefore rapidly and extensively metabolized in all three species, and a total of 20 biotransformation products [(2)–(21)] were characterized. The identified metabolites of (1) in urine and bile together accounted for *ca.* 75% of the dose administered to rats and dogs, and those in mouse urine accounted for about half of the dose. The principal radioactive component in rat plasma at 1 hour after administration was (2), which possessed an intact dihydropyridine ring; the corresponding metabolite formed by cleavage of the methyl ester was not observed in any sample analysed. Metabolite (2) was also abundant in rat and mouse urine and in the perfusate, but was only present in minor quantities in dog urine and bile. Other metabolites detected in rat plasma were (3)–(7). *In vitro* studies with liver microsomes and rat plasma and pig liver esterases showed that both the heterocyclic ring-dehydrogenation and ester-hydrolysis biotransformations were mediated by cytochrome P-450. Ten metabolites of (1) were found in mouse urine, the most important being (4), (5), (7), and (8); the last of these metabolites was markedly more abundant in samples from male animals than in those from females. At least 23 metabolites were separated in rat urine, the more important being (4)–(7), and more than 28 metabolites were observed in rat bile, where (9) represented 30% of the sample radioactivity; other biotransformation products identified in rat bile included (4), (5), and (7). Dog urine contained at least 22 metabolites, the more abundant being (4)–(7) and (13), and in dog bile, which contained more than 24 metabolites, about half of the sample radioactivity was in the form of the glucuronide conjugates (10)–(12). The ether glucuronide (14) and both diastereoisomers of the acyl glucuronide (9) were also present in dog bile, and the other minor components therein corresponded to the urinary metabolites. Interestingly, when a 1:1 mixture of the diastereoisomeric glucuronides (9) was incubated with β-glucuronidase/arylsulfatase from *Helix pomatia*, neither the corresponding aglycone (2) nor any other product bearing an intact dihydropyridine ring was observed, only the pyridine derivative (4) and ring-modified degradation products. The metabolism of (1) in these species therefore followed well established biotransformation pathways for dihydropyridine calcium antagonists, namely dehydrogenation of the heterocyclic ring, oxidative ester hydrolysis, hydroxylation of the C-2 and C-6 methyl groups, reduction of the aromatic nitro group (which was important only in mice) and subsequent acetylation (which was only observed in dogs), and ether- and acyl-glucuronide formation.

Reference

D. Scherling, W. Karl, H. J. Ahr, A. Kern, and H.-M. Siefert, Biotransformation of nitrendipine in rat, dog and mouse, *Arzneim.-Forsch.*, 1991, **41**, 1009.

Felodipine

Use/occurrence:	Antihypertensive drug
Key functional groups:	Dihydropyridine
Test system:	Rat, dog, and human liver microsomes

Structure and biotransformation pathway

(1) X = [3,4-dichlorophenyl] ; Y = H

(2) X = H; Y = [3,4-dichlorophenyl]

In general, the predominant biotransformation pathway of the dihydropyridine calcium antagonists involves oxidation of the heterocyclic ring to the inactive pyridine analogue. This process, which is catalysed by cytochrome P-450 in liver microsomes, results in high first-pass metabolism and consequently low oral bioavailability. Previous studies in dogs and humans have indicated that for felodipine this first-pass metabolism is stereoselective and species-dependent *in vivo*, as the bioavailability of the S-enantiomer (1) of felodipine was lower than that of the R-enantiomer (2) in the former species but greater in the latter. The purpose of the present study was to investigate the separate conversion of (1) and (2) into the pyridine derivative (3) *in vitro* using liver microsomes from male rats, dogs, and humans. The results showed that for the former two species (1) was metabolized slightly faster than (2), but no differences were found in V_{max} or K_m. In human liver microsomes, on the other hand, (2) was metabolized more readily than (1), and the mean intrinsic clearance (V_{max}/K_m) was about two-fold greater for the R-enantiomer (2), owing largely to its smaller K_m value. These results are consistent with those found *in vivo*. Thus, in dogs, the bioavailability of the R-enantiomer was slightly higher than that of the S-enantiomer, whereas in humans, plasma C_{max} and AUC values were about two-fold greater for the S-enantiomer.

Reference

U. G. Eriksson, J. Lundahl, C. Bäärnhielm, and C. G. Regardh, Stereoselective metabolism of felodipine in liver microsomes from rat, dog and human, *Drug Metab. Dispos.*, 1991, **19**, 889.

Isopropyl methyl 2-carbamoyloxymethyl-4-(2,3-dichlorophenyl)-1,4-dihydro-6-methyl-3,5-pyridinedicarboxylate

Use/occurrence:	Calcium channel antagonist
Key functional groups:	iso-Alkyl ester, carbamate, dihydropyridine, methyl ester, methylpyridine
Test system:	Rat (oral, 10 mg kg^{-1}), rat liver microsomes

Structure and biotransformation pathway

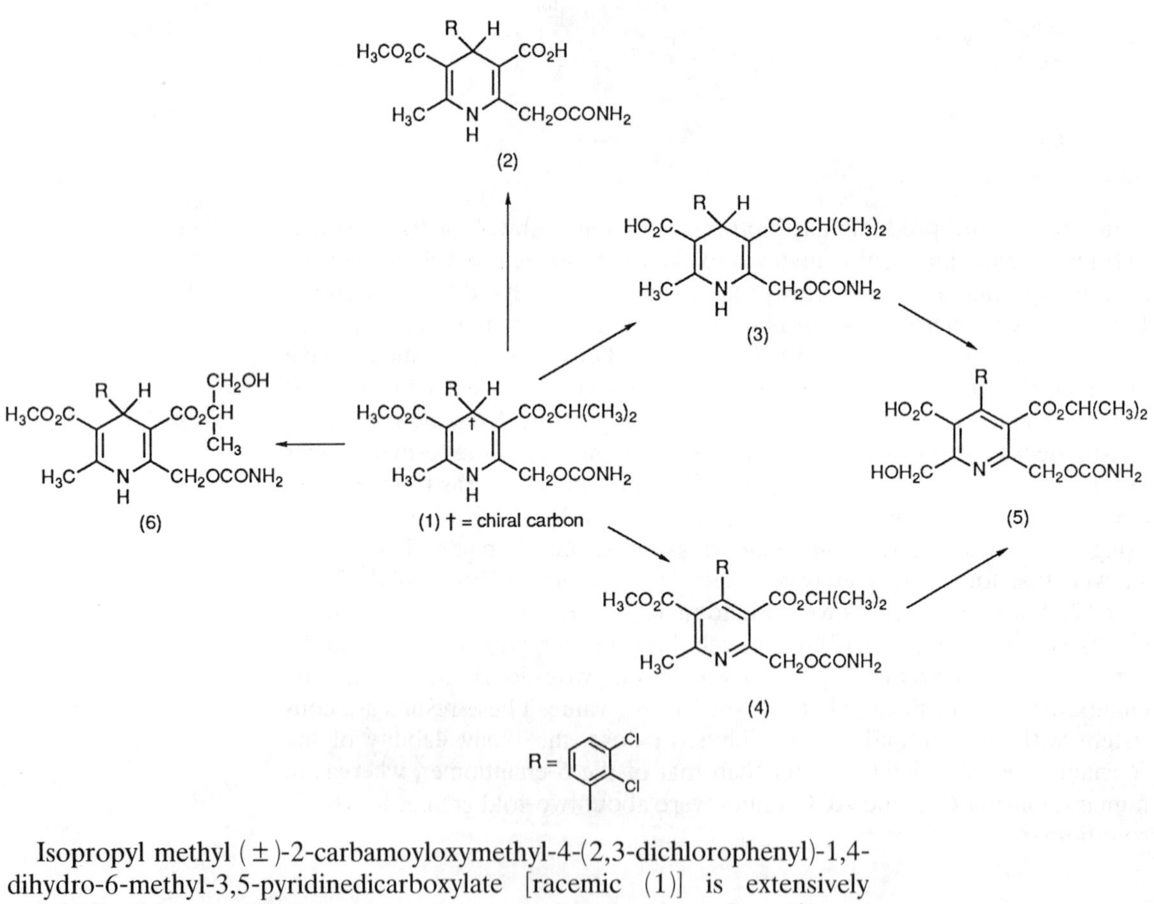

Isopropyl methyl ($\pm$)-2-carbamoyloxymethyl-4-(2,3-dichlorophenyl)-1,4-dihydro-6-methyl-3,5-pyridinedicarboxylate [racemic (1)] is extensively metabolized by rats. The present study was designed to investigate the sex difference in the stereoselective metabolism of racemic (1) in this species. Metabolites were identified following separation by HPLC and comparison with authentic synthetic standards.

Following oral administration of (±)-(1) to male and female rats, plasma levels of (+)-(1) were higher than those of the (−)-enantiomer in both sexes. However, higher levels of each enantiomer were found in female rats than in the males. A similar pattern of metabolites was seen in both sexes after dosing with racemic (1). The 3-desisopropyl metabolite (2) and the 5-desmethyl metabolite (3) were the major metabolites seen in plasma. The pyridine metabolite (4) and the corresponding hydroxymethylpyridine (5) were present at much lower concentrations.

Plasma levels of metabolite (2) after oral dosing with (−)-(1) were much higher than those after dosing with (+)-(1) in both male and female rats. These data together with those from a previous study suggested that the difference in plasma levels was due to stereoselective metabolism and sexual dimorphism in liver metabolism.

In vitro metabolic studies of enantiomers of (1) showed a pronounced sex difference in the basal rates of stereoselective metabolism of the enantiomers. Male animals were 2–3 times more efficient in metabolizing (1) enantiomers than females. The major metabolite seen *in vitro* was the pyridine metabolite (4) and a marked sex difference was seen in the formation of (4) and (6), the 3-(2'-hydroxy-1'-methylethyl) metabolite (males > females). The results suggested that the sex differences in the stereoselective metabolism of racemic (1) in rats are initially due to the preferential removal of either the dihydropyridine metabolites [(2), (3), and (6)], by hydroxylation of the iso-propyl group [(6)] and oxidative cleavage of the carboxylic esters [(2), (3)], or to the removal of the pyridine metabolite (4) by oxidation of the dihydropyridine moiety to the pyridine form.

Experiments with liver microsomes from animals treated with phenobarbitone or methylcholanthrene suggested that several forms of cytochrome P-450 were involved in the metabolism of (1). Formation of metabolite (6) was decreased in male rats and that of metabolites (2) and (4) was increased by phenobarbitone pretreatment.

Reference

F. Takayama, K. Saito, Y. Ishii, K. Shiratori, and M. Ohtawa, Sex difference in the stereoselective metabolism of a new dihydropyridine calcium channel blocker, in rat studies *in vivo* and *in vitro*, *Xenobiotica*, 1991, **21**, 557.

Tripelennamine

Use/occurrence:	Antihistamine drug
Key functional groups:	Dialkyl aryl amine, dimethylaminoalkyl, pyridine
Test system:	Human volunteers (intramuscular, 100 mg kg^{-1})

Structure and biotransformation pathway

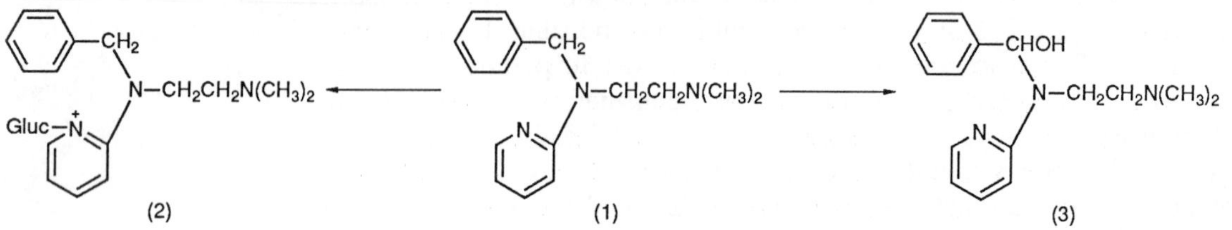

(2)   (1)   (3)

Urinary metabolites were determined by GC–MS assay, in 0–24 hour urine samples. Total amounts of unchanged (1) accounted for 1.24% of the administered dose. Total amounts of conjugated (1) (which were determined indirectly following enzyme hydrolysis) excreted in the urine in the same period accounted for 4.45% of the dose. One of the conjugated metabolites was believed to be the *N*-glucuronide (2). Total amounts of the hydroxylated compound (3) together with an unidentified metabolite accounted for *ca.* 23% of the dose. (See also Vol. 1, p. 212 and Vol. 2, p. 224.)

Reference

S. Y. Yeh, Metabolic profile of tripelennamine in humans, *J. Pharm. Sci.*, 1991, **80**, 815.

Triclopyr

Use/occurrence:	Herbicide
Key functional groups:	Aryloxyacetic acid, chloropyridine
Test system:	Crayfish (11 day static exposure to 1 and 2.5 mg l^{-1} aquarium water)

Structure and biotransformation pathway

(1) $* = {}^{14}C$

(2)

The metabolic fate of $[{}^{14}C]$triclopyr (1) in the crayfish was investigated as this edible freshwater crustacean may be raised in rice fields and so be exposed to various pesticides. The uptake, distribution, and persistence of radioactivity in crayfish tissues were determined at aquarium water concentrations of 1 and 2.5 mg l^{-1}, which simulate potential field applications. After 11 days of exposure, the majority of the residues in whole crayfish were found in the non-edible portion (shell and haemolymph), and only a small proportion was present in the edible part (hepatopancreas and tail muscle). HPLC analysis indicated the presence of two major radioactive components in organic extracts of the hepatopancreas. The most abundant of these, which represented *ca.* 80% of the sample radioactivity, was identified by co-chromatography and GC–MS as unchanged (1), and the other was identified by co-chromatography and CI-MS as the taurine conjugate (2). Several minor metabolites (<0.1 p.p.m.) were also detected, but in levels too low to allow identification. Bioconcentration factors estimated from uptake and elimination rate constants indicated a low potential for accumulation of (1) and its metabolites in this species.

Reference

M. G. Barron, S. C. Hansen, and T. Ball, Pharmacokinetics and metabolism of triclopyr in the crayfish (*Procambus clarki*), *Drug Metab. Dispos.*, 1991, **19**, 163.

HI6, Pyrimidoxime

Use/occurrence: Acetylcholinesterase reactivators

Key functional groups: Aldoxime, pyridinium

Test system: Rat (intramuscular, 30 mg kg^{-1})

Structure and biotransformation pathway

HI6 (1) and pyrimidoxime (2) are two drugs used to treat organophosphate intoxication. The mechanism of action of (1) and (2) is to reactivate acetylcholinesterases, the enzymes which are inhibited by organophosphate treatment.

The metabolism of $[^{14}C]$-(1) and $[^{14}C]$-(2) was investigated following intramuscular administration to normal rats and rats poisoned by organophosphates. The radioactivity was essentially eliminated in the urine, with more than 85% of the dose excreted by this route within 24 hours. Faecal excretion accounted for *ca.* 4% of the dose over the 24 hour period.

Compound (2) was not metabolized and was shown to be excreted primarily as the unchanged drug in the urine (identified by NMR and HPLC retention time). In contrast, (1) was degraded to pyridine-2-aldoxime (3), an inactive metabolite (with respect to reactivation of acetylcholinesterase) formed by cleavage of the bond between carbon and the quaternary nitrogen. Thus, two radiolabelled components corresponding to (1) and (3) were separable by HPLC in plasma extracts. Compound (3) was identified by co-chromatography on HPLC and NMR. In the urine, but not in the plasma, (3) was partially hydrolysed to pyridine-2-aldehyde (4).

It was thought that degradation of (1) may take place in the plasma and not the liver. Compound (1) was shown to have little affinity for the liver and catabolism of (1) was observed in plasma *in vitro* (8.6% within a 1 hour period following incubation at a concentration of 0.5 mg ml^{-1} in rat plasma). Cleavage of the quaternary nitrogen bond in buffer has also been demonstrated at pH greater than 7.

Reference

H. Garrigue, J. C. Maurizis, C. Nicolas, J. C. Madelmont, D. Godenche, T. Hulot, X. Morge, P. Demerseman, H. Sentenac-Roumanou, and A. Veyre, Disposition and metabolism of two acetylcholinesterase reactivators, pyrimidoxime and HI6, in rats submitted to organophosphate poisoning, *Xenobiotica*, 1990, **20**, 699.

Dithiopyr

Use/occurrence: Herbicide

Key functional groups: Methylthio ester

Test system: Rat liver enzymes

Structure and biotransformation pathway

Dithiopyr (1) is metabolized in rats to two major products, the nicotinic acids (3) and (5). These *in vitro* studies using male Long Evans rat liver enzymes (S9, cytosolic, and microsomal preparations) have elucidated the pathways of biotransformation of (1). The methylthioester functional groups in (1) are cleaved via rat hepatic microsomal oxygenases, and not esterases. The formation of (3) and (5) requires NADPH (or NADH) and is via oxidative and not hydrolytic processes. Thus, activation is by sulfoxide formation to (2) and (4), and these unstable intermediates are subject to hydroxide ion nucleophilic substitution to give (3) and (5) (ratio 1:5) respectively. Alternatively, (4) may form a glutathione conjugate (6) which is then hydrolysed to the corresponding dipeptide cysteinylglycine conjugate (7). Other minor metabolites were present, but they were not identified. Structural assignments followed from HPLC, positive and negative ion FAB-MS, and comparison with authentic samples. Oxidation of the sulfur atoms in dithiopyr (1) is a reaction commonly catalysed by flavin mono-oxygenases. *S*-Demethylation of (1), catalysed by cytochrome P-450 and forming the corresponding thio-acids, was detected *in vivo*, but not *in vitro*.

Reference

P. C. C. Feng and R. T. Solsten, *In vitro* transformation of dithiopyr by rat liver enzymes: conversion of methylthioesters to acids by oxygenases, *Xenobiotica*, 1991, **21**, 1265.

Minaprine

Use/occurrence: Psychotropic agent

Key functional groups: Alkyl aryl amine, methylpyridazine, morpholine

Test system: Rat and human hepatocytes, rat, rabbit, dog, baboon, and human liver microsomes

Structure and biotransformation pathway

Minaprine (1) is a psychotropic agent with antidepressant properties, used in the treatment of Alzheimer's disease and multi-infarct dementia. The biotransformation of minaprine has been extensively studied *in vivo* but the present work reports the *in vitro* hepatic metabolism of the radiolabelled compound by several species including man. Metabolites were separated by TLC and detected by a radio-TLC scanner or by autoradiography; metabolite identification was by co-chromatography with synthetic standards.

Incubation of (1) with rat hepatocytes produced a similar metabolic profile to that seen *in vivo* in the rat. The major metabolite was (3), *p*-hydroxy-(1), corresponding to 55% of the solvent-extractable metabolites. Metabolite (3) was present both as the free compound and as its glucuronide conjugate (6). The other major metabolite seen was a ring-cleavage product of the morpholino ring (5) which corresponded to 22.6% of the total metabolites. Metabolites (2), the amine, and (4), the *N*-oxide of (1) each accounted for 8% and unchanged (1) for 6% of the total metabolites. The lactam derivative of minaprine (7) and four unidentified polar metabolites were also observed.

283

The major metabolite of (1) produced by human hepatocytes was metabolite (3); its glucuronide (6) appeared in smaller amounts and small amounts of (5) and (7) were also observed. Other metabolites seen in the rat were not detected in the human hepatocyte incubations.

The metabolism of the hydroxylated metabolite (3) was also investigated in rat and human hepatocyte suspensions. The major metabolite seen in both cases was the corresponding glucuronide (6) but in addition two minor metabolites [seen previously in the rat incubations with (1)] were present in the rat but not in the human incubations with (3).

A further series of experiments studied the metabolism of (1) by hepatic microsomes from rat, rabbit, dog, baboon, and man in the presence of NADPH. In all species the major metabolite detected was (3). Metabolite (5) was formed in large amounts by rat, dog, and baboon, whereas this was a minor metabolite in rabbit and human. Metabolite (4), the *N*-oxide, was an important metabolite in dog (26% of the solvent-extractable radioactivity), compared with baboon (7%), rabbit (5%), rat (4%), or man (1–4%).

These studies show that the metabolism of (1) *in vivo* was accurately reflected by the use of hepatic *in vitro* systems. A marked interindividual variability was seen in the human *in vitro* data and this may relate to the genetic polymorphism of minaprine postulated *in vivo*.

Reference

B. Lacarelle, F. Marre, A. Durand, H. Davi, and R. Rahmani, Metabolism of minaprine in human and animal hepatocytes and liver microsomes — prediction of metabolism *in vivo*, *Xenobiotica*, 1991, **21**, 317.

Propylthiouracil

Use/occurrence:	Anti-thyroid agent
Key functional groups:	Arylalkyl, aryl thiol, phenol, pyrimidine
Test system:	Activated human polymorphonuclear leukocytes

Structure and biotransformation pathway

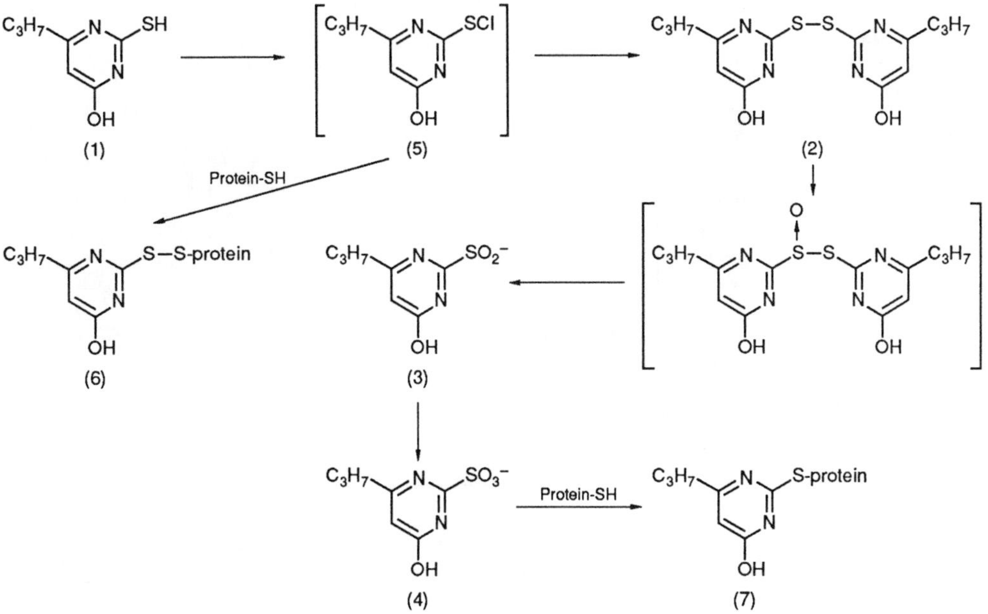

The use of propylthiouracil (1) in the treatment of Graves's disease is associated with idiosyncratic agranulocytosis, a serious depletion of polymorphonuclear leukocyte (which are mainly neutrophils). The purpose of the present study was to investigate the oxidative biotransformation of (1) during incubation with activated human neutrophils in order to determine whether any such metabolites are formed that could covalently bind to protein and so induce agranulocytosis. HPLC separated three oxidized metabolites, which were identified as (2)–(4) by chromatographic comparison with synthetic standards. The same compounds were also formed during incubation of (1) with myeloperoxidase in the presence of H_2O_2 and chloride ion (the principal oxidizing and chlorinating system responsible for the oxidative functions of neutrophil phagocytosis), but the yields were reduced in the absence of chloride ion. Accordingly, biotransformation did not occur when non-activated cells were used, and was inhibited in the presence of sodium azide (an inhibitor of myeloperoxidase) or catalase (which catalyses the decomposition of H_2O_2). These results suggest that the sulfenyl chloride (5) was an intermediary metabolite in the oxidative biotransformation pathway. As (4) and (5) were shown to react chemically with sulfhydryl-containing compounds such as *N*-acetylcysteine, it was further postulated that such reactive metabolites, generated by activated neutrophils, may be involved in propylthiouracil-induced agranulocytosis via covalent binding to sulfhydryl-containing proteins, as in (6) or (7).

Reference

L. Waldhauser and J. Uetrecht, Oxidation of propylthiouracil to reactive metabolites by activated neutrophils, *Drug Metab. Dispos.*, 1991, **19**, 354.

Phenobarbital

Use/occurrence: Anticonvulsant drug

Key functional groups: *N*-Alkylimide, arylalkyl, barbiturate

Test system: Mouse, guinea-pig, rat, and rabbit (intraperitoneal, 30–100 mg kg^{-1}), dog, cat, and pig (intravenous, 6–16 mg kg^{-1}), monkey (intramuscular, 12 mg kg^{-1})

Structure and biotransformation pathway

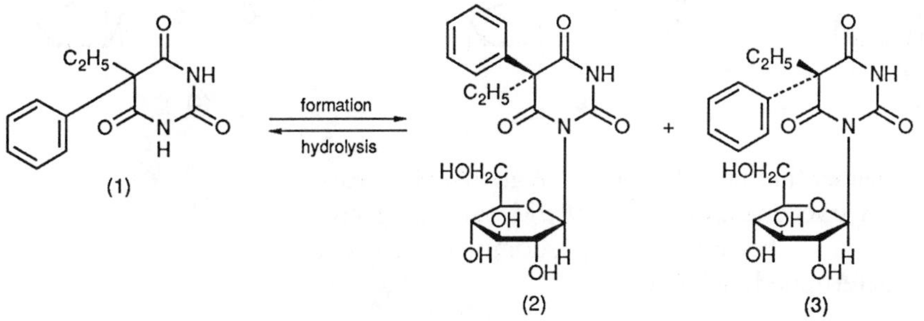

The results of a species screen (mouse, guinea-pig, rat, rabbit, dog, cat, pig, and monkey) found that only mice excreted the *N*-glucosides (2) and (3) as urinary metabolites of phenobarbital (1). The major diastereoisomer (2) excreted by mouse had the *R* configuration at the C-5 position of the barbiturate ring. The *N*-glucoside metabolites accounted for a small percentage, *ca.* 0.5%, of the administered dose. Following intraperitoneal administration of the *N*-glucoside metabolites (2) and (3) to mice free phenobarbital (1) could be detected in the urine. Metabolites were identified and quantitated by thermospray LC–MS.

Reference

W. H. Soine, P. J. Soine, T. M. England, J. W. Ferkany, and B. E. Agriesti, Identification of phenobarbital *N*-glucosides as urinary metabolites of phenobarbital in mice, *J. Pharm. Sci.*, 1991, **80**, 99.

2′,3′-Dideoxycytidine

Use/occurrence:	Anti-retroviral drug
Key functional groups:	Nucleoside
Test system:	U937 human monoblastoid cell line (10–500 μM substrate)

Structure and biotransformation pathway

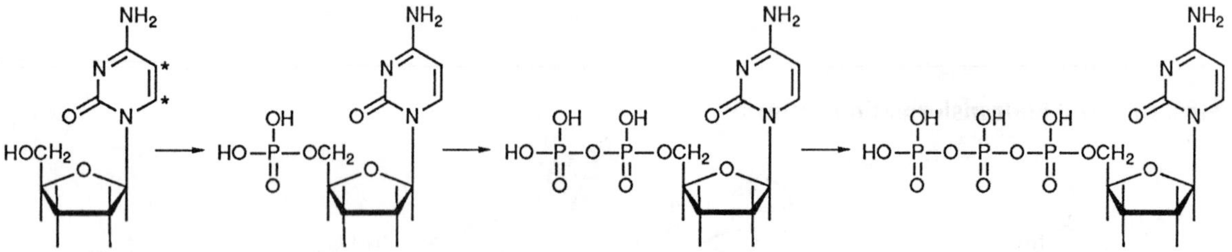

(1) $* = {}^3$H   (2)   (3)   (4)

The predominant labelled nucleotide was the triphosphate (4) after incubation of (1) up to extracellular concentrations of 150 μM. At higher concentrations the disphosphate (3) became the predominant labelled nucleotide. Metabolites were determined by HPLC.

Reference

G. Brandi, L. Rossi, G. F. Schiavano, L. Salvaggio, A. Albano, and M. Magnani, *In vitro* toxicity and metabolism of 2′,3′-dideoxycytidine, an inhibitor of human immunodeficiency virus infectivity, *Chem. Biol. Interact.*, 1991, **79**, 53.

288

E-5-(2-Bromovinyl)-2,2′-anhydrouridine

Use/occurrence:	Antiviral agent
Key functional groups:	Bromoalkene, nucleoside
Test system:	Rat (oral and intra-peritoneal, 100 mg kg^{-1})

Structure and biotransformation pathway

E-BVANUR (1) R^1 = Br, R^2 = H
Z-BVANUR (2) R^1 = H, R^2 = Br

BVANUR = 5-(2-bromovinyl-2,2′-anhydro-1-(β-D-arabinofuranosyl)uracil
   or
5-(2-bromovinyl)-2,2′-anhydrouridine

E-5-(2-Bromovinyl)-2,2′-anhydrouridine (1) is a close analogue of the potent antiviral agent *E*-5-(2-bromovinyl)-2′-deoxyuridine (BVDU). Compound (1) was synthesized to increase the stability of BVDU *in vivo*, since BVDU is rapidly metabolized by pyrimidine nucleoside phosphorylase enzymes to the inactive pyrimidine base. The present study showed that (1), in contrast to BVDU, was not a substrate of pyrimidine nucleoside phosphorylases but is an inhibitor of uridine phosphorylase.

Following oral or intraperitoneal administration of the *E*- or *trans*-isomer of [^{14}C]-(1) to male rats, two radioactive compounds were identified in the urine, namely (1) and its geometric isomer, *Z*-(1), the *cis*-isomer (2). The identity of the two molecules was confirmed by co-chromatography on HPLC with authentic standards, by comparative EI–MS, and by NMR spectroscopy. Conversion of (1) into (2) was not extensive, with 48% of the dose being excreted as unchanged (1) and 2.6% as (2) in the urine collected over a 24 hour period after the intraperitoneal dose. Elimination of (1) and (2) was essentially complete after 48 hoursm, with 62% of the total radiolabel being eliminated in the urine and 37% in the faeces. Compound (1) was shown to be poorly absorbed after oral administration by whole-body autoradiography; the radiocarbon was present mainly in the gastrointestinal tract and not in the tissues.

In summary, (1) shows markedly enhanced stability to enzymatic cleavage compared with BVDU. However, (1) is converted into its *Z*-isomer *in vivo* to a small extent. As the Z-isomer of BVDU has been shown to be a less efficient inhibitor of herpes simplex 1 virus, it may be inferred that the metabolic conversion of (1) into (2) may lead to a decrease in antiviral activity.

Reference

I. Szinai, Zs. Veres, A. Szabolcs, E. Gacs-Baitz, K. Ujszaszy, and G. Denes, *cis–trans*-Isomerization of [*E*]-5-(2-bromovinyl)-2,2′-anhydrouridine *in vivo* in rats, *Xenobiotica*, 1991, **21**, 359.

Amobarbital

Use/occurrence:	Sedative
Key functional groups:	Barbiturate
Test system:	Cat liver microsomes

Structure and biotransformation pathway

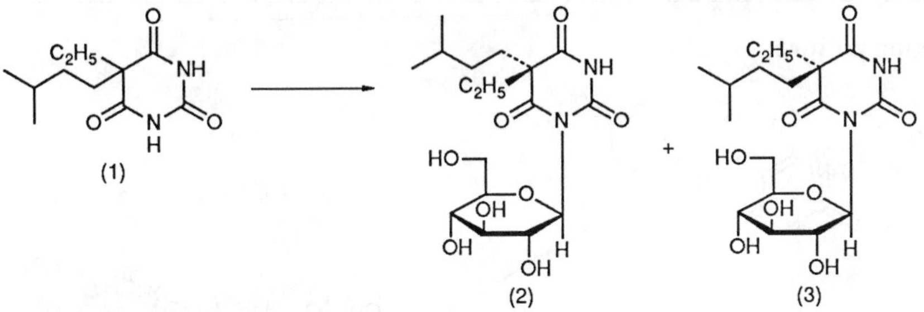

(1)   (2)   +   (3)

Glucosylation appears to be an important pathway of biotransformation of barbiturates in man. The objective of the work described in this short communication was to investigate the potential of the cat as a model for this conjugation reaction, and in particular whether it would exhibit product enantioselectivity in the formation of amobarbital *N*-glucoside. Amobarbital (1) was incubated with cat liver microsomes in the presence of UDP-D-[6-³H]-glucose and formation of the ³H-labelled glucosides (2) and (3) was monitored by radio-HPLC.

The results showed that cat liver microsomes were capable of catalysing the *N*-glucosylation of (1). Product selectivity was 2:1 in favour of the *R*-glucoside (2). The same authors had previously observed selectivity in favour of the *S*-isomer (3) in the excretion of amobarbital glucosides in humans.

Reference

S. E. Mongrain and W. H. Soine, Product enantioselectivity in the *N*-glucosylation of amobarbital by cats, *Drug Metab. Dispos.*, 1991, **19**, 1012.

Merbarone

Use/occurrence: Anticancer agent

Key functional groups: Aryl amide, thiobarbiturate

Test system: Human patients (continuous intravenous infusion, 96–1250 mg m^{-2} day^{-1})

Structure and biotransformation pathway

Merbarone (1) was administered to male human cancer patients as a continuous intravenous infusion over five days during which time the dose was increased from 96 to 1250 mg m^{-2} day^{-1}. Urine was collected on day 5 of the infusion. Attempts at isolation of separate urine metabolites were unsuccessful, but three major urinary metabolites (2)–(4) were identified by a combination of techniques applied to a crude extract obtained from XAD-2 resin clean-up. Initial characterization was obtained from CI-MS of the metabolite mixture, following which reference standards were synthesized and compared with urinary metabolites by HPLC using a diode array detector. Further structural evidence was obtained from high-resolution (500 MHz) ^{1}H NMR of the urine extract, and for the phenolic metabolites by acetylation and isolation for CI-MS of the resulting derivatives. The identified pathways of biotransformation involved aromatic hydroxylation and oxidative desulfuration. Phase II metabolites were not apparently present in urine but several minor metabolites were not identified.

Reference

J. G. Supko and L. Malseis, Characterisation of the urinary metabolites of merbarone in cancer patients, *Drug Metab. Dispos.*, 1991, **19**, 263.

Lamotrigine

Use/occurrence:	Anti-epileptic drug
Key functional groups:	Aminotriazine, aryl amine, chlorophenyl
Test system:	Man (oral, 800 mg day^{-1}), guinea-pig (intravenous, 10 mg kg^{-1}), guinea-pig hepatocytes and liver microsomes

Structure and biotransformation pathway

(1)   (2)

This study was designed to characterize fully the structure of the *N*-glucuronide conjugate (2) of lamotrigine (1), which is known to be the major human and monkey urinary metabolite of the drug. Urine was collected during a 24 hour period from a patient undergoing chronic treatment with (1), then concentrated on XAD-2 resin and purified by HPLC. Enzyme hydrolysis experiments with β-glucuronidase showed that this metabolite was indeed a glucuronic acid conjugate of (1), and high-resolution FAB-MS confirmed the expected molecular formula. ^{1}H NMR indicated through analysis of the pertinent coupling constants that the C-1' configuration of the sugar moiety was that of the β-anomer and that it was attached directly to the triazine ring. Furthermore, the ^{13}C NMR evidence strongly suggested that the point of attachment was the N-2 position, as shown (2). The metabolite was unstable at neutral and basic pH (under which conditions it produced different unidentified degradation products), but it was unusually stable at acidic pH. In a separate study, ion-pair HPLC was used to show that (2) accounted for 60% of an intravenous dose of (1) to guinea-pigs and that less than 6% of the dose was excreted as unchanged drug. The glucuronide (2) was also formed during incubation of (1) with isolated guinea-pig hepatocytes and Triton X-100-activated liver microsomes.

References

M.W. Sinz and R. P. Remmel, Isolation and characterization of a novel quaternary ammonium-linked glucuronide of lamotrigine, *Drug Metab. Dispos.*, 1991, **19**, 149.

R. P. Remmel and M. W. Sinz, A quaternary ammonium glucuronide is the major metabolite of lamotrigine in guinea pigs, *Drug Metab. Dispos.*, 1991, **19**, 630.

293

Epitiostanol

Use/occurrence:	Anti-oestrogen
Key functional groups:	sec-Alkyl alchol, thiirane
Test system:	Rat (intramuscular, 0.5 mg kg^{-1}; intravenous, 0.1 mg kg^{-1}; intraportal, 0.1 mg kg^{-1}; intrajejunal, 0.5 mg kg^{-1})

Structure and biotransformation pathway

(1) * = ^{14}C

(2)

(3)

Epitiostanol (1) has been shown to have weak pharmacological activity following administration via the oral route, but not after intramuscular (i.m.) injection. After i.m. administration of [^{14}C]-(1), 37.9% of the dose was excreted in the urine and 56.2% of the dose in the faeces, but no unchanged (1) was excreted. Analysis of the liver ^{14}C content directly after intraportal administration showed that 36% of the total was unchanged (1) and *ca.* 60% consisted of metabolites [23% as 2α,3α-epithio-5α-androstan-17-one (2), 13% as 5α-androst-2-en-17-one (3), and 17% as other unidentified metabolites]. Metabolites were identified following separation by TLC and co-chromatography with synthetic compounds.

Direct intravenous (tail vein) injection of (1) also suggested that the liver was the major site of metabolism. 30% of the administered ^{14}C was present in this tissue 2 minutes after dosing, most of which was present as metabolites. After intrajejunal administration of (1), 83% of the absorbed (plasma) ^{14}C was due to (2) and (3) with the remainder being present as unchanged (1). After oral treatment, all plasma ^{14}C was due to polar metabolites. Thus, the

294

low biological activity of (1) after oral dosing was due to first-pass metabolism by both the intestinal mucosa and the liver.

Reference

T. Ichihashi, H. Kinoshita, K. Shimamura, and Y. Yamada, Absorption and disposition of epithiosteroids in rats (1): Route of administration and plasma levels of epitiostanol, *Xenobiotica*, 1991, **21**, 865.

Monocrotaline

Use/occurrence:	Natural product (phytotoxin)
Key functional groups:	Alkyl ester, pyrrolizidine
Test system:	Rat (intravenous, 60 mg kg^{-1})

Structure and biotransformation pathway

(1)   (2)

Following intravenous administration of [^{14}C]monocrotaline (1) to rats with cannulated bile ducts, 83% of the dose was excreted in urine during 7 hours and 12% dose in the bile. Radioactive components in both excreta samples were separated and quantified by HPLC and, where possible, identified by FAB-MS and/or chromatographic comparison with reference compounds. One of the two major components in urine, which accounted for nearly two-thirds of the radioactivity therein, was shown to be unchanged (1) and the other as a collection of polar metabolites which included compounds that gave a positive Ehrlich's test for pyrroles. One of the two minor urinary components was positively identified as monocrotaline *N*-oxide (2), but the other was not identified. By contrast, less than 5% of the biliary radioactivity was in the form of parent compound and the *N*-oxide was not detected at all; the vast majority of the biliary radioactivity was present as the collection of polar Ehrlich-positive metabolites (one of which has been previously identified as a glutathione-conjugated pyrrole). Retention of radioactivity in the red blood cells was found to be relatively high in rats treated with (1), and it was proposed that transport of metabolites by this tissue from the liver to the lungs may play a role in its pulmonary toxicity.

Reference

J. E. Estep, M. W. Lamé, D. Morin, A. D. Jones, D. W. Wilson, and H. J. Segall, [^{14}C]Monocrotaline kinetics and metabolism in the rat, *Drug Metab. Dispos.*, 1991, **19**, 135.

Monocrotaline

Use/occurrence:	Plant alkaloid
Key functional groups:	Alkyl ester, pyrrolizidine
Test system:	Isolated perfused rat liver

Structure and biotransformation pathway

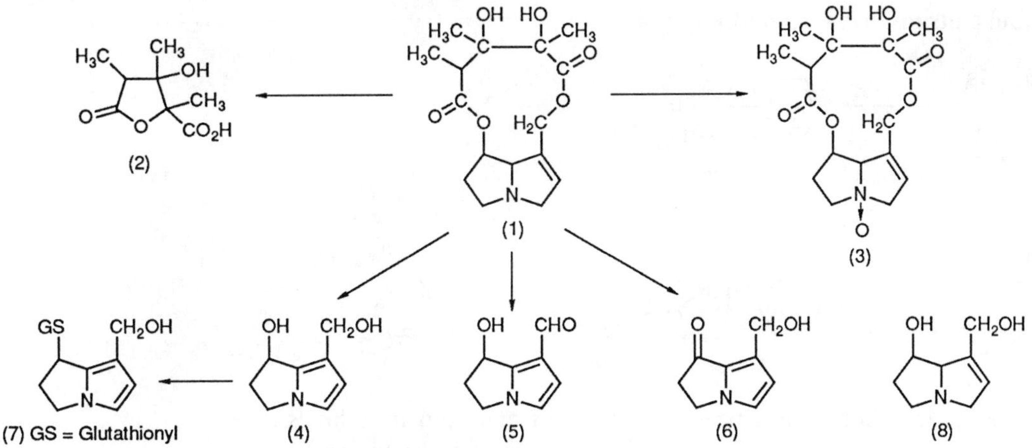

Monocrotaline (1) uniformly labelled with ^{14}C was isolated from plants grown in an atmosphere containing $^{14}CO_2$. Monocrotalic acid (2) was reported to be the major acidic metabolite of (1) in both bile and perfusate. Metabolites (4)–(6) were present in trace amounts in the perfusate. The metabolites were identified by GC–MS, as TMS derivatives, but the assignment of structure (6) was tentative. Monocrotaline N-oxide (3), identified by HPLC co-chromatography, was also a minor perfusate metabolite. A large proportion of both bile and perfusate radioactivity was not partitionable into organic solvents under acidic or basic conditions. Using FAB-MS–MS, part of this material was identified as the glutathione conjugate (7). The necine base (8) was not detected in the perfusate, leading the authors to propose that pyrrole formation rather than hydrolysis was the major pathway of metabolism for (1).

Reference

M. W. Lamé, A. D. Jones, D. Morin, and H. J. Segall, Metabolism of [^{14}C]monocrotaline by isolated perfused rat liver, *Drug Metab. Dispos.*, 1991, **19**, 516.

Levamisole

Use/occurrence:	Anti-colon cancer, anthelmintic drug, immunostimulant
Key functional groups:	Imidazoline, thiazoline
Test system:	Human intestinal flora

Structure and biotransformation pathway

Levamisole (1) metabolism has been described in man and in animals: extensive metabolism *in vivo* is by aromatic ring hydroxylation or thiazoline ring scission to the cyclic urea (2). This 2-oxoimidazolidine derivative (2) may be responsible for the pharmacological effects of (1). As therapy with (1) combined with 5-fluorouracil decreases the risk of colorectal carcinoma, the metabolism of (1) in the colon has been investigated. The major metabolites are *N*-hydroxylated (3) and its corresponding disulfide (21%). The major active metabolite in animal plasma and urine (2) was not detected in this *in vitro* system. However, a minor bacterial metabolite was the mixed disulfide (4) (1.8%) which contains (2). There was no interconversion between (2) and (3) or between their disulfides. The formation of the hydroxamic lactam (3) from (1) must involve an oxidation step, of particular interest in the anaerobic conditions required for bacterial viability. The metabolites (2) and (3), their disulfides, and (4) were all dextrorotary. Racemization of the asymmetric carbon is unlikely in the ring-opening process, and the metabolites were therefore all assigned the 5*S*-configuration as in (1). Full synthetic and analytical details are reported, including: specific rotation, IR, UV, ^{1}H and ^{13}C NMR, EI-, CI-, and FAB-MS, TLC, and HPLC. Strong metabolizers include *Bacteroides* and *Clostridium* species, but the metabolism includes, in part, a non-enzymatic process. The thiazoline (1) may be a pro-drug for the thiols (2) and (3).

Reference

Y.-Z. Shu, D. G. I. Kingston, R. L. van Tassell, and T. D. Wilkins, Metabolism of levamisole, an anti-colon cancer drug, by human intestinal bacteria, *Xenobiotica*, 1991, **21**, 737.

6-Benzoylbenzoxazolinone

Use/occurrence: Analgesic drug

Key functional groups: Aryl ketone, benzoxazolinone

Test system: Human (oral, 1 g)

Structure and biotransformation pathway

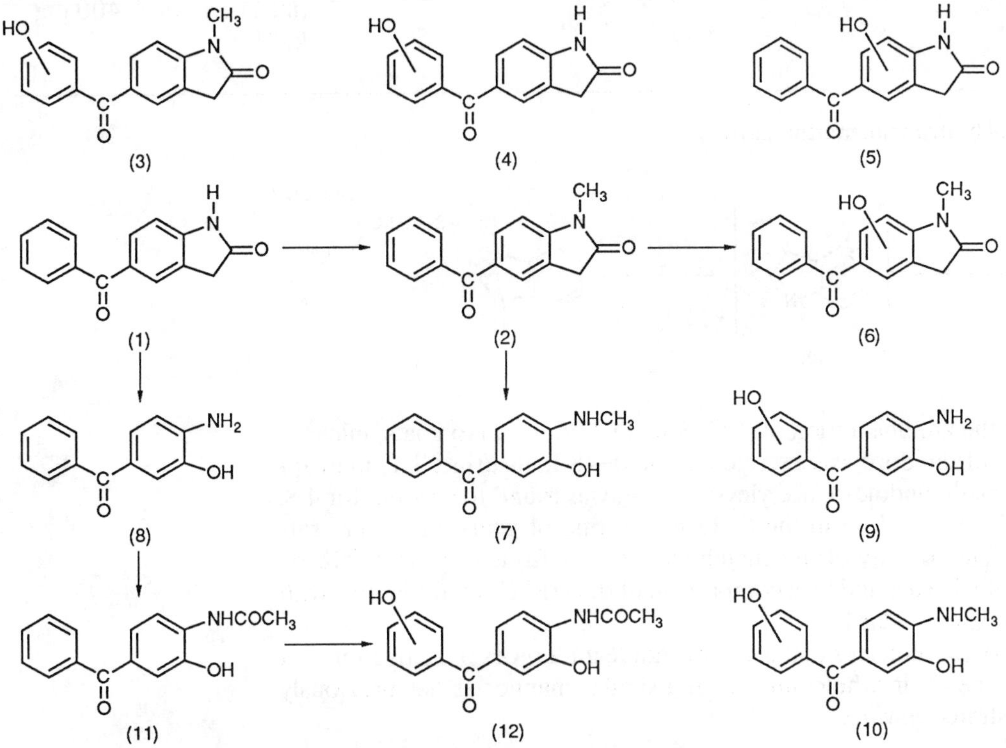

Urinary metabolites of (1) were purified by TLC and analyed by GC–MS or by direct insertion probe MS. Metabolites were identified by reference to the mass spectra of synthesized standards and/or deduced from the spectra obtained. The four metabolites that were definitely identified were (2), (7), (8), and (11). Hydrolytic cleavage of the benzoxazolinone ring leads to (8) which is subsequently acetylated to give (11). Other tentatively identified metabolites [(3), (4), (6), (9), (10), (12)] appear to arise from the hydroxylation of the other metabolites.

Reference

M. Bastide, J. Chabard, C. Lartigue, H. Bargnoux, J. Petit, J. Berger, H. Mansour, D. Lesieur, and N. Busch, Identification of several human urinary metabolites of 6-benzoylbenzoxazolinone by gas chromatography/mass spectrometry, *Biol. Mass Spectrom.*, 1991, **20**, 484.

3-Methylindole

Use/occurrence: By-product of bacterial fermentation of dietary tryptophan; cigarette smoke

Key functional groups: Indole

Test system: Goats (intravenous, 15 mg kg^{-1}), mice and rats (intraperitoneal, 400 mg kg^{-1})

Structure and biotransformation pathway

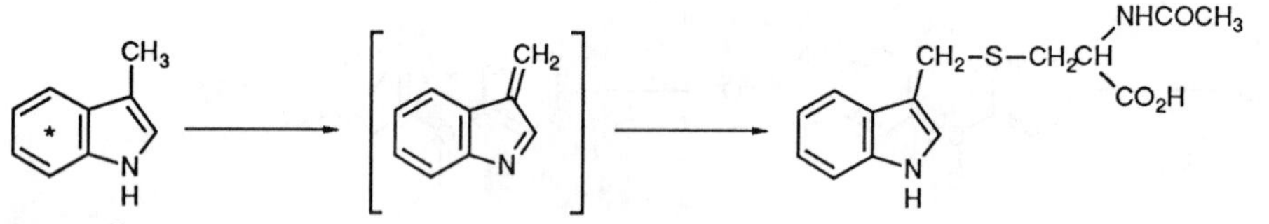

(1) * = ^{14}C   (2)   (3)

Following the administration of [^{14}C]-3-methylindole (1) to goats, mice, or rats, *ca.* 68% of the dose was excreted in the 0–48 hour urine. The mercapturic acid, 3-methylindole-*N*-acetylcysteine (3), was found to account for 4.8, 2.6, and 7.3% of the dose in the 0–48 hour urine of goats, mice, and rats respectively. The identity of this metabolite was confirmed by ^{1}H NMR, by MS of the methyl ester, and by a comparison of the HPLC retention time with that of a synthetic standard.

Excretion of (3) indicates that 3-methylindole undergoes bioactivation to a methylene imine (2) in whole animals in a similar manner as has previously been demonstrated *in vitro*.

Reference

G. L. Skiles, D. J. Smith, M. L. Appleton, J. R. Carlson, and G. S. Yost, Isolation of a mercapturate adduct produced subsequent to glutathione conjugation of bioactivated 3-methylindole, *Toxicol. Appl. Pharmacol.*, 1991, **108**, 531.

1-Methoxyindole-3-carboxylic acid

Use/occurrence:	Model compound
Key functional groups:	*N*-Methoxy
Test system:	Rat liver $15\,000\,g$ supernatant

Structure and biotransformation pathway

(1) OCH_2OH HCHO   (2)

The formation of *N*-hydroxyindoles from their *N*-methoxy derivatives *in vivo* could result in unforeseen biological effects, as a number of *N*-hydroxy compounds have been shown to have appreciable toxicity. In view of this possibility, the *O*-demethylation of 1-methoxyindole-3-carboxylic acid (1) was studied. The method involved determination of the product formaldehyde, which gave an indirect estimate of the metabolite, 1-hydroxyindole-3-carboxylic acid (2). Formaldehyde was assayed spectrophotometrically using the Nash reaction following incubation of (1) with rat liver $15\,000\,g$ supernatant in the presence of an NADPH regenerating system.

The *O*-demethylation of a heterocyclic *N*-methoxyindole may be considered to be a novel metabolic reaction. In an earlier investigation, the pathway of metabolism of (1) was considered to involve the direct reduction of the N—O bond to give indole-3-carboxylic acid. The present results indicate that the formation of the indole analogue may proceed predominantly by the *O*-demethylation pathway, after initial formation of the unstable intermediate (2).

Reference

J. O. Nwankwo, *In vitro O*-demethylation from a heterocyclic nitrogen: a novel metabolic reaction?, *Xenobiotica*, 1991, **21**, 569.

6,7-Dichloro-5(*N*,*N*-dimethylsulfamoyl)-2,3-dihydrobenzofuran-2-carboxylic acid (DBCA)

Use/occurrence:	Uricosuric diuretic
Key functional groups:	Chiral carbon, dihydrobenzofuran
Test system:	Rat liver $100\,000\,g$ supernatant

Structure and biotransformation pathway

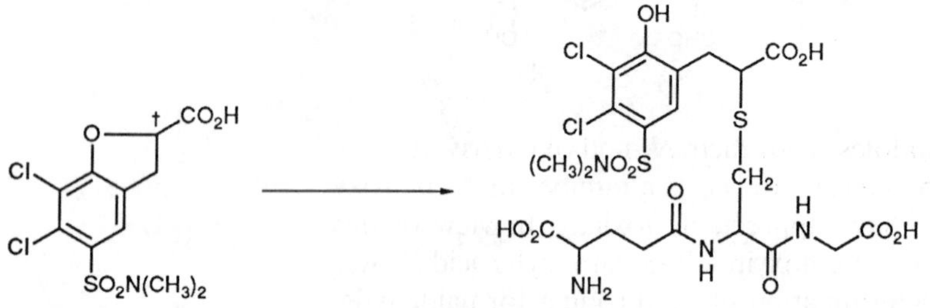

(1) † = chiral carbon
(2)

The $(-)$-enantiomer of DBCA (1) was incubated at a concentration of 1 mM in the presence of glutathione (5 mM) and rat liver and $100\,000\,g$ supernatant for 2 hours. Analytes were separated on TLC and further purified by HPLC. The glutathione conjugate of DBCA (2) was identified by SI-MS and NMR. The conjugation of (1) is apparently unique in that the glutathione is linked to a carbon which is not rendered electrophilic by adjacent halide or epoxide substituents.

DBCA was designed to be originally administered as the racemate and the metabolic rates of the R-(+)- and S-(−)-enantiomers have previously been examined in isolated rat hepatocytes. It was found that the R-(+)-enantiomer inhibited the glutathione conjugation of the S-(−)-enantiomer (1). Furthermore the K_i for the R-(+)-enantiomer was 30-fold smaller than the K_m for the glutathione conjugation of (1) in rat liver cytosol.

Reference

J. Higaki, J. Kikuchi, Y. Ikenishi, and M. Hirata, Structure determination of diuretic-glutathione conjugate by mass and n.m.r. spectrometry, *Xenobiotica*, 1991, **21**, 47.

Albendazole

Use/occurrence:	Anthelmintic
Key functional groups:	Aryl thioether
Test system:	Rat (oral, 10 mg kg^{-1}), dog (oral, 10 mg kg^{-1}), human volunteers (oral, 10 mg kg^{-1})

Structure and biotransformation pathway

(1)   (2)   (3)

The oxidation of the prochiral thioether (1) to the chiral sulfoxide (2), and subsequently to the achiral sulfone (3), was compared in male Sprague–Dawley rats, male beagle dogs, and male human volunteers. HPLC with a chiral stationary phase was used to monitor the enantioselective oxidation. Plasma samples were subjected to preparative HPLC on a C_{18} column followed by rechromatography over α_1-glycoprotein immobilized on silica. The racemate of (2) was initially formed in all species; subsequently the (+)/(−) ratio changed and reached 13.1 and 9.3 in man and dogs respectively. However, the (+)/(−) ratio decreased to 0.6 in male rats. The (+)-enantiomer (structure not shown) represented 80% of the total sulfoxide plasma AUC in man. There was essentially no albendazole (1) in plasma; the major metabolic products were (2) and (3). The (+)/(−) ratio changing with time is hypothesized to be due to enantioselective cytochrome P-450 oxidation of sulfoxide (2) to its corresponding sulfone (3), a phenomenon which has been previously described. In this instance, the biotransformation data obtained from dogs are certainly more similar to those from man than are the data from rats, where the opposite (−)-enantiomer predominated.

Reference

P. Delatour, E. Benoit, S. Besse, and A. Boukraa, Comparative enantioselectivity in the sulfoxidation of albendazole in man, dogs and rats, *Xenobiotica*, 1991, **21**, 217.

Mebendazole

Use/occurrence:	Anthelmintic drug
Key functional groups:	Aryl ketone, methyl carbamate
Test system:	Isolated perfused rat jejunum

Structure and biotransformation pathway

(1) benzoyl-benzimidazole—NHCO$_2$CH$_3$ → (2) hydroxy(CHOH)-benzimidazole—NHCO$_2$CH$_3$

(1) → (3) benzoyl-benzimidazole—NH$_2$

(2) → (4) OGluc-benzimidazole—NHCO$_2$CH$_3$ (4) Gluc = Glucuronyl

The intestinal metabolism of (1) has been investigated using isolated perfused jejunal segments from the rat. Significant absorption and metabolism of (1) was observed in this system resulting in the formation of both oxidized and conjugated metabolites. The oxidized metabolites were identified by co-chromatography with authentic standards and the conjugates identified following β-glucuronidase treatment. Metabolites were separated by HPLC. The metabolites seen in the resorbate were the hydroxy metabolite (2) and its glucuronide conjugate (4) present at similar concentrations (0.3% and 0.2% of the dose respectively). Unchanged (1) accounted for 2.1% of the dose. In the perfusate 70% of the dose was present as (1), 0.9% as (2), and 2.1% as the decarbamoylated amino metabolite (3). When a portion of gut from a rat treated with 3-methylcholanthrene (3-MC) was used, the rate of metabolism of (1) to metabolite (3) was markedly increased; 40% of the dose in the perfusate was present as (3) with 58% as unchanged (1). Metabolites (2) and (4) were not detected in either the perfusate or the resorbate of the gut segment from 3-MC treated animals.

The present results indicate a significant first-pass metabolism of (1) by the gut wall and the pattern of metabolites was similar to that observed *in vivo* in this species.

Reference

H. Gottmanns, R. Kroker, and F. R. Ungemach, Investigations on the biotransformation of mebendazole using an isolated perfused rat gut system, *Xenobiotica*, 1991, **21**, 1431.

Benzothiazole

Use/occurrence:	Model compound
Key functional groups:	Thiazole
Test system:	Guinea-pig (intraperitoneal, 30 mg kg^{-1})

Structure and biotransformation pathway

Following intraperitoneal administration of (1) to guinea-pigs, four major metabolite peaks were present in a gas chromatogram of the 0–24 hour urine sample not subjected to enzyme hydrolysis. These peaks were identified as unchanged (1), 2-methylmercaptoaniline (2), and the S-oxidized products of (2), 2-methylsulfinylaniline (3) and 2-methylsulfonylaniline (4). Identification of these four compounds was by ^{1}H NMR, IR, and EI-MS together with comparison with authentic chemical standards. Treatment of the urinary excretion products with β-glucuronidase and sulfatase provided evidence that the identified metabolites (2)–(4), particularly (3), were partially excreted as conjugates. In addition sulfate hydrolysis resulted in two new peaks in the chromatogram, which were identified as the hydroxylated metabolites, 2-methylsulfinylphenylhydroxylamine (5) and 2-methylsulfonylphenyl-hydroxylamine (6) on the basis of their mass spectra. No authentic chemical standards were available for metabolites (5) and (6).

Thus the preferred route of benzothiazole metabolism in the guinea-pig involves thiazole ring scission, with no evidence of metabolism yielding intact carbocyclic ring products. Metabolites (2)–(4) were present in conjugated and unconjugated forms; metabolites (5) and (6) were identified only after hydrolysis with sulfatase. The mechanism of ring cleavage remains unclear, but it could involve base-catalysed attack at C-2 of the thiazole ring.

Reference

K. Wilson, H. Chissick, A. M. Fowler, F. J. Frearson, M. Gittins, and F. J. Swinbourne, Metabolism of benzothiazole I. Identification of ring-cleavage products, *Xeno-biotica*, 1991, **21**, 1179.

6-Hydroxy-2-(4-sulfamoylbenzylamino)-4,5,7-trimethylaminobenzothiazole (E-6080)

Use/occurrence:	5-Lipoxygenase inhibitor
Key functional groups:	Arylmethyl, phenol, sulfonamide
Test system:	Rat, guinea-pig, dog, rhesus monkey (oral, 10 mg kg^{-1}), bile duct-cannulated rat and guinea-pig (intravenous, 0.2 and 5 mg kg^{-1}, rat and guinea-pig; 20 mg kg^{-1}, rat)

Structure and biotransformation pathway

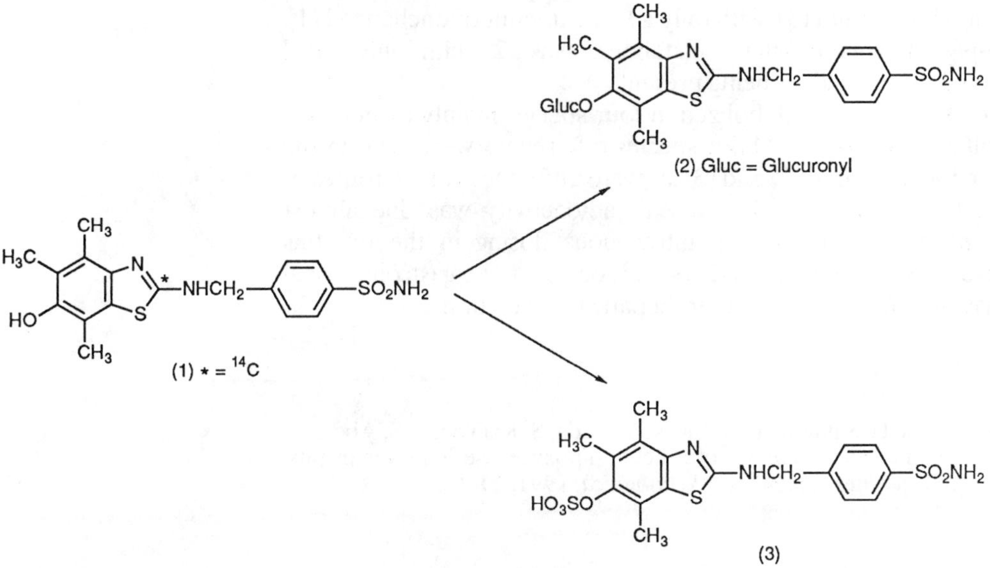

(2) Gluc = Glucuronyl

(1) * = ^{14}C

(3)

5-Lipoxygenase inhibitors such as (1) block the formation of leucotrienes which are important chemical mediators in allergic and inflammatory response. These inhibitors are thus potentially useful for treatment of disease states such as asthma. The purpose of the present study was to study the metabolic pathway and excretion of (1) in laboratory animals.

Following a single oral administration of [^{14}C]-(1), over 94% of the dose was excreted within 3 days in three of the four species studied (rat, guinea-pig, and dog); excretion was essentially complete in the monkey after 7 days. Species differences were seen in the extent of urinary excretion; 2.8% of the dose was excreted by the rat, 46.9% by the guinea-pig, 2.6% by the dog, and 68.2% by the monkey. The remainder of the dose was excreted in the faeces in all cases. Metabolites isolated from urine, bile, and faeces were separated by TLC and HPLC. Metabolites were identified by chemical and enzymic hydrolysis, by comparison with standards synthesized *in vitro* with rat liver preparations, and by ^{1}H NMR and FAB-MS. After oral dosing in rhesus monkeys, the major urinary metabolite was (2), the glucuronide conjugate of (1) accounting for 63% of the dose with the corresponding sulfate (3) present at only 0.25% of the dose. Metabolite (2) was also the dominant urinary metabolite in the guinea-pig [33% (2) and 3% (3)]. In contrast, the major rat urinary metabolite was (3), 1.2% of the dose, whereas (2) accounted for 0.5%. In dogs, approximately equal amounts of (2) and (3) were present in the urine (0.3% of each metabolite). Metabolites present in gall bladder bile of monkey, guinea-pig, and dog were predominantly (2) and (3): unchanged (1) accounted for less than 1% and unknown metabolites 1–4% of the dose. In monkeys and guinea-pigs, over 96% and 91% respectively of the biliary ^{14}C was due to the glucuronide (2). In dogs, the sulfate (3) was more predominant and a value of 1.7 was calculated for the (2):(3) ratio.

After intravenous dosing of (1) to bile duct-cannulated rats at three dose levels, 85–91% of the dose was recovered in the 0–24 hour bile. In contrast, only 60% of the dose was excreted in the bile of guinea-pigs following intravenous administration. Almost all of the radioactivity present in rat bile was present as metabolites (2) and (3), with only a trace amount of unchanged (1). In the guinea-pig, the major biliary metabolite was (2) with only small amounts of (3) and unchanged (1) being present.

Thus (1) was shown to be metabolized in four species mainly by glucuronide and/or sulfate conjugation. Major species differences were seen in the extent of conversion of (1) into (2) and (3), as were differences in the routes of elimination for the metabolites. The faecal radioactivity was due almost entirely to unchanged (1), even after intravenous dosing in the rat; this suggests that hydrolysis of conjugates (2) and (3) occurs in the gastrointestinal tract and (1) may therefore undergo enterohepatic recirculation.

Reference

T. Yoshimura, S. Tanaka, H. Sakurai, Y. Nagai, K. Yamada, S. Katayama, S. Abe, and I. Yamatsu, Metabolites and excretion of a new 5-lipoxygenase inhibitor in rats, guinea pigs, beagles and rhesus monkeys, *Xenobiotica*, 1991, **21**, 627.

Thiabendazole

Use/occurrence:	Anthelmintic drug
Key functional groups:	Benzimidazole, thiazole
Test system:	Pregnant mouse (oral, 1000 mg kg^{-1}), mouse embryo, rat (oral, 800 mg kg^{-1})

Structure and biotransformation pathway

(1) * = ^{14}C   (2)   (3)

(5)   (6)   (4)

Thiabendazole (1) is known to be mainly metabolized in laboratory animals to the 5-hydroxy derivative (2) and glucuronide and sulfate conjugates (3) and (4). Besides these known metabolites, two novel metabolites were identified from rat and mouse urine in this study using EI- and CI-MS and ^{1}H NMR. These were the 4-hydroxy metabolite (5) and 2-acetylbenzimidazole (6). Metabolites (2) and (6) were also detected, in addition to unchanged (1), in embryos of mice given [^{14}C]-(1) orally on day 10 of gestation, and were formed when (1) was incubated in a whole-embryo culture or embryo homogenate. However, the amounts of metabolites formed *in vitro* were low in comparison with amounts detected *in vivo* leading the authors to propose that most of the metabolites in the embryos *in vivo* came from the dam.

Reference

T. Fujitani, M. Yoneyama, A. Ogata, T. Ueta, K. Mori, and H. Ichikawa, New metabolites of thiabendazole and the metabolism of thiabendazole by mouse embryo *in vivo* and *in vitro*, *Food Chem. Toxicol.*, 1991, **29**, 265.

Tiospirone

Use/occurrence:	Antipsychotic drug
Key functional groups:	*N*-Alkylpiperazine, benzoisothiazole, cycloalkyl, piperidinedione
Test system:	Human (oral, 60 mg)

Structure and biotransformation pathway

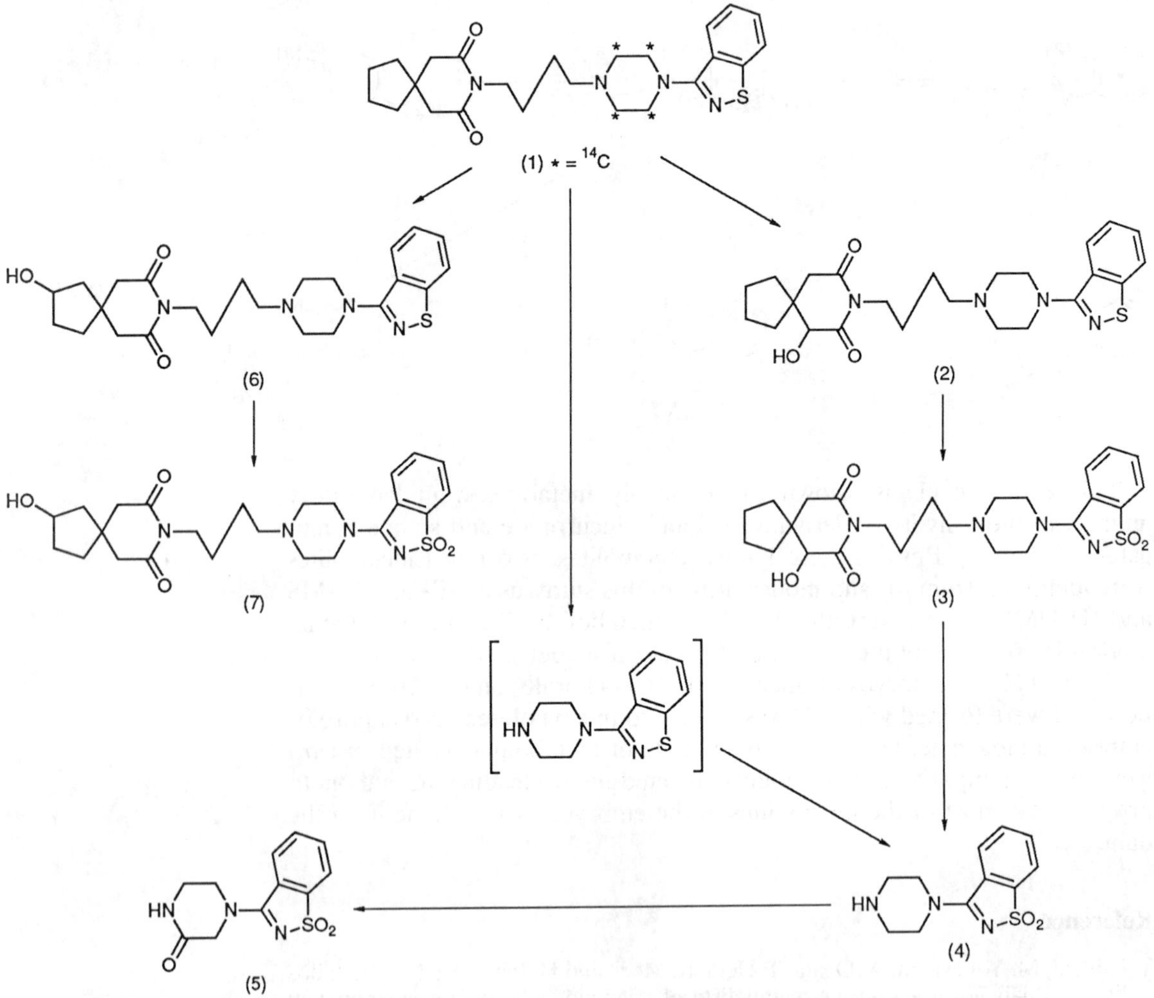

(1) * = ^{14}C

(6)

(7)

(2)

(3)

(4)

(5)

After single oral doses of [^{14}C]tiospirone (1) to healthy male volunteers, *ca.* 39% of the dose was excreted in urine within 24 hours, increasing to 46% dose after 5 days. Radioactive components in pooled 0–24 hour urine were separated and purified by HPLC, and identified by DCI-MS and chromatographic comparison with reference compounds. Incubation experiments with

310

β-glucuronidase/sulfatase showed that few, if any, of the urinary metabolites were present as conjugates. In addition to small amounts of unchanged drug, five metabolites [(2)–(6)] were characterized unequivocally and one, (7), tentatively. The most important of these was (4), and together the identified components accounted for about half of the urinary radioactivity (20% of the dose). The major identified primary routes of biotransformation of (1) in man were therefore: N-dealkylation of the butyl side-chain, hydroxylation of the piperidinedione ring α to the glutarimidyl carbon, hydroxylation at position-3 of the cyclopentyl ring, oxidation of the benzoisothiazole sulfur to the sulfone, and oxidation α to the nitrogen of the piperazine ring yielding a lactam; most of these pathways have been found previously to occur *in vitro* using rat liver microsomes. Interestingly, all the identified S-oxidized metabolites of (1) in man were sulfones, and no sulfoxides were detected, although several metabolites were not identified.

Reference

R. F. Mayol, H. K. Jajoo, L. J. Klunk, and I. A. Blair, Metabolism of the antipsychotic drug tiospirone in humans, *Drug Metab. Dispos.*, 1991, **19**, 394.

ABC-99

Use/occurrence:	Antibronchospastic drug
Key functional groups:	Dithiolane, xanthine
Test system:	Rat liver microsomes

Structure and biotransformation pathway

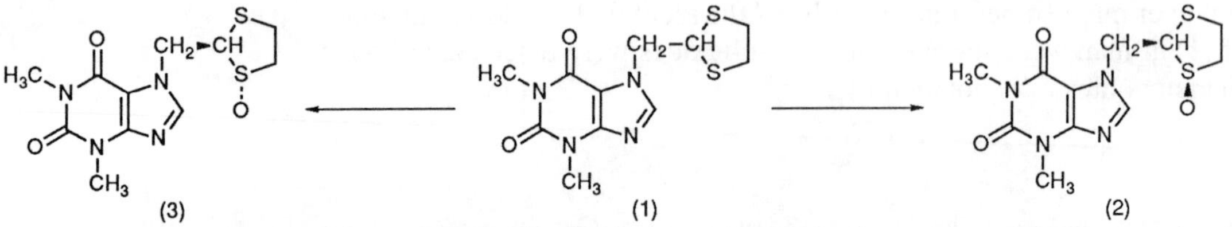

(3)   (1)   (2)

Rat liver microsomes transformed ABC-99 (1) into two isomeric sulfoxides (2) and (3). Production of the *trans*-isomer (3) was favoured over that of the *cis*-isomer (2) in the ratio 7:3. The reaction was apparently catalysed by flavin-dependent mono-oxygenases. The metabolites were identified by chemical reduction to the parent compound and ^{1}H NMR, MS, IR, and UV in comparison with synthetic standards. No *N*-dealkylation to theophylline was observed.

Reference

G. Grosa, O. Caputo, M. Ceruti, G. Biglino, J. S. Franzone, and C. Cravanzola, Metabolism of 7-(1,3-dithiolan-2-ylmethyl)-1,3-dimethylxanthine by rat liver microsomes, *Drug Metab. Dispos.*, 1991, **19**, 454.

1,2,3,4-Tetrahydroisoquinoline, 1-Methyl-1,2,3,4-tetrahydroisoquinoline

Use/occurrence:	Parkinsonism-related compounds
Key functional groups:	Tetrahydropyridine
Test system:	Rat (oral, 50 mg kg^{-1})

Structure and biotransformation pathway

(2a) R = H
(2b) R = CH$_3$

(1a) R = H, * = ^{14}C
(1b) R = ^{14}CH$_3$

(3a) R = H
(3b) R = CH$_3$

(4b)

(5a)

After oral administration of [^{14}C]-1,2,3,4-tetrahydroisoquinoline (1a) or [^{14}C]-1-methyltetrahydroisoquinoline (1b) most of the dose was excreted unchanged. Unchanged compound accounted for more than 70% of the 24 hour urinary excretion in both cases. The main metabolites were the 4-hydroxy compounds, which accounted for 2.7% (2a) or 8.7% (2b) urinary radioactivity. *N*-Methylation also occurred with both compounds but the products (3a) and (3b) each represented less than 1% urinary radioactivity. In the case of (1a), isoquinoline (5a) was found as a metabolite (2.5%); the intermediate dihydro compound was not detected. In the case of (1b) the fully oxidized isoquinoline was not detected, while the dihydro metabolite (4b) was present in urine at a low level (1%). *N*-Methyl quaternary ion metabolites, implicated in the neurotoxicity of other similar compounds, were not detected at a level of 0.03% dose. Radioactivity in brain was mainly (>90%) associated with unchanged compound. Identification of urinary metabolites rested on TLC comparison with standards, except for (2b) where GC–MS was also used.

Reference

K. Kikuchi, Y. Nagatsu, Y. Makino, T. Mashino, S. Ohta, and M. Hirobe, Metabolism and penetration through the blood–brain barrier of Parkinsonism-related compounds, *Drug Metab. Dispos.*, 1991, **19**, 257.

Sparfloxacin

Use/occurrence:	Antibacterial drug
Key functional groups:	Aryl carboxylic acid, quinolone
Test system:	Rat (oral, 4.4, 10, 44, and 500 mg kg^{-1}; intravenous, 4.4 mg kg^{-1})

Structure and biotransformation pathway

(1) R = H, * = ^{14}C
(2) R = Gluc

During 96 hours after single oral doses of [^{14}C]sparfloxacin (1) (4.4 mg kg^{-1}) to rats, 17% and 81% dose were excreted in urine and faeces respectively, compared with 22% and 77% dose respectively after intravenous administration at the same dose level. Urinary excretion gradually became less extensive as the oral dose level was increased until at 500 mg kg^{-1} it represented 13% dose and faecal excretion 86% dose. In bile duct-cannulated rats treated intraduodenally with (1), 46% of the dose was excreted in bile during 48 hours and 21% dose in the urine; enterohepatic circulation of drug-related material was demonstrated by administering this bile to other cannulated rats. The extent of absorption of an oral dose of (1) was estimated to be *ca.* 70% in this species. TLC analysis of plasma, urine, bile, and milk samples obtained from treated rats indicated the presence of only two radioactive components in each case. The less polar of these in urine was identified as unchanged (1) by MS and chromatographic comparison with the authentic compound, and the more polar was characterized as (2), the glucuronic acid conjugate of parent drug, on the basis of its ability to liberate (1) during treatment with β-glucuronidase and with aqueous alkali. During and after 14 consecutive daily oral doses of (1) (10 mg kg^{-1}) to rats, radioactivity excretion profiles were similar to those found after single oral doses, and TLC showed that the proportions of (1) and (2) in urine were not statistically different from those after a single administration.

References

Y. Matsunaga, H. Miyazaki, Y. Oh-e, K. Nambu, H. Furukawa, K. Yoshida, and M. Hashimoto, Disposition and metabolism of [^{14}C]sparfloxacin in the rat, *Arzneim.-Forsch.*, 1991, **41**, 747.

Y. Matsunaga, H. Miyazaki, N. Nomura, and M. Hashimoto, Disposition and metabolism of [^{14}C]sparfloxacin after repeated administration in the rat, *Arzneim.-Forsch.*, 1991, **41**, 760.

Ofloxacin

Use/occurrence:	Antibacterial drug
Key functional groups:	Aryl carboxylic acid, N-methylpiperazine, quinolone
Test system:	Rat (intravenous, 10 and 20 mg kg^{-1}), rat liver microsomes

Structure and biotransformation pathway

(1) R = H
(3) R = Gluc

(2) R = H
(4) R = Gluc

It is known that the major biotransformation pathway of ofloxacin in rats is glucuronidation of the carboxyl group, and that biliary excretion of the S-($-$)-conjugate (3) in rats treated with S-($-$)-ofloxacin (1) was considerably greater (*i.e.* 31% *versus* 7% dose) than that of the R-($+$)-conjugate (4) in rats treated with R-($+$)-ofloxacin (2). The purpose of the present study was to investigate this apparent stereoselective glucuronidation process *in vitro*. Thus, when [^{14}C]-(1) and -(2) were separately incubated *in vitro* with rat liver microsomes fortified with UDPGA, TLC showed that (3) was produced with a V_{max} that was seven-fold greater than that of its optical antipode (4). Both reactions obeyed Michaelis–Menten kinetics with similar K_m values, resulting in a V_{max}/K_m ratio (a measure of the intrinsic hepatic clearance) that was eight-fold greater for formation of (3). Other studies in rat hepatic microsomes confirmed that R-($+$)-ofloxacin competitively inhibited glucuronidation of the S-($-$)-enantiomer ($K_i = 2.92$ mM), and accordingly AUC values for serum levels of S-($-$)-ofloxacin after intravenous administration of racemic ofloxacin (20 mg kg^{-1}) to rats were 1.7-fold greater than those after administration of the S-($-$)-enantiomer (10 mg kg^{-1}) itself. Hence, the stereoselectivity in the metabolism and disposition of ofloxacin in this species may be explained in terms of enantioselective glucuronidation.

Reference

O. Okazaki, T. Kurata, H. Hakusui, and H. Tachizawa, Stereoselective glucuronidation of ofloxacin in rat liver microsomes, *Drug Metab. Dispos.*, 1991, **19**, 376.

MDL 74270

Use/occurrence:	Free radical scavenger
Key functional groups:	Alkyl quaternary ammonium, arylmethyl, benzopyran, methyl carboxylate
Test system:	Rat (oral and intravenous, 1 mg kg^{-1})

Structure and biotransformation pathway

(1) * = ^{12}C; † = ^{14}C
(2) * = ^{14}C; † = ^{12}C

(3)

(5) * = ^{14}C

(4)

After intravenous administration of [N-*methyl*-^{14}C]-MDL 74270 (1) to rats, 40% of the dose was excreted in urine and 45% dose was excreted in faeces during 4 days, whereas the corresponding values after oral administration of (1) were 1% and 79% respectively, demonstrating poor oral absorption in this species. HPLC analysis of 0–24 hour urine indicated, by chromatographic comparison with the reference compounds, that the deacetylated metabolite (3) and the quinone (4) each accounted for *ca.* 15% of the sample radioactivity, and that unchanged drug represented at least 20%. Comparisons were made between the concentrations of radioactivity in selected tissues of rats following intravenous administration of [^{14}C]-MDL 74270 (2) and its tertiary amine analogue, [^{14}C]-MDL 74366 (5), both of which were labelled in the ethanamine side chain. Concentrations in blood were similar for both compounds, but the heart:blood radioactivity concentration ratio was > 20 for (2) during 6 hours post-dose, compared with < 2 for (5); radioactivity levels in skeletal muscle were less than those in heart for both compounds at all sampling times. These results demonstrated a marked cardioselectivity for the quaternary amine, an observation which is consistent with previous findings for other quaternary antiarrhythmic drugs whose long duration of action is ascribed to this cardiac uptake. Rat heart homogenates were analysed by HPLC in order to determine the nature of the drug-derived material therein, and the results showed that most of this was in the form of the phenol metabolite (3), which is considered to be the active drug. As *in vitro* studies showed that (2) was rapidly hydrolysed to (3) in rat blood, it is likely that the same process operating *in vivo* preceded distribution into myocardial tissue.

Reference

J. Dow, M. A. Petty, J. M. Grisar, E. R. Wagner, and K. D. Haegele, Cardioselectivity of α-tocopherol analogues, with free radical scavenger activity, in the rat, *Drug Metab. Dispos.*, 1991, **19**, 1040.

($\pm$)-5-[(3,4-Dihydro-2-phenylmethyl-2*H*-1-benzopyran-6-yl)methyl]thiazolidine-2,4-dione (CP-68,722)

Use/occurrence:	Antidiabetic
Key functional groups:	Dihydropyran, thiazolidine
Test system:	Bile duct-cannulated rat (intravenous, 6.5 mg kg^{-1}), rat liver microsomes and cytosol

Structure and biotransformation pathway

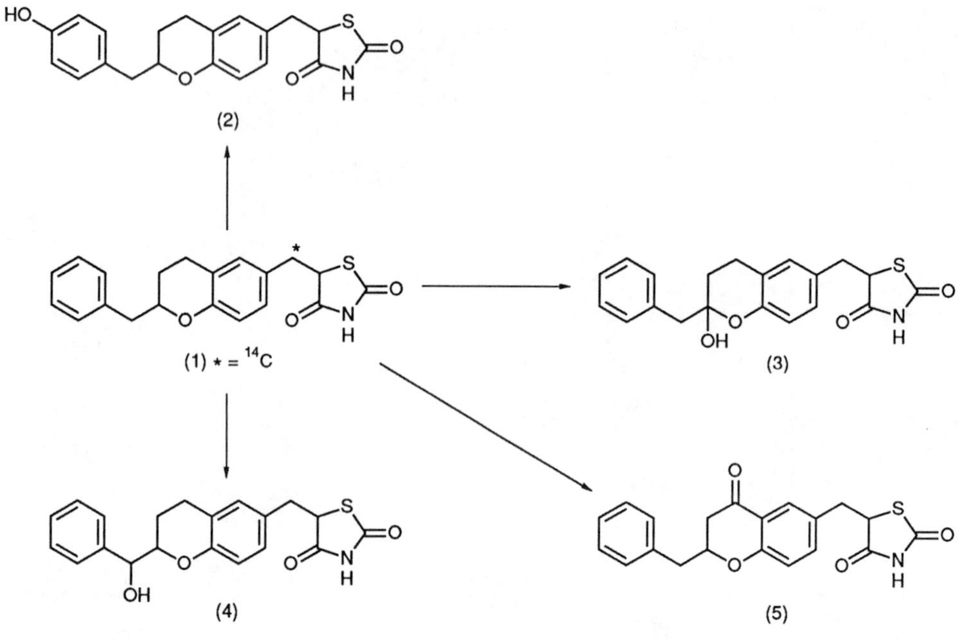

CP-68,722 (1) is a new antidiabetic agent which is expected to benefit patients with non-insulin-dependent diabetes. Previous studies have shown that the bile was the major route of excretion of (1). Following intravenous administration of [^{14}C]-(1) to bile duct-cannulated rats, four metabolites were excreted in the bile as glucuronide conjugates and accounted for 75–85% of the biliary radioactivity. β-Glucuronidase treatment yielded four aglycones which were separable by HPLC. These metabolites were identified (following derivatization to ^{2}H-labelled or non-labelled compounds) by co-chromatography, GC–MS, and comparison with synthetic standards. The metabolites were the 4'-hydroxy- (2), one 2-hydroxy- (3), and the diastereoisomers of the 11-hydroxy- (4) derivative. The incubation of [^{14}C]-(1) with rat liver microsomes in the presence of NADPH elicited the formation of the metabolites (2)–(4) formed *in vivo*, and one additional metabolite not formed in the *in*

318

vivo study, (5), the 4-keto metabolite of (1). Structures for two other metabolites (which were formed only using the *in vitro* metabolizing system) were tentatively proposed on the basis of their mass spectra. These were thought to be a metabolite resulting from hydroxylation in the thiazolidine-dione ring and a phenol resulting from opening of the dihydrobenzopyran ring.

Reference

H. G. Fouda, J. Lukaszewicz, D. A. Clark, and B. Hulin, Metabolism of a new thiazolidinedione hypoglycemic agent CP-68,722 in rat: metabolite identification by gas chromatography mass spectrometry, *Xenobiotica*, 1991, **21**, 925.

Zenarestat

Use/occurrence:	Aldose reductase inhibitor
Key functional groups:	Alkyl carboxylic acid, bromophenyl, chlorophenyl
Test system:	Rat (oral, intravenous, intraperitoneal, 10 mg kg^{-1})

Structure and biotransformation pathway

(4) Gluc = glucuronyl

(1) * = ^{14}C

(2) R^1 = H, R^2 = OH
(3) R^1 = OH, R^2 = H

Zenarestat, [3-(4-bromo-2-fluorobenzyl)-7-chloro-2,4-dioxo-1,2,3,4-tetra-hydroquinazolin-1-yl]acetic acid (1), is being developed for the treatment of diabetic neuropathy. Recently a sex difference in the urinary excretion of another compound of this class (ponarestat) in the rat was reported. This study was designed to study the sex difference in metabolism and excretion of (1) in the male and female rat.

Following the administration of [^{14}C]-(1) to rats by these routes, there was a marked difference in urinary excretion between the male and female rats. Only about 1% of the radioactivity was excreted in the urine of males after 72 hours, whereas 45% of the ^{14}C was excreted in the female rat urine. The majority (95%) of the radiolabel was excreted in the faeces of male rats whereas only 51% was eliminated via this route by females. The mode of excretion did not differ with the route of administration.

After intravenous dosing of [^{14}C]-(1), a similar sex difference in excretion was seen in urine and biliary samples collected over a 48 hour period from bile duct-cannulated rats. There was a marked sex difference in urinary excretion, with only 0.7% of the radioactivity being excreted in the male rat urine compared with 18.2% in female rat urine. Biliary excretion was 96% of the radioactivity in the males and 68% in the females.

Metabolites in faecal, bile, and urine samples were separated by TLC and detected by autoradiography. Quantitation was achieved by scraping appropriate areas from the TLC plate and measuring radioactivity by LSC. Besides the unchanged drug (39–47% in males and females), the major faecal metabolites were thought to be aromatic hydroxylated metabolites of (1). These metabolites, (2) and (3), each accounted for ca. 21–23% of the faecal radiolabel in both male and female rats. Three minor unidentified metabolites each accounted for ≤ 2% of the faecal radiolabel. In bile, unchanged (1),

20–26% of the biliary ^{14}C, and zenarestat-1-O-glucuronide (4) as two isomers (33–38%) were the major components. Three other major (8–12%) and four minor (1–3%) metabolites were also detected in the bile. There were no metabolites common to bile and faeces, and when biliary metabolites were mixed with caecal contents, only faecal metabolites could be detected. This indicates that biliary metabolites are probably conjugates of faecal metabolites. In the urine, more than 90% of the radioactivity was present as unchanged (1).

Thus, no sex differences in the metabolism of (1) were detected. The sex difference in the urinary excretion of ^{14}C after administration of $[^{14}C]$-(1) was thought to be due to an active renal tubular secretion mechanism in the female animals which was absent or present at a much lower level in the males. Probenecid, an inhibitor of tubular secretion, caused a marked decrease in cumulative renal excretion of $[^{14}C]$-(1) in female but not in male rats.

Reference

Y. Tanaka, Y. Deguchi, I. Ishii, and T. Terai, Sex differences in excretion of zenarestat in rat, *Xenobiotica*, 1991, **21**, 1119.

Piritrexim

Use/occurrence:	Folate antagonist
Key functional groups:	Methoxyphenyl, pyridopyrimidine
Test system:	Dog (intravenous, oral, 1.8 mg kg^{-1})

Structure and biotransformation pathway

Following either oral or intravenous administration of [^{14}C]piritrexim (1) urinary and faecal excretion of radioactivity accounted for *ca.* 20% and 70% of the dose respectively. About 10% of an intravenous dose and 15% of an oral dose was excreted unchanged in urine and faeces. The *O*-demethylated compounds (3) and (4) were the major excretory metabolites of (1). They were excreted mainly in faeces but were also found in urine. Their respective glucuronide and sulfate conjugates (5) and (6) were only detected in urine. Demethylation at the 2′-position was the more favoured route of biotransformation. The 2′-hydroxy metabolite (3) and its glucuronide (5) together accounted for 29–33% dose, while metabolite (4) and the sulfate (6) accounted for 14–15% dose. The 3′-hydroxy compound (2) was a very minor metabolite detected only in urine after oral dosing. ^{1}H NMR provided the main evidence of metabolite structure, together with enzymatic deconjugation experiments for the conjugated metabolites.

Reference

J. L. Woolley, D. V. Deangelis, M. E. Grace, S. H. T. Liao, R. C. Crouch, and C. W. Sigel, The disposition and metabolism of [^{14}C]piritrexim in dogs after intravenous and oral administration, *Drug Metab. Dispos.*, 1991, **19**, 1139.

Piritrexim

Use/occurrence:	Anticancer/antipsoriasis drug
Key functional groups:	Aminopyrimidine, methoxyphenyl, methylpyridine, pyridopyrimidine
Test system:	Rat (oral, 5, 10, and 20 mg kg^{-1}; intravenous, 5 and 10 mg kg^{-1})

Structure and biotransformation pathway

After oral administration of [^{14}C]piritrexim (1) (10 mg kg^{-1}) to rats, 84% and 9% of the dose respectively were excreted in faeces and urine during 24 hours, compared with 57% and 32% dose respectively during 48 hours after intravenous administration. Quantitative TLC determination of (1) in serial blood samples showed that at all times parent drug accounted for less than half of the plasma radioactivity; the remaining drug-related material therein was present as metabolites that showed substantial dihydrofolate reductase-binding activity. Radioactive components in urine and faeces were separated and quantified by TLC, then purified by HPLC for structural identification by NMR and LC–MS comparison with synthetic reference compounds (where available). In this way it was shown that unchanged (1) was present in both biofluids after oral and intravenous dosing, and that the O-desmethyl (2) and (3) and hydroxylated (4) metabolites accounted for most of the remaining faecal radioactivity. Only small amounts of intact piritrexim were found in urine, where (2) and (3), mainly as their sulfate conjugates, were the principal radioactive components therein. The O-desmethyl metabolites were shown to be potent inhibitors of dihydrofolate reductase, and were cytotoxic to cells in culture.

Reference

J. L. Woolley, D. V. Deangelis, R. C. Crouch, J. P. Shockcor, and C. W. Sigel, The disposition and metabolism of [^{14}C]piritrexim in rats after intravenous and oral administration, *Drug Metab. Dispos.*, 1991, **19**, 600.

Prazepam, Halazepam

Use/occurrence:	Anxiolytic drugs
Key functional groups:	Benzodiazepine
Test system:	Human liver microsomes

Structure and biotransformation pathway

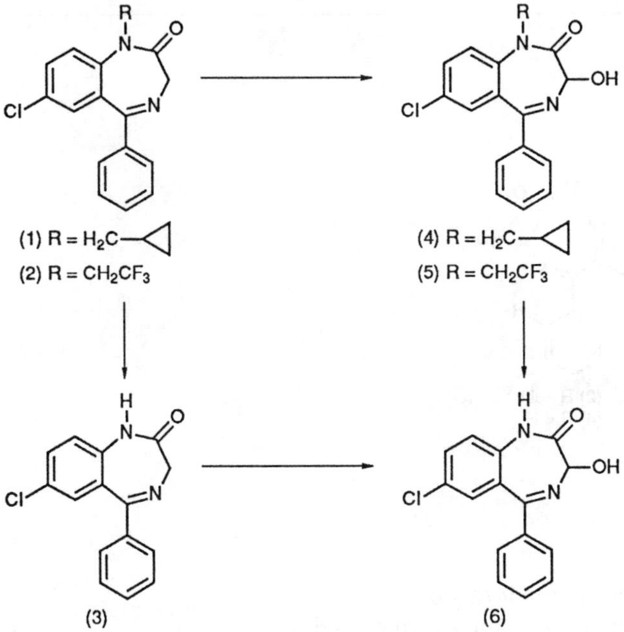

(1) R = H₂C—△

(2) R = CH₂CF₃

(4) R = H₂C—△

(5) R = CH₂CF₃

(3)

(6)

The metabolism of prazepam (1) and halazepam (2) was studied in microsomes prepared from livers of two male and one female subject who had died of head injuries. Metabolites were analysed by normal-phase HPLC and chiral stationary-phase HPLC. Both compounds were metabolized by *N*-dealkylation and 3-hydroxylation but there were differences in the relative amounts of products formed. With prazepam, *N*-dealkylation was favoured whereas 3-hydroxylation was very much the predominant pathway for halazepam. The 3-hydroxylation reaction was stereoselective with both (1) and (2), resulting in the formation of a product enriched (60–70%) in the 3*R*-enantiomer. The *N*-dealkylations of (4) and (5) were substrate enantio-selective; 3*S*-(4) and 3*R*-(5) were each dealkylated slightly faster than their opposite enantiomers.

Reference

X.-L. Lu, F. P. Guengerich, and S. K. Yang, Stereoselective metabolism of prazepam and halazepam by human liver microsomes, *Drug Metab. Dispos.*, 1991, **19**, 637.

Lorazepam 3-acetate

Use/occurrence:	Anxiolytic drug
Key functional groups:	Chiral carbon, methyl carboxylate
Test system:	Human and rat liver microsomes, rat brain S9 fraction

Structure and biotransformation pathway

† = chiral carbon
(1) R = Cl
(3) R = H

(2) R = Cl
(4) R = H

The rates of hydrolysis of racemic and enantiomeric lorazepam 3-acetate (1) to lorazepam (2) by esterases in rat and human liver microsomes and in the S9 fraction of rat brain were determined with the aid of chiral stationary-phase HPLC. With *rac*-(1) as substrate, *R*-(1) was hydrolysed by rat and human liver microsomes 2.7- and 6.8-fold faster respectively than *S*-(1), whereas in the rat brain S9 fraction hydrolysis of *S*-(1) was 2.3-fold faster than that of *R*-(1). When the pure enantiomers were used as substrates, *R*-(1) was hydrolysed at a rate 8.4-fold greater than that of *S*-(1) in rat liver microsomes and 166-fold greater in a human liver microsomal preparation; in rat brain S9 fraction *S*-(1) was hydrolysed twice as fast as *R*-(1). Enantiomeric interactions occurred in the hydrolysis of the enantiomeric acetates. Thus, the presence of *R*-(1) stimulated the hydrolysis rate of *S*-(1) in all esterase preparations, whereas the presence of the latter enantiomer enhanced the hydrolysis rate of its optical antipode in the rat brain S9 fraction but inhibited it in both liver microsomal preparations.

Similar results to these have been reported previously for hydrolysis of the structurally related compound oxazepam 3-acetate (3) to oxazepam (4). Comparison of the results indicated that esterases in rat brain S9 fraction hydrolysed *rac*-(1) faster than *rac*-(3), but the reverse situation obtained for rat and human liver microsomes. Similarly, *R*-(1) was hydrolysed faster than *R*-(3) in rat liver microsomes, bu more slowly in human liver microsomes, while *S*-(1) was hydrolysed faster than *S*-(3) in both liver microsomal preparations.

Reference

K. Liu, F. P. Guengerich, and S. K. Yang, Enantioselective hydrolysis of lorazepam 3-acetate by esterases in human and rat liver microsomes and rat brain S9 fraction, *Drug Metab. Dispos.*, 1991, **19**, 609.

Diltiazem

Use/occurrence:	Antihypertensive drug
Key functional groups:	Dimethylaminoalkyl, methoxyphenyl, methyl carboxylate
Test system:	Rat liver microsomes and mitochondria

Structure and biotransformation pathway

(1)

(2) R = CH$_3$
(3) R = H

(4) R^1 = CH$_3$, R^2 = CH$_3$CO
(5) R^1 = CH$_3$, R^2 = H
(6) R^1 = H, R^2 = CH$_3$CO
(7) R^1 = H, R^2 = H

Acidic metabolites of diltiazem (1) formed by deamination of the dimethyl-aminoalkyl side-chain have previously been identified as major metabolites in rat, dog, and man. In this study the formation of metabolites via this pathway has been investigated *in vitro* using rat liver preparations. Metabolites were extracted, isolated by HPLC and identified by GC–MS. Both neutral and acidic metabolites were formed by microsomal and mitochondrial fractions. Two neutral metabolites were identified as the aldehydes (2) and (3), being precursors to the carboxylic acids. Four carboxylic acid metabolites were identified, two major components (4) and (6) and two minor components (5) and (7). The rates of formation of the aldehydes were about 4–5-fold greater than those of the acids. NADPH was necessary for the formation of these metabolites, indicating the participation of cytochrome P-450.

Reference

S. Nakamura, Y. Ito, T. Fukushima, Y. Sugawara, and M. Ohashi, Metabolism of diltiazem III. Oxidative deamination of diltiazem in rat liver microsomes, *J. Pharmacobio-Dyn.*, 1990, **13**, 612.

N-Methylcarbazole

Use/occurrence:	Present in tobacco smoke
Key functional groups:	*N*-Methylindole
Test system:	Rat hepatocyte cultures

Structure and biotransformation pathway

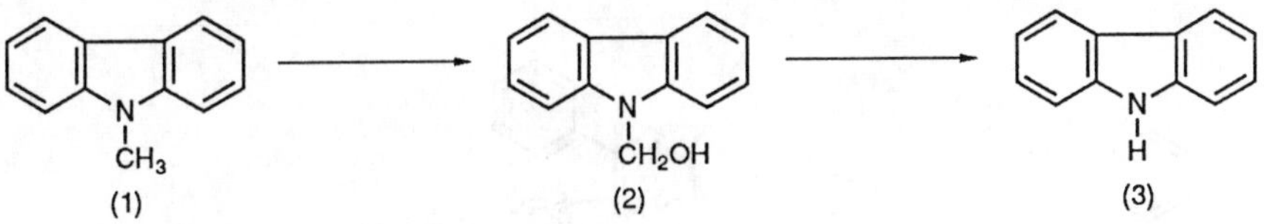

(1)   (2)   (3)

N-Methylcarbazole (1) was incubated with cultures of primary rat hepatocytes and the metabolites were extracted with ethyl acetate before being analysed by HPLC. The chromatogram showed that (1) had been oxidized to *N*-hydroxymethylcarbazole (2) which could be further degraded to carbazole (3). The structure of (2) was confirmed by GC–MS of the trimethylsilyl derivative. Analysis of cytotoxicity showed that the formation of (2) produced a more toxic metabolite.

Reference

W. Yang, T. Jiang, P. J. Davis, and D. Acosta, *In vitro* metabolism and toxicity assessment of *N*-methylcarbazole in primary cultured rat hepatocytes, *Toxicology*, 1991, **68**, 217.

Dibenzothiophene

Use/occurrence: Environmental contaminant

Key functional groups: Thiophene

Test system: Rat hepatic microsomes

Structure and biotransformation pathway

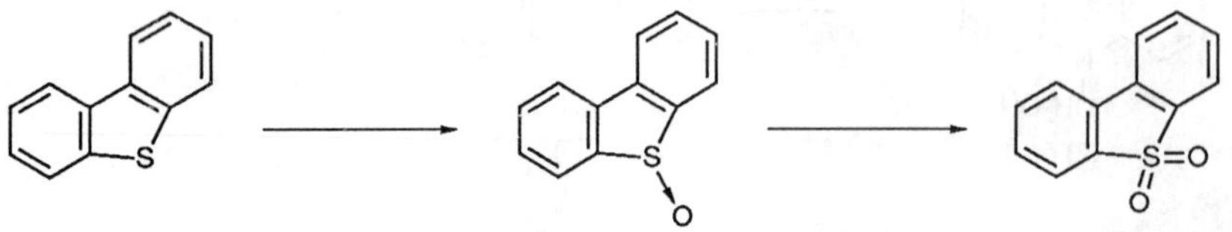

(1)   (2)   (3)

Following incubation with rat liver microsomes dibenzothiophene was found to be converted into the corresponding sulfoxide (2) and sulfone (3). The products were analysed by GC and to reference standards. Pretreatment of the rats with phenobarbital or aroclor increased the formation of sulfone five-fold. The metabolism of a series of thiaarenes was also examined, with the thiophene ring in a central or a peripheral position. In general compounds with peripheral thiophene rings more readily underwent oxidation of the carbon skeleton, whereas in those with a central thiophene ring S-oxidation was the major route of metabolism.

Reference

J. Jacob, A. Schmoldt, C. Augustin, G. Raab, and G. Grimmer, Rat liver microsomal ring and S-oxidation of thiaarenes with central or peripheral thiophene rings, *Toxicology*, 1991, **67**, 181.

330

Abecarnil

Use/occurrence:	Anxiolytic, anticonvulsant drug
Key functional groups:	Alkyl ester, benzyl ether
Test system:	Human volunteers (intravenous, 1.88 mg followed, 14 days later, by oral, 10.3 mg)

Structure and biotransformation pathway

(1) * = ^{14}C

(2)

(3)

New chemical classes which interact with the benzodiazepine–GABA-receptor complex include β-carbolines. Abecarnil (1) is a potent β-carboline which is extensively metabolized by benzyl ether cleavage to the phenol (2) and subsequent glucuronidation or sulfation. A minor metabolic pathway is ester hydrolysis to the carboxylic acid (3) and then acyl glucuronide formation. The bioavailability of (1) was 40%, the terminal half-life was four hours, and the total clearance was 11 ml min^{-1} kg^{-1}. The main route for rapid and complete excretion was in the faeces (60% of the dose) rather than in the urine (25%). Analysis was by EI- or CI-FAB-MS and NMR after incubation with glucuronidase or sulfatase, using HPLC with fluorescence detection, liquid scintillation counting, and comparison with authentic standards. Two unknown metabolites were detected by radio-HPLC, together with unchanged (1). The isolation of the metabolites from different animal species and from man will be reported separately.

Reference

W. Krause, T. Duka, and H. Matthes, Pharmacokinetics and biotransformation of the anxiolytic abecarnil in healthy volunteers, *Xenobiotica*, 1991, **21**, 763.

2-Amino-3,8-dimethylimidazole[4,5-*f*]-quinoxaline (MeIQX)

Use/occurrence:	Food mutagen
Key functional groups:	Aminoimidazole, *N*-methylimidazole, methylquinoxaline
Test system:	Rat (oral, 0.01, 0.1, 0.2, and 20 mg kg^{-1})

Structure and biotransformation pathway

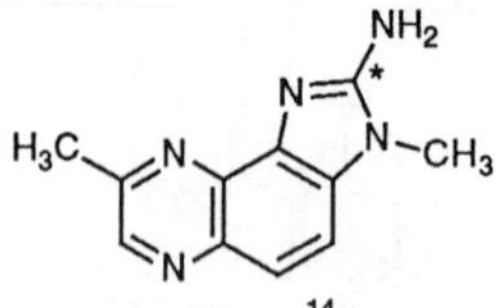

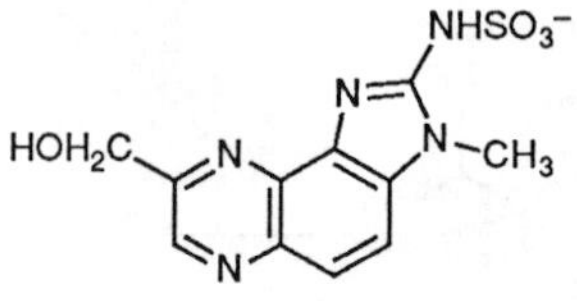

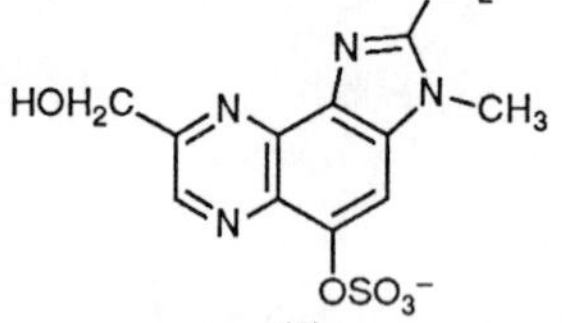

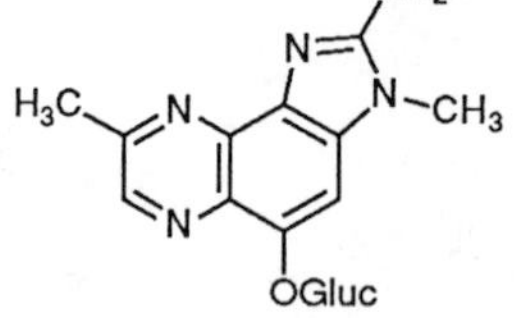

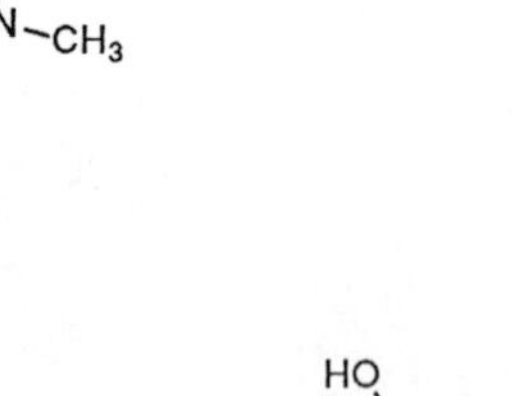

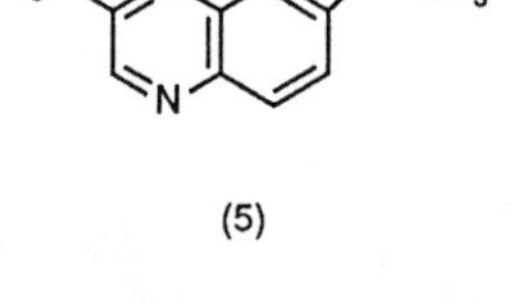

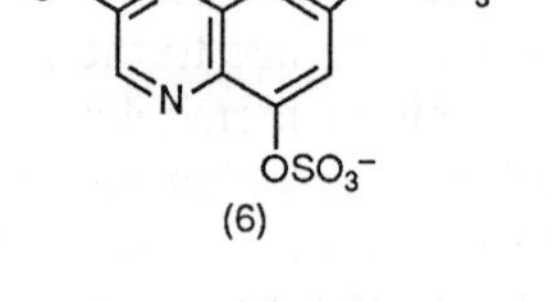

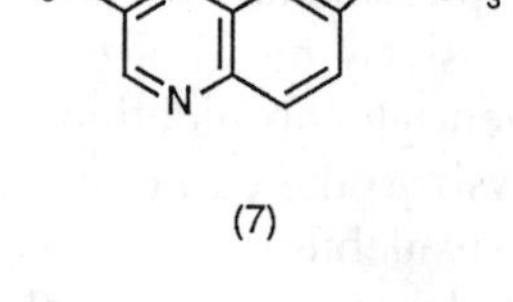

Urine and faeces were collected for up to 72 hours following an oral dose of [^{14}C]-(1). Recovery of radioactivity in excreta averaged 90% or greater for each dose group. Faecal elimination was generally a more important route than urinary elimination; however, *ca*. 45% of the dose was eliminated in the urine of non-induced rats following a 20 mg kg^{-1} dose. Significant amounts of unchanged (1) were recovered in the urine of non-induced rats accounting for *ca.* 20% and 6% of the dose respectively following the 20 and 0.2 mg kg^{-1} doses. Following the 0.01 mg kg^{-1} dose the amount of unchanged (1) excreted in urine represented less than 2% of the dose. At the 20 mg kg^{-1} dose compounds (8) and (5) were the major urinary metabolites in non-induced rats. In polychlorinated biphenyl induced rats given the same dose compound (4) was the major urinary metabolite. The corresponding sulfate metabolite (6) was the major biliary metabolite in both induced and non-induced rats at both the 0.1 and 20 mg kg^{-1} doses. Metabolites were analysed by a radiochemical HPLC assay and/or FAB-MS.

Reference

R. J. Turesky, J. Markovic, I. Bracco-Hammer, and L. B. Fay, The effect of dose and cytochrome P450 induction on the metabolism of the food-borne carcinogen 2-Amino-3,8-dimethylimidazole[4,5-*f*]quinoxaline (MeIQX) in the rat, *Carcinogenesis*, 1991, **12**, 1839.

Chlorpromazine *N*-oxide

Use/occurrence:	Model compound
Key functional groups:	Dibenzothiazine, dimethylaminoalkyl
Test system:	Rat (intraperitoneal, 20 mg kg^{-1}), bile duct-cannulated rat (oral, intraperitoneal, intravenous, 20 mg kg^{-1})

Structure and biotransformation pathway

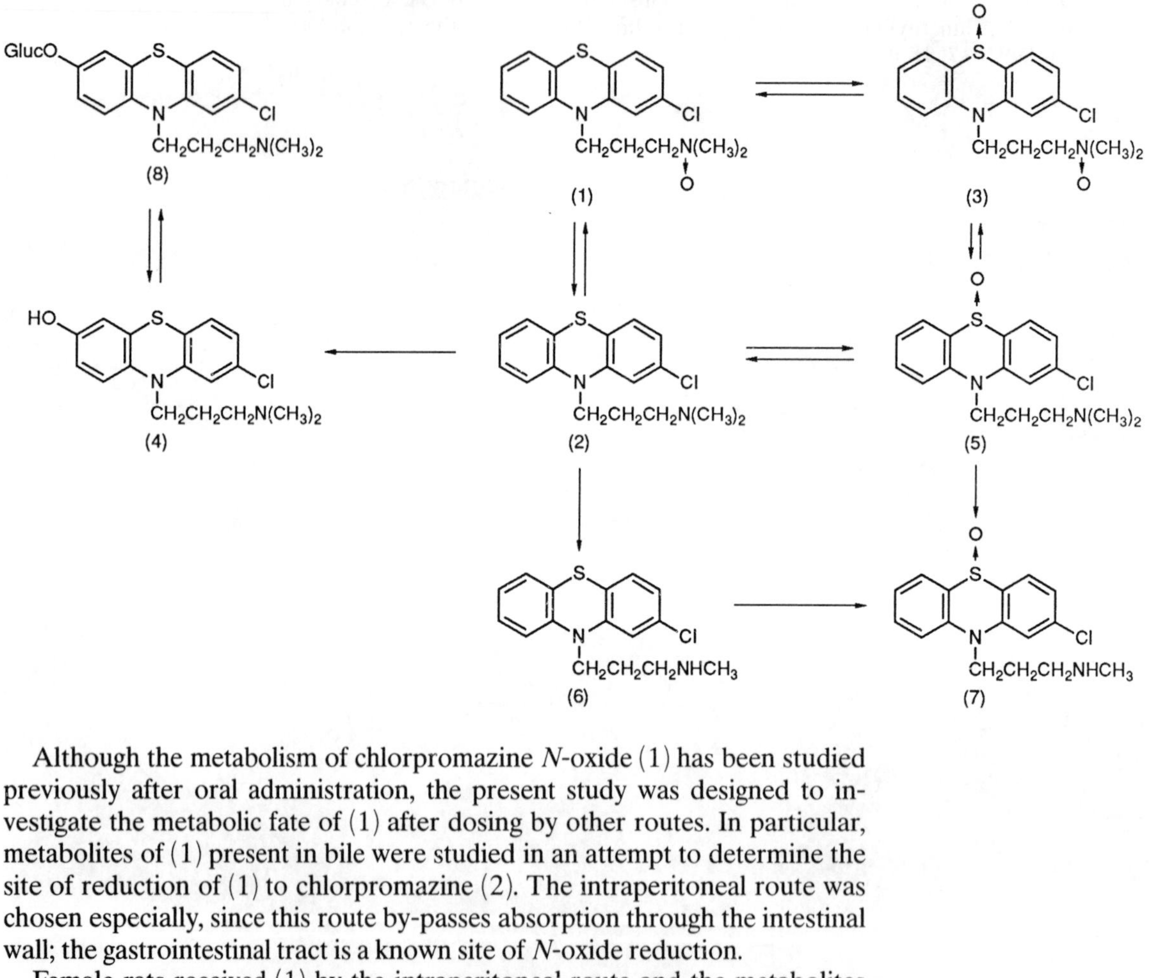

Although the metabolism of chlorpromazine *N*-oxide (1) has been studied previously after oral administration, the present study was designed to investigate the metabolic fate of (1) after dosing by other routes. In particular, metabolites of (1) present in bile were studied in an attempt to determine the site of reduction of (1) to chlorpromazine (2). The intraperitoneal route was chosen especially, since this route by-passes absorption through the intestinal wall; the gastrointestinal tract is a known site of *N*-oxide reduction.

Female rats received (1) by the intraperitoneal route and the metabolites present in extracts of urine and faeces were separated by HPLC. Metabolites were identified by comparison of retention times with authentic standards. The five metabolites identified in urine were chlorpromazine (2), 7-hydroxychlorpromazine (4), chlorpromazine sulfoxide (5), *N*-desmethylchlorpromazine (6), and *N*-desmethylchlorpromazine sulfoxide (7). Metabolites (2) and (4)–(6) inclusive [but not (7)] were identified in the faeces.

For each of the three routes of administration used in the bile duct-cannulated animals there were no differences in the qualitative metabolic profiles seen in bile samples. Following administration of (1), unchanged (1), (2), (4), (5), and chlorpromazine *N,S*-dioxide (3) were identified in biliary extracts following separation by HPLC using an isocratic solvent system. Identification of metabolites was by MS using CI or FAB ionization and comparison with authentic standards. After direct analysis of bile using a gradient HPLC system, three peaks were seen which were not present in control bile. These metabolites were 7-hydroxychlorpromazine-*O*-glucuronide (8) together with (3) and unchanged (1). Identification of metabolite (8) was considered to be the first unequivocal identification of this intact glucuronide in any species.

In contrast to the metabolites seen in urine and faeces, the biliary metabolites of (1) included those containing an *N*-oxide group, namely (1) and (3). This suggests that after oral or intraperitoneal administration (1) is absorbed as the intact molecule into the body. The fact that neither (1) nor (3) was present in faeces suggests that these compounds are excreted in the bile and that the *N*-oxide group is subsequently reduced in the intestinal tract.

Reference

T. J. Jaworski, E. M. Hawes, J. W. Hubbard, G. McKay, and K. K. Midha, The metabolites of chlorpromazine *N*-oxide in rat bile, *Xenobiotica*, 1991, **21**, 1451.

Fluphenazine

Use/occurrence:	Neuroleptic drug
Key functional groups:	Alkyl alcohol, phenol
Test system:	Immobilized rabbit hepatic microsomes

Structure and biotransformation pathway

(1)

(2)

(3) Gluc = glucuronyl

(4)

Fluphenazine (1) is a potent neuroleptic drug which is widely used in the treatment of schizophrenia. To date, there has only been indirect evidence for the presence of Phase II metabolites of (1) and the present study was designed to biosynthesize and characterize the glucuronides of (1). The products were identified by various means following separation by TLC and HPLC, including FAB-, EI-, and CI-MS and MS–MS. The major metabolite was further characterized by ^1H and ^{13}C NMR.

The system used to generate the glucuronide metabolites was an immobilized rabbit hepatic microsomal enzyme. The substrates used were (1) or its 7-hydroxy derivative (2). Conversion of (1) into its aliphatic hydroxy ether glucuronide metabolite (3) in the presence of UDPGA was very poor. However, (2) was metabolized very efficiently by the immobilized enzyme system (60% conversion) and the metabolite was identified as (4), the phenolic ether β-D-glucuronide of (2).

It was thought that the low yield (< 1%) of (3) was due to the relatively low activity of the rabbit glucuronyl transferase enzyme in catalysing the formation of the primary alcohol glucuronide. Human glucuronyl transferase may be more efficacious in this reaction but it is also possible that sulfation is a major Phase II reaction in man.

Reference

C.-J. C. Jackson, J. W. Hubbard, and K. K. Midha, Biosynthesis and characterization of glucuronide metabolites of fluphenazine: 7-hydroxyfluophenazine glucuronide and fluphenazine glucuronide, *Xenobiotica*, 1991, **21**, 383.

Fluphenazine

Use/occurrence:	Neuroleptic drug
Key functional groups:	Alkyl alcohol, alkyl tert-amine, phenothiazine, piperazine
Test system:	Rat (intraperitoneal, 20 mg kg^{-1})

Structure and biotransformation pathway

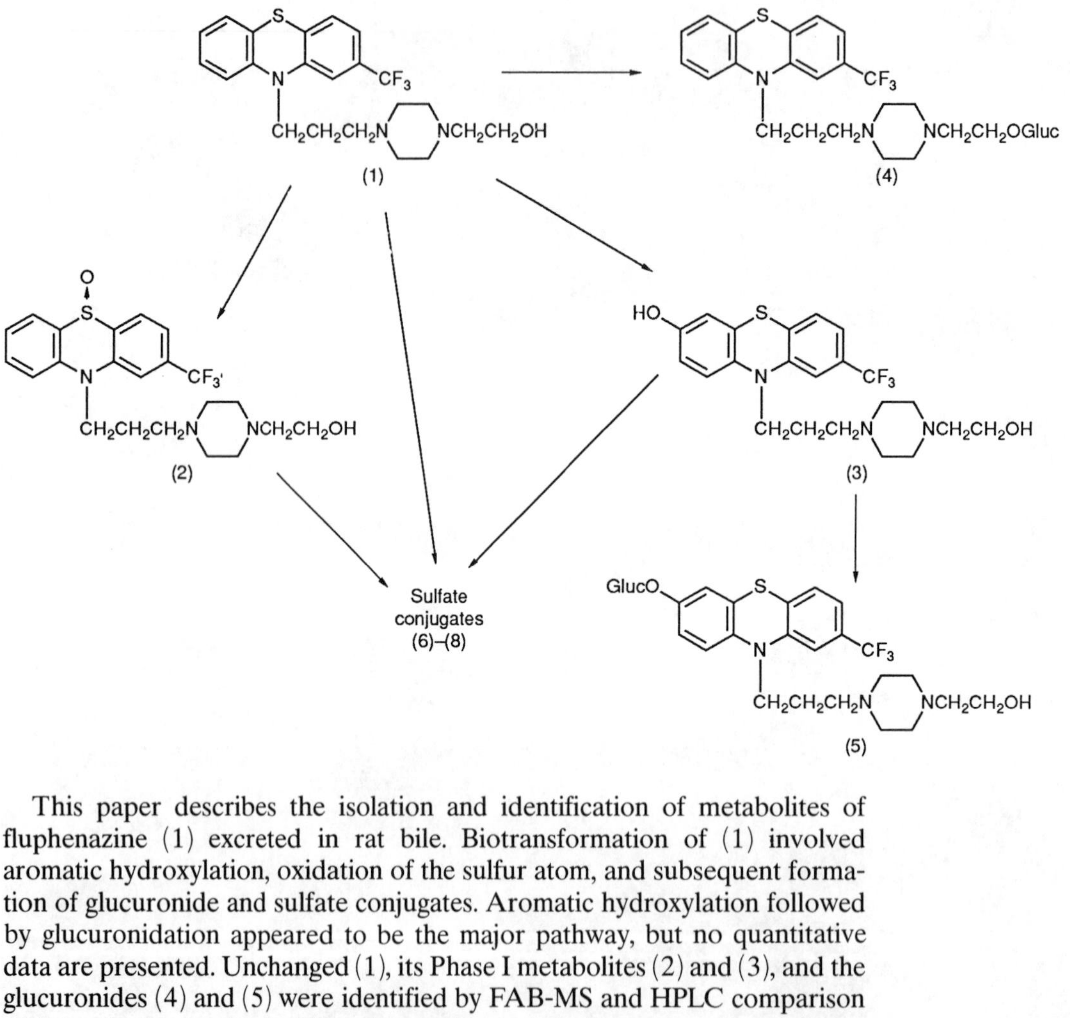

This paper describes the isolation and identification of metabolites of fluphenazine (1) excreted in rat bile. Biotransformation of (1) involved aromatic hydroxylation, oxidation of the sulfur atom, and subsequent formation of glucuronide and sulfate conjugates. Aromatic hydroxylation followed by glucuronidation appeared to be the major pathway, but no quantitative data are presented. Unchanged (1), its Phase I metabolites (2) and (3), and the glucuronides (4) and (5) were identified by FAB-MS and HPLC comparison with synthetic standards. The glucuronide standards were prepared by an immobilized enzyme technique. MS–MS was used to provide additional structural evidence for these conjugates. Some FAB-MS and MS–MS data were obtained indicating the presence of sulfate conjugates in bile, but

338

evidence for the formation of sulfates from (1), (2), and (3) rested mainly on identification of the exocons after sulfatase hydrolysis. An indication was also obtained for the presence of a further Phase II metabolite, a monoglucuronide of a dihydroxy derivative of (1), in bile.

Reference

C.-J. Jackson, J. W. Hubbard, G. McKay, J. K. Cooper, E. M. Hawes, and K. K. Midha, Identification of Phase I and Phase II metabolites of fluphenazine in rat bile, *Drug Metab. Dispos.*, 1991, **19**, 188.

Quinelorane

Use/occurrence:	Dopamine (D_2) agonist
Key functional groups:	*N*-Alkylpiperidine, aminopyrimidine
Test system:	Rat (oral, 0.1 and 40 mg kg^{-1}), mouse (oral, 1 and 50 mg kg^{-1}), dog (intravenous, 0.1 mg kg^{-1}), rhesus monkey (oral, 0.25 mg kg^{-1})

Structure and biotransformation pathway

In rats, dogs, and monkeys treated with [^{14}C]quinelorane (1), most of the administered radioactivity was excreted in the urine, whereas in mice the extents of urinary and faecal excretion were similar. [*N.B.* Owing to the sensitivity of dogs to the emetic effect of (1), the intravenous dose of labelled drug to these animals was preceded by 22 consecutive daily oral doses of non-labelled drug to induce tachyphylaxis to this effect.] Biliary excretion was not an important feature of the disposition of (1) in the rat, as only 5% of an oral dose to cannulated animals was excreted by this route. Radioactive components in urine were separated and quantified by TLC and the major components in rat and monkey urine were subsequently identified by a variety of mass spectroscopic techniques (*i.e.* EI-, CI-, FD-, FAB-, and/or MS–MS). In addition to unchanged (1), four major metabolites (2)–(5) were identified in this way, although not all were formed by each species. Since the hydroxylated metabolites (2) and (4) reacted with the diazomethane, the hydroxy group was presumed to be located in the aromatic (pyrimidine) ring. Metabolite (2) was essentially only produced by monkeys, (3) was produced by all four species, and (4) and (5) were only produced by mice and monkeys.

340

The *N*-despropyl metabolite (3) was a major urinary metabolite in all species, but biotransformation was generally more extensive in the mouse and monkey than in the rat or dog.

Reference

B. M. Manzione, J. R. Bernstein, and R. B. Franklin, Observations on the absorption, distribution, metabolism and excretion of the dopamine (D_2) agonist, quinelorane, in rats, mice, dogs and monkeys, *Drug Metab. Dispos.*, 1991, **19**, 54.

cis-Flupentixol

Use/occurrence:	Neuroleptic drug
Key functional groups:	Alkene, *N*-alkylpiperazine, arylalkene, diaryl thioether
Test system:	Rat (intraperitoneal, 48 mg kg^{-1})

Structure and biotransformation pathway

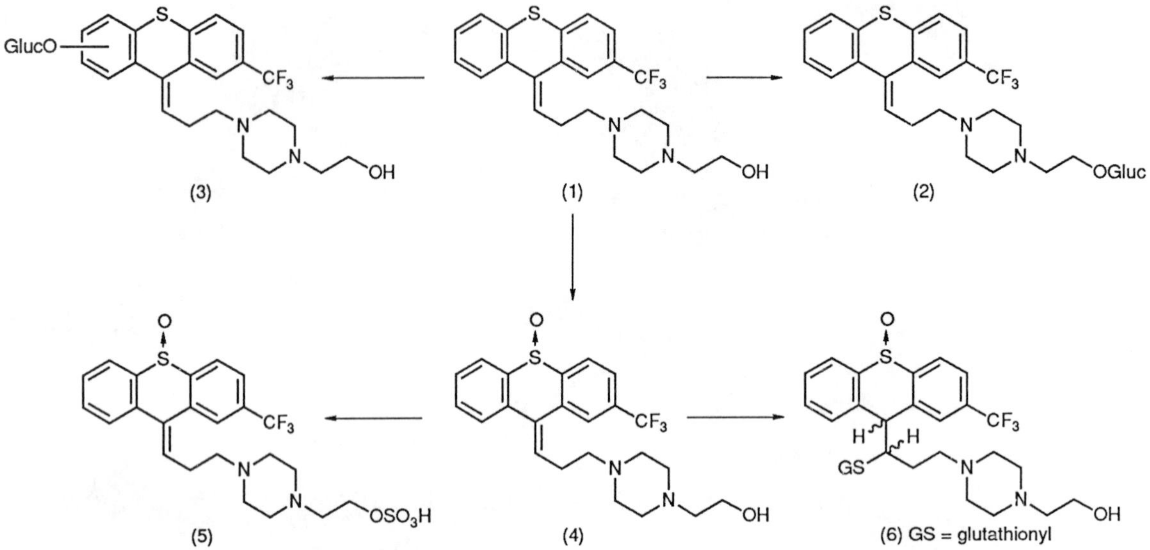

Plasma and bile samples obtained from bile duct-cannulated rats treated intraperitoneally with *cis*-flupentixol (1) were analysed by HPLC, using isocratic elution conditions to separate Phase I metabolites and gradient conditions to separate Phase II metabolites. In each case, metabolites were isolated and identified by FAB-MS, ^{1}H NMR, and (where possible) co-chromatography with reference compounds. Phase II metabolites were also subjected to hydrolysis with β-glucuronidase/sulfatase and the liberated aglycones isolated for structural analysis. In this way it was shown that the sulfoxide (4) was the only important Phase I metabolite of the drug in plasma and bile. Analysis of bile under the Phase II metabolite conditions indicated the presence of five components, which were identified as (2)–(6). The most interesting of these was the unstable glutathione adduct (6), which could also be synthesized non-enzymatically from glutathione and (4); the corresponding adducts of several other psychotropic drugs containing a tricyclic ring system and an exocyclic double bond were also prepared similarly. It is believed that this is the first example of such a glutathione adduct formed *in vivo* from thioxanthene antipsychotic or tricyclic antidepressant drugs.

Reference

Y.-Z. Shu, J. W. Hubbard, G. McKay, E. M. Hawes, and K. K. Midha, *cis*-Flupentixol metabolites in rat plasma and bile, *Drug Metab. Dispos.*, 1991, **19**, 154.

Mianserin

Use/occurrence:	Antidepressant drug
Key functional groups:	N-Methylpiperazine
Test system:	Microsomes from mouse, rat, guinea-pig, rabbit, and human liver

Structure and biotransformation pathway

(3)   (1) * = ³H   (2)

The formation of stable, reactive, and cytotoxic metabolites of [³H]mianserin (1) was assessed in microsomes prepared from each species under identical conditions. HPLC analysis indicated that desmethylmianserin (3) was the principal stable metabolite from four of the five species but was either not formed or was completely metabolized further in rabbit liver microsomes. 8-Hydroxymianserin (2) was apparently detected (HPLC) in incubations from all five species, but was significant only in the mouse. The formation of chemically reactive metabolites, as evidenced by irreversible binding of radioactivity to the protein precipitate from each incubation, showed less variability between species. In contrast, the results of an *in vitro* lymphocyte cytotoxicity assay showed that in the presence of NADPH human liver microsomes produced a far higher level of cell death than microsomes from the other species. In the absence of NADPH there were no significant differences between species.

Reference

P. Roberts, N. R. Kitteringham, and B. K. Park, Species differences in the activation of mianserin to a cytotoxic metabolite, *Drug Metab. Dispos.*, 1991, **19**, 841.

(−)-Sparteine, (+)-Sparteine (pachycarpine), 2,3-Didehydrosparteine

Use/occurrence:	Alkaloids
Key functional groups:	Quinolizidine
Test system:	Rat [oral, 0.094 mmol kg^{-1} (1) and (2), 0.04 mmol kg^{-1} (3)], rat liver microsomes

Structure and biotransformation pathway

The metabolism of sparteine (1) has been studied intensively in animals and man following the observation that the formation of its major metabolites was under genetic control. However, the metabolism of its (+)-enantiomer (2) has not been investigated. After administration of (1) and (2) to rats, marked differences in metabolism were observed. Metabolites were separated by GC and identified by comparison with authentic standards, MS, and ^{13}C or ^{2}H NMR. For (1), two metabolites accounted for the majority of the dose excreted in the urine, a total of 25% of the dose of [^{2}H]-(1): metabolites (3), 2,3-didehydrosparteine (16.9%), and (4), lupanine (6.3%). Unchanged (1) represented 2.1% of the urinary material. The formation of (3) from (1) was shown to occur via stereospecific abstraction of the axial 2β hydrogen atom. When synthetic (3) was given to rats, 10–13% of the urinary content was unchanged (3) and *ca.* 16% was metabolite (4).

Following administration of [^{2}H]-(2), 61.6% of the dose was recovered in the urine, 58.6% of which was present as (5), 4S-hydroxypachycarpine. Unchanged (2), 2,3-didehydropachycarpine (6), and 5,6-didehydropachycarpine (7) accounted for 2.4%, 0.3%, and 0.3% of the dose respectively. These investigations were thought by the authors to be the first reported case of metabolic reaction at an aliphatic methylene group of the sparteine molecule not adjacent to nitrogen.

Inhibitions studies *in vitro* with rat liver microsomes in the presence of NADPH showed that both sparteine enantiomers (1) and (2) were metabolized by the same cytochrome P-450 isozyme (cytochrome P-450 IID1). Thus this enzyme showed marked substrate and product stereoselectivity for the metabolism of (1) and (2).

Reference

T. Ebner, C. O. Meese, and M. Eichelbaum, Regioselectivity and stereoselectivity of the metabolism of the chiral quinolizidine alkaloids sparteine and pachycarpine in the rat, *Xenobiotica*, 1991, **21**, 847.

Sparteine

Use/occurrence:	Alkaloid, oxytocic agent
Key functional groups:	Quinolizidine
Test system:	Human extensive metabolizer

Structure and biotransformation pathway

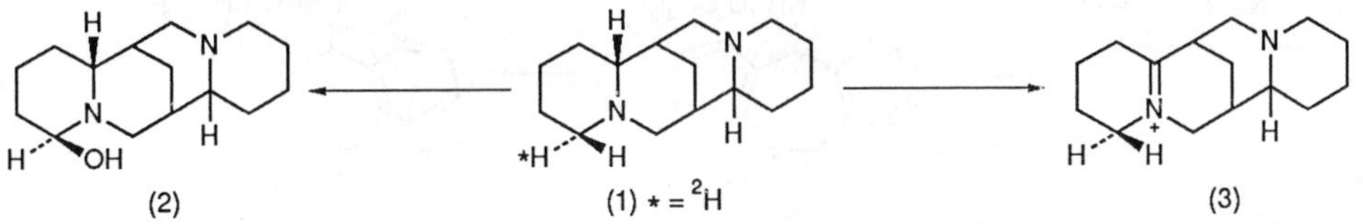

Sparteine (1) has been shown to exhibit genetic polymorphism in man. Enamine structures (not shown) had previously been proposed for the two metabolites found in the urine of extensive metabolizers. Evidence for those structures came from GC–MS after a clean-up procedure in which the urine was made alkaline. In this study ^{2}H NMR of urine was used to show that, as excreted, these metabolites had the carbinolamine (2) and iminium structures (3) shown in the scheme. HPTLC comparison of urine with standards was used to provide further evidence of these structures. Quantitatively (2) and (3) were described as the major and minor metabolites respectively.

Reference

T. Ebner, C. O. Meese, P. Fischer, and M. Eichelbaum, A nuclear magnetic resonance study of sparteine delta metabolite structure, *Drug Metab. Dispos.*, 1991, **19**, 955.

AJ-2615

Use/occurrence:	Antihypertensive drug
Key functional groups:	Dibenzothiepine
Test system:	Rat (oral, 30 mg kg^{-1})

Structure and biotransformation pathway

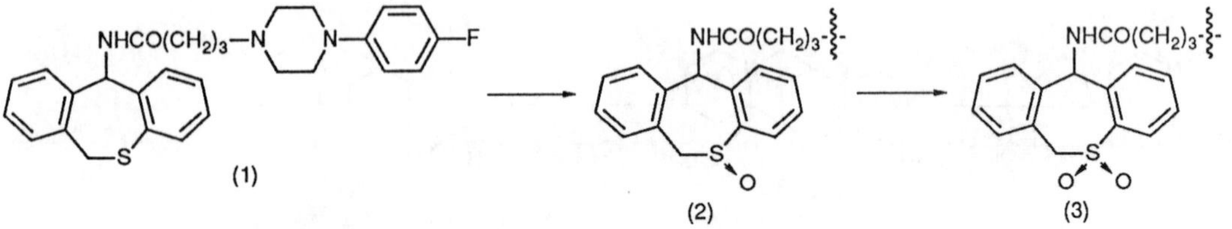

Plasma samples were collected from rats 3 hours after a single oral administration of (1). Parent compound (1) and its metabolites (2) and (3) were identified, without chromatographic separation, in a crude plasma extract by FAB-MS–MS. Two metabolites were identified as the *S*-oxide (2) and the *S,S*-dioxide (3). The presence of two diastereoisomeric *S*-oxides in the plasma extract was ascertained by HPLC. Their relative configurations were determined by IR and ^{1}H NMR.

Reference

M. Kurono, A. Itogawa, K. Yoshida, S. Naruto, P. Rudewicz, and M. Kanai, Identification of AJ-2615 and its *S*-oxidized metabolites in rat plasma by use of tandem-mass spectrometry, *Biol. Mass Spectrom.*, 1991, **21**, 17.

Functional Nitrogen Compounds

Cyanamide

Use/occurrence:	Chemical intermediate
Key functional groups:	Nitrile
Test system:	Human (oral and dermal, 0.25 mg kg^{-1}), rat (oral, 10 mg kg^{-1})

Structure and biotransformation pathway

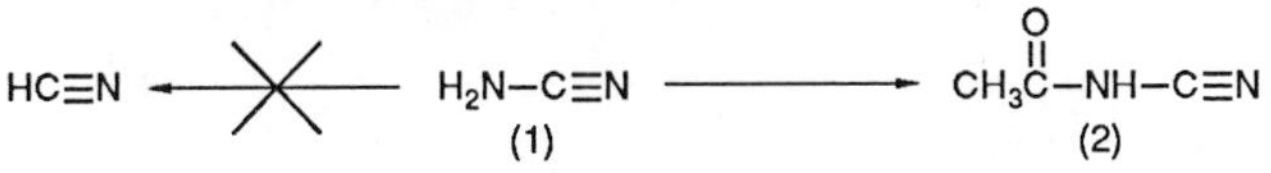

$$HC\equiv N \longleftarrow\!\!\!\times\!\!\!\longrightarrow H_2N-C\equiv N \longrightarrow \underset{(2)}{CH_3\overset{\overset{O}{\|}}{C}-NH-C\equiv N}$$

(1)

An HPLC method was developed to determine acetylcyanamide (2) in urine. In rats 45.6% of an oral dose of cyanamide (1) was recovered in urine as acetylcyanamide (2). In human volunteers 40% of an oral dose of cyanamide (1) was recovered in urine as acetylcyanamide (2). Following dermal administration of cyanamide (1) to the volunteers only 7.7% of the available dose was recovered in urine as acetylcyanamide (2). Specific assay of urinary thiocyanate or blood cyanide failed to demonstrate any treatment-related increase in concentration. It was concluded that *in vivo* metabolism to cyanide is irrelevant in man.

Reference

B. Mertschenk, W. Bornemann, J. G. Filser, L. von Meyer, U. Rust, J. C. Schneidr, and Ch. Gloxhuber, Urinary excretion of acetylcyanamide in rat and human after oral and dermal application of hydrogen cyanamide (H$_2$NCN), *Arch. Toxicol.*, 1991, **65**, 268.

p-Nitrophenyl azide, *p*-Cyanophenyl azide, *p*-Chlorophenyl azide, *p*-Methoxyphenyl azide, Phenyl azide, Phenethyl azide

Use/occurrence:	Model compounds
Key functional groups:	Azide
Test system:	Mouse liver microsomes, mouse hepatocytes

Structure and biotransformation pathway

Substrate	Metabolite (anaerobic)	Metabolite (aerobic)
R = H (1)	ND	(10)
Cl (2)	(7)	
OMe (3)	ND	(10)
CN (4)	(8)	
NO$_2$ (5)	(9)	
(6)	ND	

ND = not detected

Little is known about the metabolism of the azide group, which has recently attracted interest in drug design following the development of 3′-azido-3′-deoxythymidine as an antiviral agent. Compounds (1)–(6) were incubated in the dark with either mouse hepatic microsomes (in the presence of a NADPH generating system) or mouse hepatocytes under an atmosphere of either air or nitrogen.

Metabolites were separated by HPLC. On incubation of (2), (4), and (5) with microsomes under anaerobic (but not aerobic) conditions, the corresponding aromatic amine metabolites, (7), (8), and (9) respectively, were formed. Formation of the amine (9) from *p*-nitrophenyl azide (5) was *ca.* 20-fold more rapid than the metabolism of the corresponding cyano- (4) or chloro- (2) substrates to their respective amines. Under similar anaerobic conditions, compounds (1), (3), and (6) did not produce any detectable metabolite in the presence of hepatic microsomes. When (1) and (3) were incubated with hepatocytes or microsomes under aerobic conditions, *p*-hydroxyphenyl azide (10) was formed.

The results of this study showed that the reduction to the corresponding amine was predominantly enzyme-mediated (heat-treated microsomes did not catalyse the reaction) and required the presence of NADPH and the absence of oxygen. The data further suggest that for azido-substituted compounds where the azido group is not subject to an electron-withdrawing environment, bioreductive metabolism is unlikely to be a major route of metabolism of azido-substituted compounds.

Reference

D. Nicholls, A. Gescher, and R. J. Griffin, Medicinal azides. Part 8. The *in vitro* metabolism of *p*-substituted phenyl azides, *Xenobiotica*, 1991, **21**, 935.

Phenelzine

Use/occurrence:	Antidepressant drug
Key functional groups:	Hydrazine
Test system:	Bovine adrenal methyl transferase

Structure and biotransformation pathway

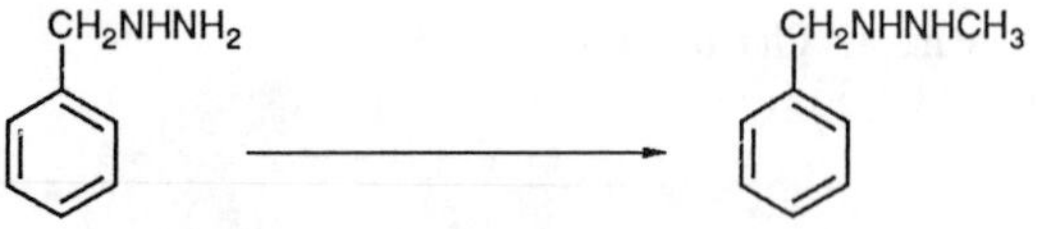

CH$_2$NHNH$_2$

(1)

CH$_2$NHNHCH$_3$

(2)

Phenelzine (1) was found to be methylated by enzymes obtained from bovine adrenal and some rat tissues in the presence of S-adenosylmethionine as methyl group donor. Methylation occurred at the terminal nitrogen and the metabolite (2) was identified by TLC and HPLC comparison with a chemically synthesized standard and by GC–MS as its pentafluoropropionyl derivative.

Reference

P. H. Yu, B. A. Davis, and D. A. Durden, Enzymatic N-methylation of phenelzine catalysed by methyltransferases from adrenal and other tissues, *Drug Metab. Dispos.*, 1991, **19**, 836.

Cyclohexanone oxime

Use/occurrence:	Industrial chemical
Key functional groups:	Cyclohexyl, oxime
Test system:	Rat (oral, 1, 10, 30 mg kg^{-1}; dermal, 30 mg kg^{-1}), rat liver S9 and microsomes

Structure and biotransformation pathway

Following oral administration of [^{14}C]cyclohexanone oxime (1), 68–87% dose was recovered in urine in the first 24 hours and 5–9% in faeces. About 2% dose was excreted as $^{14}CO_2$ in expired air. Urine contained three major metabolites, which were characterized, using ^{1}H NMR and enzymatic deconjugation, as cyclohexyl glucuronide (7) and monoglucuronide conjugates (8) and (6) of *cis-* and *trans-*cyclohexane-1,2-diol. Cyclohexanone (2) was detected by GC analysis of a dichloromethane extract of urine. After dermal application of (1) a large proportion of the dose volatilized from the skin. About 4% dose was excreted in urine, which contained the same profile of conjugated metabolites observed after oral dosing.

In the *in vitro* experiments, components with the same retention times as cyclohexanone (2), cyclohexanol (4), and the 1,2-diols (3) and (5) were detected by GC in extracts of S9 preparations, but only cyclohexanone was detected when microsomes were used. Cyclohexanone could evidently be formed by chemical hydrolysis under the incubation conditions.

Reference

D. Parmer and L. T. Burka, Metabolism and disposition of cyclohexanone oxime in male F-344 rats, *Drug Metab. Dispos.*, 1991, **19**, 1101.

N-Methyl-*N*-formylhydrazine

Use/occurrence:	Component of edible mushrooms
Key functional groups:	Formyl, hydrazine
Test system:	Rat liver microsomes

Structure and biotransformation pathway

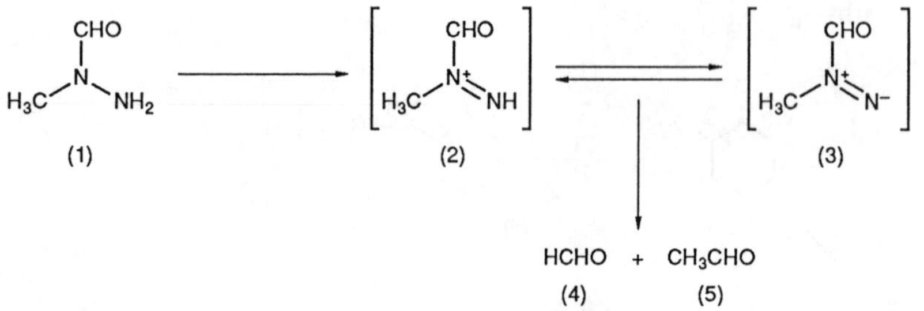

To obtain an initial insight into the possible metabolism of *N*-methyl-*N*-formylhydrazine (1) the authors studied its oxidative chemistry using principally mercuric oxide as the oxidizing agent. Both chemical and microsomal oxidation of (1) yielded formaldehyde (4) and acetaldehyde (5) although the chemical reaction yielded other products as well. The aldehydes were quantified by HPLC after derivatization with 2,4-dinitrophenylhydrazine. The authors proposed that the aldehydes were formed via an intermediate diazenium ion (2) or diazene (3) which could fragment to yield formyl and methyl radicals. It was suggested that these radical intermediates could be involved in the carcinogenesis of (1).

Reference

P. M. Gannett, C. Garrett, T. Lawson, and B. Toth, Chemical oxidation and metabolism of *N*-methyl-*N*-formylhydrazine. Evidence for diazenium and radical intermediates, *Food Chem. Toxicol.*, 1991, **29**, 49.

2-Nitropropane, Propane-2-nitronate

Use/occurrence:	Model compounds
Key functional groups:	Nitroalkane
Test system:	Mouse hepatocytes and liver microsomes (0–10 mM substrate)

Structure and biotransformation pathway

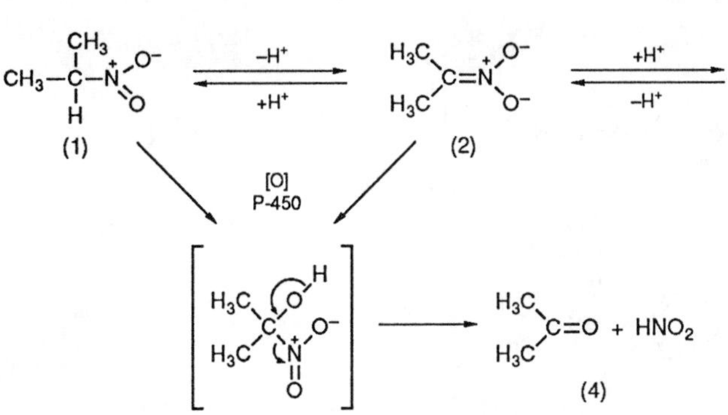

Substrate binding of (1) and (2) by mouse liver microsomes was followed by recording optical difference spectra. A colorimetric assay was used to determine nitrate, the oxidative denitrification of (2) was measured using an ion pair HPLC method, and acetone (4) was determined by a GC assay. Metabolism of either (1) or (2) was dependent on the presence of viable microsomes and of NADPH, and was inhibited by P-450 inhibitors. The cytotoxicity of both (1) and (2) was determined in mouse liver hepatocytes; both had low toxicity as concentrations of 20 mM were required to elicit a response.

Reference

R. Dayal, B. Goodwin, I. Linhart, K. Mynett, and A. Gescher, Oxidative denitrification of 2-nitropropane and propane-2-nitronate by mouse liver microsomes: Lack of correlation with hepatotoxic potential, *Chem. Biol. Interact.*, 1991, **79**, 103.

1-(Pyridin-3-yl)-3,3-dimethyltriazene,
1-(2-Methoxypyridin-5-yl)-3,3-
dimethyltriazine, 1-(Pyridin-3-yl)-3-
ethyl-3-methyltriazene

Use/occurrence:	Model compounds
Key functional groups:	Methyltriazine
Test system:	Rat hepatocytes

Structure and biotransformation pathway

	X	R^1	R^2
(1) a	H	CH_3	CH_3
b	H	CD_3	CD_3
c	H	CH_3	CD_3
(2)	CH_3O	CH_3	CH_3
(3)	H	CH_3	C_2H_5

Reaction kinetics for the *N*-demethylation of (1)–(4) were followed by HPLC analysis of the triazenes (4). Formaldehyde (5) [and/or acetaldehyde for (3)] and/or its deuterated analogues were identified and measured as the corresponding 2,4-dinitrophenylhydrazones. The K_m and V_{max} were determined for each substrate, (1), (2), and (3). As expected a clear deuterium isotope effect was demonstrated. Interestingly for the ethyl analogue (3) loss of the ethyl group was favoured over the loss of the methyl group. Depending on substrate concentration the loss of the ethyl group was 2–4 times greater than loss of the methyl group.

Reference

J. Iley and G. Ruecroft, Mechanism of the microsomal demethylation of 1-aryl-3,3-dimethyltriazines, *Biochem. Pharmacol.*, 1990, **40**, 2123.

Nitrosamines

N-Nitrosodimethylamine

Use/occurrence:	Model compound
Key functional groups:	Nitrosamine, *N*-methylnitrosamine
Test system:	Rat liver microsomes (0.09–2.4 mM substrate)

Structure and biotransformation pathway

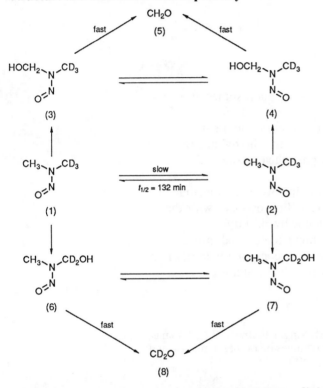

The possibility that *N*-nitrosodimethylamine [(1) and (2)] might be metabolized preferentially at either the *syn*- (relative to the nitroso group) or the *anti*-methyl group was examined by studying the metabolism of the mono-C^2H_3 isomers or the fully deuterated analogue (not shown) by comparing rates of formaldehyde production [(5) and (8)]. The formaldehyde was analysed for its deuterium content by MS of the dimedone derivative. Although a strong preference for attack at C^1H_3 *versus* C^2H_3 regardless of stereochemistry was observed there was little evidence of regioselectivity in the metabolism of (1).

Reference

L. K. Keefer, M. B. Kroeger-Koepke, H. Ishizaki, C. J. Michejda, J. E. Saavedra, J. A. Hrabie, C. S. Yang, and P. P. Roller, Stereoselectivity in the microsomal conversion of *N*-nitrosodimethylamine to formaldehyde, *Chem. Res. Toxicol.*, 1990, **3**, 540.

N-Nitrosodimethylamine, *N*-Nitrosodiethylamine

Use/occurrence:	Environmental carcinogens
Key functional groups:	Nitrosamine
Test system:	Rat nasal mucosa microsomes

Structure and biotransformation pathway

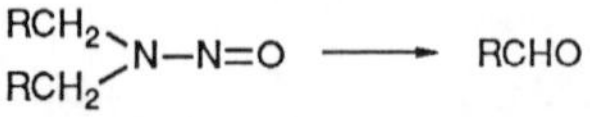

$$\begin{array}{c} RCH_2 \\ RCH_2 \end{array}\!\!>\!N\!-\!N\!=\!O \longrightarrow RCHO$$

(1) R = H or CH₃   (2)

N-Nitrosodialkylamines (1) were incubated with microsomes prepared from rat nasal mucosa. The levels of formaldehyde (2; R = H) produced from the dimethyl analogue were determined using a radiometric assay, and the levels of acetaldehyde (2; R = CH₃) produced from the diethyl analogue by derivatization to the corresponding 2,4-dinitrophenylhydrazone followed by HPLC.

The results showed a higher rate of metabolism in 9 week old rats than in 4 week old animals, but no sex-related differences. Pretreatment with diallyl sulfide decreased the metabolism of the nitrosamine by 60–80%.

Formaldehyde and acetaldehyde have been shown to be nasal carcinogens. The ability of rat nasal microsomes to produce these compounds *in situ* may be central in the carcinogenic effect of the *N*-nitrosodialkylamines.

Reference

J.-Y. Hong, T. Smith, M.-J. Lee, W. Li, B.-L. Ma, S. M. Ning, J. F. Brady, P. E. Thomas, and C. S. Yang, Metabolism of carcinogenic nitrosamines by rat nasal mucosa and the effect of diallyl sulfide, *Cancer Res.*, 1991, **51**, 1509.

Azoxymethane, Methylazoxymethanol, N-Nitrosodimethylamine

Use/occurrence: Model compounds

Key functional groups: Azoxy, nitrosamine, N-methylnitrosamine

Test system: Rat liver microsomes and reconstituted P-450 IIE1 (1 mM substrate)

Structure and biotransformation pathway

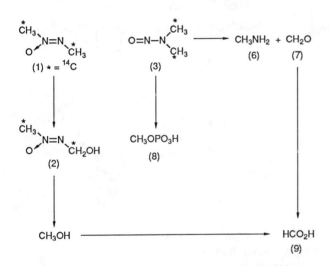

Methylazoxymethanol (2) was detected as a product of microsomal metabolism of azoxymethane (1). In a separate experiment methanol (4) and formic acid (5) were detected as a product of microsomal metabolism of methylazoxymethanol (2). Metabolites of the nitrosamine (3) were methylamine (6), formaldehyde (7), methanol (4), methyl phosphate (8), and formic acid (9). Incubations carried out in the presence of a monoclonal antibody to cytochrome P-450 IIE1 resulted in an 85–90% inhibition of all three reactions, thus providing evidence that all three substrates are metabolized by the same form of cytochrome P-450 in the rat. Metabolites were measured by a radiochemical HPLC assay.

Reference

O. Soon Sohn, H. Ishizaki, C. S. Yang, and E. S. Fiala, Metabolism of azoxymethane, methylazoxymethanol, and N-nitrosodimethylamine by cytochrome P450IIE1, *Carcinogenesis*, 1991, **12**, 127.

4-(Methylnitrosamino)-1-(3-pyridyl)-butan-1-one

Use/occurrence:	Tobacco carcinogen
Key functional groups:	Aryl ketone, *N*-methyl-nitrosamine, nitrosamine, pyridine
Test system:	Hamster lung tissues

Structure and biotransformation pathway

The effect of *S*-(−)-nicotine upon the metabolism of 4-(methylnitros-amino)-1-(3-pyridyl)butan-1-one (1) has been studied using hamster lung explants. *S*-(−)-nicotine was found to have no significant effect when present at equimolar concentrations with (1). As the concentration of *S*-(−)-nicotine was increased, there was an inhibition of the metabolic activation of (1) by α-carbon hydroxylation, which leads to metabolites (2)–(4), and also of pyridine *N*-oxidation leading to metabolites (5) and (6), by up to 90%. The formation of metabolite (7) by carbonyl reduction, however, was not affected by the presence of *S*-(−)-nicotine.

Administration of *S*-(−)-nicotine to hamsters also slowed the metabolism of (1) in the liver reducing the levels of the major metabolic product (4) and the other hydroxylated metabolites (2), (3), and (8).

Reference

H. M. Schuller, A. Castonguay, M. Orloff, and G. Rossignol, Modulation of the uptake and metabolism of 4-(methylnitrosamino)-1-(3-pyridyl)-1-butanone by nicotine in hamster lung, *Cancer Res.*, 1991, **51**, 2009.

N-Nitroso-N-methylaniline

Use/occurrence:	Model compound
Key functional groups:	Aryl amine, N-methylnitrosamine, nitrosamine
Test system:	Mouse liver microsomes (1 and 10 mM substrate)

Structure and biotransformation pathway

After incubation of 1 mM (1) with liver microsomes from phenobarbitone-pretreated mice for 15 minutes the main metabolite was (4) with a 15% yield in relation to the substrate concentration. Comparatively small amounts of (2) and (3) were also detected. The possible metabolite N-methyl-N-phenyl-hydroxylamine could not be detected. Owing to its chemical instability (6) could not be detected by the HPLC methods used to detect the other metabolites, but it was detected by a colorimetric method. Phenol (5) was also detected. A mechanism (not shown) was proposed involving the concurrent generation of nitric oxide and primary amines during the reductive denitrosation of (1). However, it was concluded that this study neither proved nor excluded the proposed mechanism.

Reference

T. Scheper, K. E. Appel, W. Schunack, A. Somogyi, and A. Hildebrandt, Metabolic denitrosation of N-nitroso-N-methylaniline: Detection of amine-metabolites, *Chem. Biol. Interact.*, 1991, **77**, 81.

Amino Acids and Peptides

S-Carboxymethyl-L-cysteine

Use/occurrence:	Mucolytic agent
Key functional groups:	Amino acid, cysteine
Test system:	Human volunteers (oral, 375–1500 mg)

Structure and biotransformation pathway

$HO_2C\overset{*}{C}H_2SCH_2CHCO_2H$ — $HO_2CCH_2SCH_2CHCO_2H$

NH_2 (1) * = ^{13}C

OH (4)

$HO_2CCH_2SCH_2CO_2H$ — $HO_2CCH_2SCH_2CO_2H$

(2)

O (3)

In addition to its therapeutic use, S-carboxymethyl-L-cysteine (1) has also been used as a probe drug for investigating polymorphic sulfoxidation of other sulfur-containing drugs in humans. However, recent investigations have questioned the metabolic basis of this usage. The objective of this study was to investigate the formation of nitrogen-free acidic metabolites of (1), which would escape detection by analytical methods used in sulfoxidation phenotyping. Using GC–MS, three metabolites of this nature were identified, thiodiglycolic acid (2) (mean 19.8% dose/24 hours), thiodiglycolic acid sulfoxide (3) (mean 13.3% dose/24 hours), and (3-carboxymethylthio)lactic acid (4) (mean 2.1% dose/8 hours). Trace amounts of endogenous (2) and (4) were also detected in urine from subjects not administered (1). Quantification of metabolites was based on GC–MS using ^{2}H- and ^{13}C-labelled internal standards. No significant intra-individual correlation was observed in formation of (2) and (3). In view of these results the authors questioned the existence of polymorphic sulfoxidation of (1) in humans.

Reference

U. Hofmann, M. Eichelbaum, S. Seefried, and C. O. Meese, Identification of thiodiglycolic acid, thiodiglycolic acid sulphoxide and (3-carboxymethylthio)lactic acid as major human biotransformation products of S-carboxymethyl-L-cysteine, *Drug Metab. Dispos.*, 1991, **19**, 222.

S-Benzyl-N-malonyl-L-cysteine

Use/occurrence: Model xenobiotic plant metabolite

Key functional groups: Alkyl amide, cysteine

Test system: Rat (oral, 10 mg kg^{-1})

Structure and biotransformation pathway

There has been an increasing interest in the fate in mammals of pesticide metabolites derived from plants. The aim of this study was to investigate the fate of a model phytomercapturic acid, S-benzyl-N-malonyl-L-cysteine (1), following oral administration to the male rat. The molecule was labelled with either ^{14}C or ^{13}C and metabolites were separated by TLC and HPLC. Metabolite identification was by co-chromatography with authentic standards and by EI-MS, CI-MS, and NMR.

Elimination of ^{14}C following an oral dose of (1) to rats was mainly via the urine. 66% of the dose was excreted by this route during the first 24 hours and a total of 79% within 72 hours of the dose; the remainder was excreted in the faeces, mostly in the first 24 hours. The major metabolite seen in the 0–24 hour urine was hippuric acid (2) which represented 50% of the excreted radiolabelled material in the sample. Other metabolites seen were benzyl methyl sulfoxide (3), which together with unchanged (1) accounted for 11% of urinary ^{14}C, and the corresponding sulfone (4) representing ca. 3% of urinary ^{14}C. The remaining polar metabolites (30% of urinary ^{14}C) were unidentified but the data suggested that some may have been glucuronides.

370

The results indicate that the *N*-malonyl bond in (1) is biolabile. The result-ant cysteine conjugate is thought to be metabolized by cysteine conjugate C-S-lyase to benzyl thiol. This compound would then be methylated by a methyl transferase enzyme to yield metabolites (3) and (4). However, the major route from benzyl thiol would yield the glycine conjugate, hippuric acid (2). A comparison of the results of this study with those from experiments with *N*-acetyl-*S*-benzyl-L-cysteine and *S*-benzyl-L-cysteine support the pro-posed routes of biotransformation.

Reference

K. A. Richardson, V. T. Edwards, B. C. Jones, and D. H. Hutson, Metabolism in the rat of a model xenobiotic plant metabolite *S*-benzyl-*N*-malonyl-L-cysteine, *Xeno-biotica*, 1991, **21**, 371.

Z-103

Use/occurrence: Anti-ulcer agent

Key functional groups: Amino acid, carnosine

Test system: Rat (oral, 3–100 mg kg^{-1}), rat tissue homogenates

Structure and biotransformation pathway

(1) * = ^{14}C

(2)

(3)

(4)

(5)

During 5 days after single oral doses of $[^{14}C]$-Z-103 (1) to rats at a dose level of 50 mg kg^{-1}, 4%, 13%, and 39% of the dose were excreted in urine, faeces, and expired air respectively, and a further 39% dose remained in the carcasses; biliary excretion represented 4% of the dose. The biotransformation of (1) was investigated *in vitro* by incubating it with rat liver S9 fraction and small intestine homogenates, and determining the resultant metabolite profiles by HPLC. In this way it was shown that the organic component of (1) was converted into L-carnosine (2), which was further metabolized in both of these tissues to L-histidine (3) and β-alanine (4). The cumulative excretion of radioactivity in urine, faeces, and expired air after 21 consecutive daily oral doses of $[^{14}C]$-(1) to rats were similar to those found after single doses, and it appeared that zinc did not accumulate in the body after repeated administration of (1). TLC analysis of acid-hydrolysed tissue homogenates from these rats indicated the presence of only one major radioactive component, and this was shown to be (3) by co-chromatography with the reference compound; glutamic acid (5) was similarly identified as a minor radioactive component in certain tissues. After single oral doses of $[^{65}Zn]$-Z-103 to rats, excretion in urine and faeces accounted for < 1% and 85% of the dose respectively, and 11% dose was present in the carcasses at 120 hours; biliary excretion represented < 1% of the dose. High concentrations of ^{65}Zn radioactivity were found in the gastrointestinal contents, liver, pancreas, and kidney, which was presumed to be related to the fact that zinc is an essential element. It was concluded that most of the dose of Z-103 dissociated to zinc and L-carnosine, and that the former was metabolized with the endogenous metal, whereas the latter was metabolized and utilized in the synthesis of endogenous high molecular weight compounds, such as proteins.

References

H. Sano, S. Furuta, S. Toyama, M. Miwa, Y. Ikeda, M. Suzuki, H. Sato, and K. Matsuda, Study on the metabolic fate of *catena*-(S)-$[\mu$-$[N^{\alpha}$-(3-aminopropionyl)-histidinato$(2-)$-N^1,N^2,O:$N^{\tau}]$-zinc], *Arzneim.-Forsch.*, 1991, **41**, 965.

S. Toyama, S. Furuta, M. Miwa, M. Suzuki, H. Sano, and K. Matsuda, Study on the metabolic fate of *catena*-(S)-$[\mu$-$[N^{\alpha}$-(3-aminopropionyl)histidinato$(2-)$-N^1,N^2, O:$N^{\tau}]$-zinc], *Arzneim.-Forsch.*, 1991, **41**, 976.

MDL 27210

Use/occurrence:	Angiotensin converting enzyme inhibitor
Key functional groups:	Alkyl ester, isoalkyl carboxylic acid
Test system:	Dog, monkey (intravenous, 3 mg kg^{-1})

Structure and biotransformation pathway

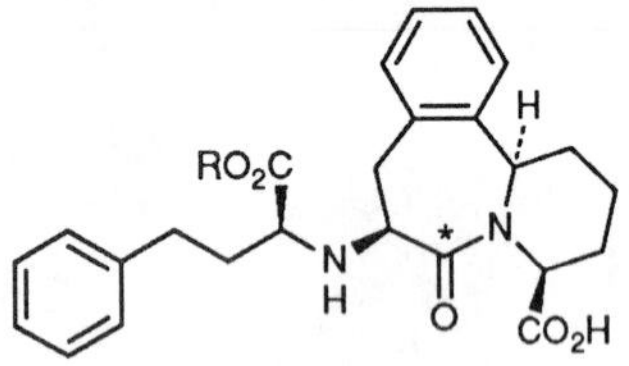

$* = {}^{14}C$
(1) R = C$_2$H$_5$
(2) R = H

After intravenous administration of [^{14}C]-MDL 27210 (1) monkeys excreted 52% dose in faeces and 41% in urine. Dogs excreted 80% dose in faeces and 14% in urine. Faecal excretion in the monkey was more prolonged, but otherwise excretion was largely complete within 24 hours of dosing. MDL 27210 was metabolized by ester hydrolysis to the acid (2) which is the active drug. In the dog, the metabolite (2) was the only radiolabelled component detected in 0–24 hour urine and faecal extracts. In the monkey, (2) was the only metabolite in faecal extracts and represented at least 90% of urinary radioactivity. Two minor unidentified metabolites were detected in monkey urine. Metabolite (2) was isolated from both monkey and dog urine and identified by CI-MS–MS.

Reference

H. Cheng, J. D. Stuhler, B. R. Dorrbecker, and W. P. Gordon, Disposition and metabolism of the angiotensin-converting enzyme inhibitor [4S-[4α,7α,(R*), 12bβ]]-7-(S-(1-ethoxycarbonyl-3-phenylpropyl)amino]-1,2,3,4,6,7,8,12b-octa-hydro-6-oxo-pyrido[2,1-a][2]benzazepine-4-carboxylic acid in monkeys and dogs, *Drug Metab. Dispos.*, 1991, **19**, 212.

Captopril

Use/occurrence:	ACE inhibitor (antihypertensive)
Key functional groups:	Alkyl thiol
Test system:	Human liver, renal cortex, renal medulla, and intestinal (ileum) mucosa microsomal fractions

Structure and biotransformation pathway

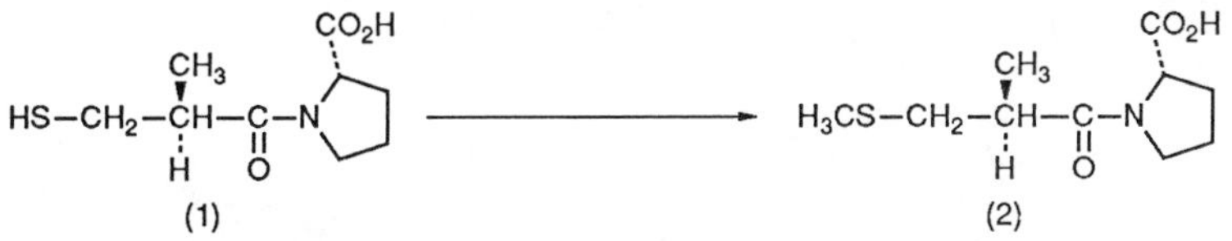

Captopril (1) is extensively metabolized in man, but little is known about the metabolism of (1) by human tissues. The methyl derivative (2) is a known human metabolite and although the contribution of the methylation pathway to the overall metabolism of (1) is unknown, S-methylation is an important pathway. Methylation of (1) by microsomal fractions from several human tissues was studied in the presence of S-adenosyl-L-[methyl-^{14}C]methionine. The results showed that liver is the primary site of methylation of (1) whereas the intestine plays only a minor role. The kidney may also contribute substantially to the hepatic methylation of (1), with renal cortex being more active than renal medulla.

Reference

G. M. Pacifici, S. Santerini, and L. Giuliani, Methylation of captopril in human liver, kidney and intestine, *Xenobiotica*, 1991, **21**, 1107.

Loxistatin

Use/occurrence: Proteolytic enzyme inhibitor

Key functional groups: Alkyl amide, alkyl ester, epoxide

Test system: Rat (oral, 50 mg kg^{-1})

Structure and biotransformation pathway

(1) * = ^{14}C

(2)

(3)

(4)

The rat urinary metabolites of [^{14}C]loxistatin (1), which is an ester pro-drug of the pharmacologically active form (2), have been investigated. Metabolites were isolated from acidified desalted urine by ethyl acetate extraction, and then treated with diazomethane to yield methyl ester derivatives. The esters were further purified by column chromatography and preparative TLC before examination by EI- and CI-MS, and by ^{1}H NMR, using synthetic reference compounds for comparison. 16% of the dosed radioactivity was excreted in urine over 24 hours, predominantly as three metabolites (2)–(4), which accounted for *ca.* 8%, 4%, and 4% of the dose respectively. Three other metabolites collectively accounted for less than 1% of the dose, and there was no detectable unchanged (1). Metabolites (3) and (4) represent the result of ω- and (ω − 1)-hydroxylation respectively of the alkyl chain in (2). Although ω-hydroxylation of (1) introduces a new chiral centre into the molecule, NMR comparison of metabolite (3) with synthetic reference compounds suggests that only one diasteromer, with the *S*-configuration, is formed. Metabolite (2) together with (3) and/or (4), was also present in plasma after administration of (1). Moreover, since (3) was about as potent as (2) in terms of papain inhibitory activity, it may contribute to the pharmacological effect observed *in vivo*.

Reference

K. Fukushima, M. Arai, M. Tamai, C. Yokoo, M. Murata, T. Suwa, and T. Satoh, Metabolic fate of loxistatin in rat, *Xenobiotica*, 1990, **20**, 1043.

Loxistatin

Use/occurrence:	Proteolytic enzyme inhibitor
Key functional groups:	Alkyl amide, alkyl ester, epoxide
Test system:	Normal and dystrophic hamster (oral, 5 or 50 mg kg^{-1})

Structure and biotransformation pathway

Loxistatin (1) is an ester pro-drug of loxistanic acid (a potent inhibitor of cysteine proteinases) which may be useful in the treatment of muscular dystrophy. The metabolism of [^{14}C]loxistatin was studied in normal hamsters after oral administration. Most of the total radioactivity (88%) was excreted after 24 hours following an oral dose of 5 mg kg^{-1} and the amounts excreted in urine, faeces, and expired air up to 120 hours were 64%, 28%, and 6% respectively. Biliary excretion of ^{14}C in bile duct-cannulated hamsters accounted for 13.2% of the dose within 24 hours.

Two major metabolites and several minor metabolites (tentatively identified by TLC and co-chromatography with reference standards) were detected in plasma, but unchanged (1) was not detected at any time point studied. The major plasma metabolite was loxistatin free acid (2) which in turn gave rise to two other metabolites (3) and (4) which were hydroxylated in the terminal isopropyl group. All metabolites had a short half-life of *ca.* 1 hour. Previous studies have demonstrated the presence of (2), (3), and (4) in the plasma and urine of rats and that the metabolites were pharmacologically active. The metabolites (2)–(4) were shown to account for the majority of the plasma radiolabel at 0.5 hour after dosing in the hamster and were presumably the form of the compound found in the target tissues (cardiac and skeletal muscle fibres) in both normal and dystrophic hamsters treated with (1) at a dose level of 50 mg kg^{-1}.

Reference

K. Fukushima, Y. Kohno, W. Osabe, T. Suwa, and T. Satoh, The disposition of loxistatin and metabolites in normal and dystrophic hamsters, *Xenobiotica*, 1991, **21**, 23.

RS-26306

Use/occurrence:	LHRH antagonist
Key functional groups:	Peptide
Test system:	Rat (intravenous, 1 mg kg^{-1}; subcutaneous, 1 and 10 mg kg^{-1}), cynomolgus monkey (intravenous, 1 mg kg^{-1}; subcutaneous, 1 mg kg^{-1})

Structure and biotransformation pathway

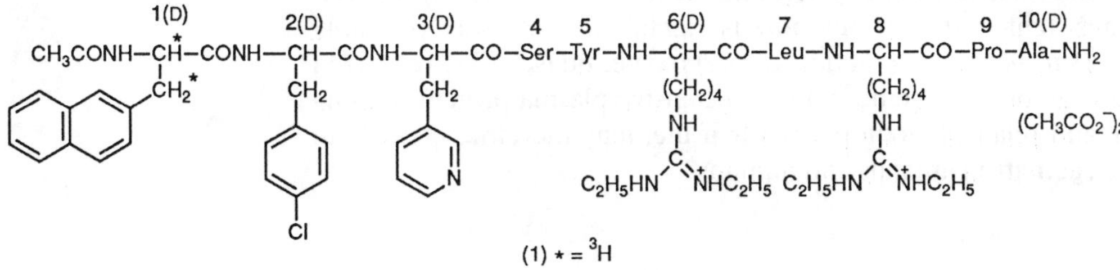

(1) * = ^{3}H

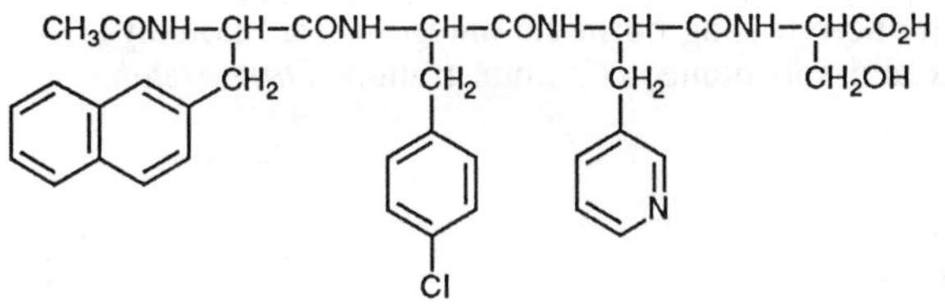

(2)

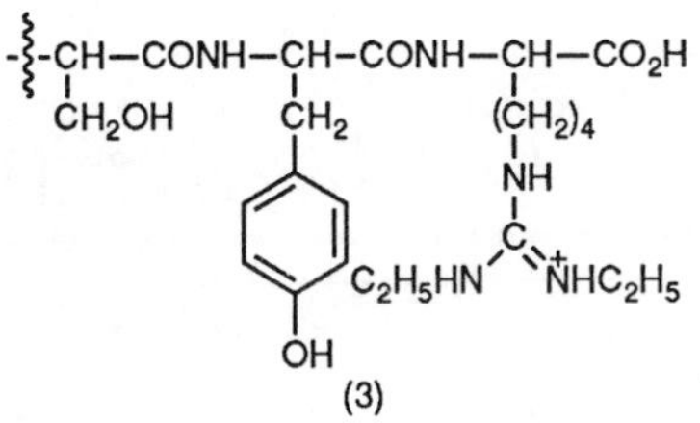

(3)

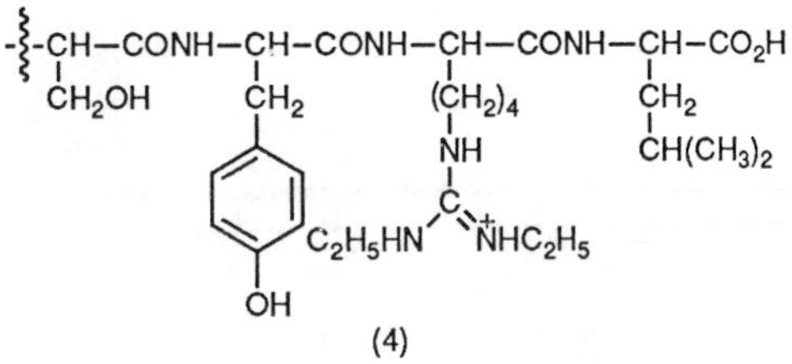

(4)

After intravenous or subcutaneous administration of the synthetic deca-peptide [^{3}H]-RS-26306 (1) to intact rats, 12–21% of the dose was excreted in urine during 7 or 10 days and 66–84% dose was excreted in the faeces, while after subcutaneous administration of (1) to bile duct-cannulated rats about half of the labelled dose was recovered in bile during 72 hours and 12% dose remained at the injection site at the end of this time. In monkeys, 25% of an intravenous dose of (1) was excreted in urine and 62% dose was excreted in the faeces during 7 days, compared with 16% and 55% dose respectively after sucutaneous administration. Radioactive components in plasma, urine, and bile were separated and quantified by radio-HPLC, and the results showed that the former two body fluids from both species contained mostly unchanged drug, although monkey plasma also contained an important metabolite. Only traces of parent compound were found in rat bile, however, where the bulk of the sample radioactivity was in the form of three metabolites, one of which co-chromatographed with the monkey plasma metabolite. The three principal rat biliary metabolites were isolated, further purified by HPLC, and identified as the truncated peptide fragments (2)–(4) by chromatographic and FAB-MS comparison with the synthetic reference compounds. Plasma kinetic data obtained indicated that, for a peptide, (1) possessed a relatively long terminal half-life in rats and monkeys, which was rationalized in terms of restricted cleavage of the five D-amino acids it contains by the endogenous peptidases. The biotransformation results, which showed that the amino acid linkages in (1) that are most susceptible to enzymatic cleavage are the Ser^4-Tyr^5 and Leu^7-$hArg(Et_2)^8$ bonds between two adjacent L-amino acids [thereby yielding metabolites (2) and (4) respectively], are consistent with this hypothesis; metabolite (3) was probably formed from (4) *in vivo* by the action of carboxypeptidase. This restricted enzymatic cleavage of (1), together with its extensive plasma protein binding (*viz.* 82–84%) and relatively long plasma half-life, may therefore contribute to its prolonged activity in animals and humans.

Reference

R. L. Chan, S. C. Hsieh, P. E. Haroldsen, W. Ho, and J. J. Nestor, Disposition of RS-26306, a potent luteinizing hormone-releasing hormone antagonist, in monkeys and rats after single intravenous and subcutaneous administration, *Drug Metab. Dispos.*, 1991, **19**, 858.

Cyclosporine

Use/occurrence:	Immunosuppressant drug
Key functional groups:	Peptide
Test system:	Human (intravenous, 1.9 mg kg^{-1})

Structure and biotransformation pathway

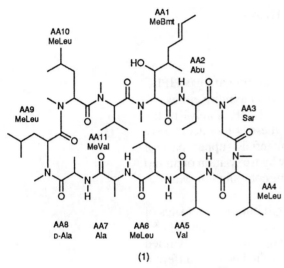

(1)

AA1 to AA11 are the component amino acids.
MeBmt = *N*-methylbut-4-enyl-4-methylthreonine

Known metabolites

Modification of amino acid				Further
1	4	6	9	modifications
+OH			+OH	
	−CH$_3$	+OH	+OH	
+OH			+OH	cyclization
			+OH	
+OH	−CH$_3$			
	−CH$_3$		+OH	
		+OH	+OH	
carboxyl				
+OH				
			+OH	
AM1c +OH				cyclization
	−CH$_3$			

381

New metabolites seen in this study

	Modification of amino acid				Further
	1	4	6	9	modifications
(2)	$+OH$	$-CH_3$			$+2OH$ ⎤
(3)	$+OH$				$+2OH$ ⎥
(4)	$+OH$	$-CH_3$			$+OH$ ⎥
(5)	$+OH$				$+2OH$ ⎥ positions of hydroxylation
(6)		$-CH_3$			$+2OH$ ⎥ undefined
(7)	$+OH$				$+2OH$ ⎥
(8)	$+OH$				$+2OH$ ⎥
(9)	carboxyl				$+OH$ ⎦
(10)	glucuronide				cyclization
(11)	carboxyl				cyclization
(12)	$+OH$	$-CH_3$			
(13)	$+OH$				$+OH (CH_3, AA1)$
(14/ 15)	aldehyde				isomers

Cyclosporine (1) metabolites were isolated by semi-preparative HPLC from human bile taken from liver-grafted patients receiving intravenous (1). Their structures were elucidated by FAB-MS and ^1H NMR. Twelve of the biliary metabolites were known products of (1) and these components were isolated and identified using authentic standard materials (upper part of Table). These metabolites were formed principally by hydroxylation of amino acids (AA) 1, 6, or 9 (in some cases at two sites) in combination with de-methylation at AA4.

Of the new metabolites a number were characterized but the exact sites of metabolism were only partially defined. Trihydroxylated metabolites (3), (5), (7), and (8) with the hydroxy groups present at AA1 and two other undefined sites were observed; the four metabolites were distinguished by their different retention times on HPLC. Other metabolites were two dihydroxylated de-methylated (AA4) metabolites (4) and (6), separable by HPLC; a trihydroxyl-ated (AA1 plus two other sites) demethylated (AA4) metabolite (2); a monohydroxylated (AA1) demethylated (AA4) metabolite (12); two carboxylated metabolites (9) and (11), one of which (9) was hydroxylated at an unknown site; a glucuronide (10) (which yielded the previously known metabolite AM1c with a single hydroxy group at AA1) following treatment with β-glucuronidase; a dihydroxylated (two positions at AA1) metabolite (13); and a metabolite with an aldehyde function at AA1 which had two isomeric forms (14/15). The isomerism of (14/15) was thought to be caused by conjugation of the aldehyde group with the double bond between C-6 and C-7 of AA1.

Reference

U. Christians, S. Strohmeyer, R. Kownatzki, H.-M. Schiebel, J. Bleck, J. Greipel, K. Kolhaw, R. Schottmann, and K.-Fr. Sewing, Investigations on the metabolic path-ways of cyclosporine: I. Excretion of cyclosporine and its metabolites in human bile — isolation of 12 new cyclosporine metabolites, *Xenobiotica*, 1991, **21**, 1185.

Steroids

3,7-Dioxo-5β-cholanic acid

Use/occurrence:	Choleretic agent
Key functional groups:	Steroid
Test system:	Guinea-pig, hamster, rabbit, rat (intraduodenal, 1, 100 mg kg^{-1})

Structure and biotransformation pathway

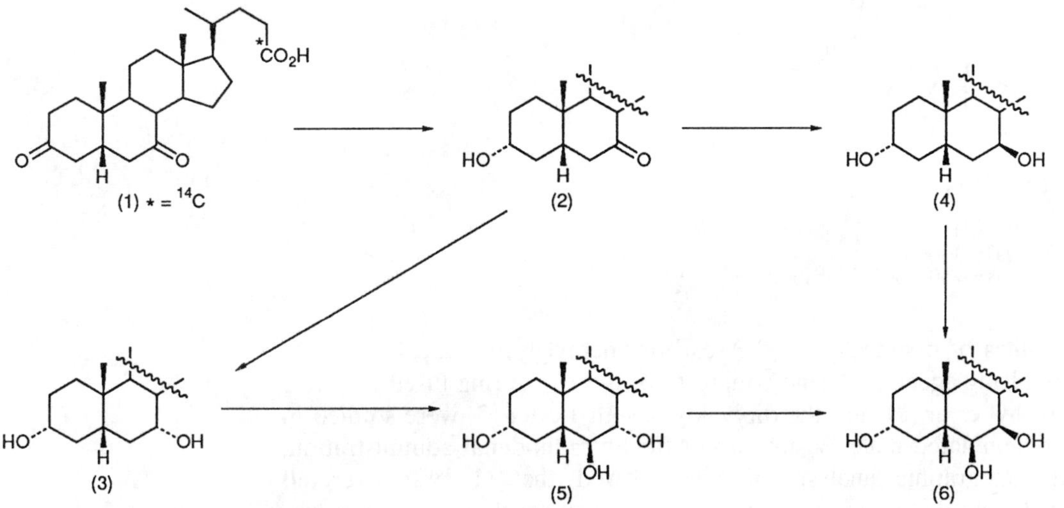

(1) * = ^{14}C   (2)   (4)

(3)   (5)   (6)

There were species differences in the excretion of conjugated metabolites of the dioxocholanic acid (1) in bile. In hamsters and guinea-pigs most of the metabolites were excreted as taurine or glycine conjugates while in rats only taurine conjugates were detected and in rabbits only glycine conjugates. The proportion of unconjugated metabolites increased at the higher dose to 17–29%. The major metabolites were ketolithocholic acid (2), chenodeoxycholic acid (3), and urodeoxycholic acid (4). In rats the dihydroxycholic acids were further metabolized to the α- (5) and β-muricholic acid (6).

Reference

S. Miki, Y. Asanuma, M. Une, and T. Hoshita, *J. Pharmacobio.-Dyn.*, 1990, **13**, 637.

Ursodeoxycholyl-*N*-carboxymethylglycine

Use/occurrence: Cholelitholytic, calcium gallstone dissolving agent

Key functional groups: Alkyl ester, alkyl carboxylate

Test system: Rat (oral 10 mg kg^{-1})

Structure and biotransformation pathway

(1) R = H
(2) R = C$_2$H$_5$
(3) R = CH(CH$_3$)CO$_2$C(CH$_3$)$_3$

Metabolites of ursodeoxycholyl-*N*-carboxymethylglycine, a calcium gallstone dissolving agent, a glycine conjugate of the *cis*-AB-ring fused bile acid (1), its diethyl ester (2), and the dipivaloyloxyethyl ester (3) were studied in bile duct-cannulated male Wistar rats. After intraduodenal administration, radiochromatographic analysis of bile showed that (1) was excreted unchanged. The diethyl ester (2) was not found intact in the bile, but was present as the monoester (80%) and the diacid (1) (20%). The dipivaloyloxyethyl ester (3) was rapidly recovered in the bile as the diacid (1) (> 80%). The intact diester (3) was not detected, but small amounts of the monoester were detected. The efficient conversion of (3) into (1) means that (3) may be a useful pro-drug in calcium gallstone dissolution therapy. The cholic acid amides (1), (2), and (3) were analysed by TLC and ^{1}H NMR.

Reference

S. Hatono, H. Yoshida, M. Matsunami, Y. Ide, K. Matsuda, T. Yatsunami, T. Fuwa, K. Kihira, T. Kuramoto, and T. Hoshita, Absorption, biliary excretion, and metabolism of a new cholelitholytic agent, ursodeoxycholyl *N*-carboxymethylglycine and its esters in rats, *J. Pharmacobio.-Dyn.*, 1991, **14**, 561.

Methenolone

Use/occurrence:	Anabolic steroid
Key functional groups:	Allylic methyl, steroid
Test system:	Human (oral, 50 mg)

Structure and biotransformation pathway

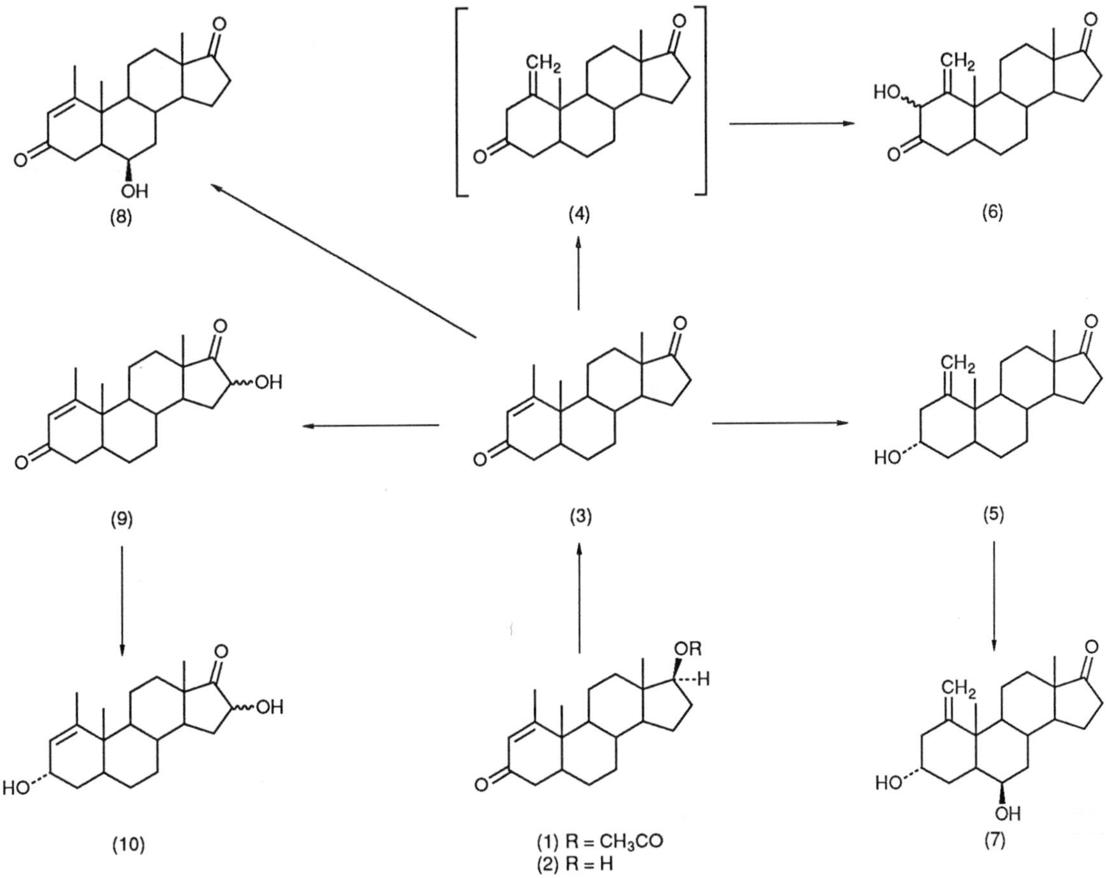

Urine samples were collected during 7 days after single oral doses of methenolone (1) and metabolites were extracted with ether before and after treatment with β-glucuronidase and sulfatase. The extracts were analysed by GC–MS after trimethylsilylation. About 1.6% of the dose was excreted as unchanged methenolone. Up to nine metabolites were detected, excreted mainly as glucuronides, while three metabolites were also excreted as sulfates. Only one metabolite (3) was identified in the unconjugated fraction. A series of metabolites (5)–(7) was derived by initial formation of an exocyclic methylene group. It is proposed that these metabolites are formed by rearrangement to (4) followed by hydroxylation (6) or reduction of the keto group (5). Other

387

metabolites (8)–(10) were also formed by oxidation of the 17β-hydroxy group and hydroxylation at positions-6 and -16. The total excretion of methenolone and the major detected metabolite accounted for less than 5% of the dose, indicating either that urinary excretion was not the major route of excretion or that more polar metabolites were not extracted.

Reference

D. Goudreault and R. Massé, Studies on anabolic steroids — 4. Identification of new urinary metabolites of methenolone acetate (Primoblan^R) in human by gas chromatography/mass spectrometry, *J. Steroid Biochem. Mol. Biol.*, 1990, **37**, 137.

Medroxyprogesterone acetate

Use/occurrence: Fertility control agent, cancer therapy

Key functional groups: Methyl ketone

Test system: Human patients

Structure and biotransformation pathway

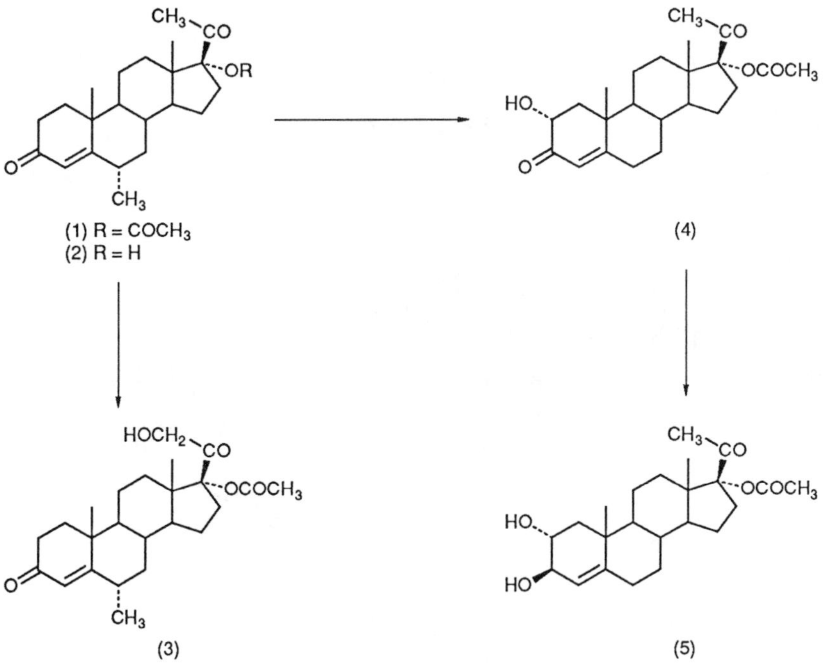

(1) R = COCH₃
(2) R = H

(3)

(4)

(5)

Plasma and urine samples were collected from long-term patients receiving high doses of medroxyprogesterone acetate (1). Plasma samples were extracted with ether and analysed by HPLC. GC–MS was performed after formation of methoxime trimethylsilyl derivatives. Structures were assigned on the basis of mass spectra and in some cases NMR. The identified metabolites were formed by hydrolysis to give (2), hydroxylation at positions-21 and -2 to give (3) and (4) respectively, and reduction of the 3-keto group (5). Other hydroxylated metabolites were also tentatively identified.

Reference

G. Sturm, H. Haberlein, T. Bauer, T. Plaum, and D. J. Stalker, Mass spectrometric and high-performance liquid chromatographic studies of medroxyprogesterone acetate metabolites in human plasma, *J. Chromatogr. Biomed. Appl.*, 1991, **562**, 351.

Boldenone

Use/occurrence:	Veterinary anabolic steroid (drug of abuse)
Key functional groups:	Steroid
Test system:	Human volunteer (oral, 0.2–1 mg kg^{-1})

Structure and biotransformation pathway

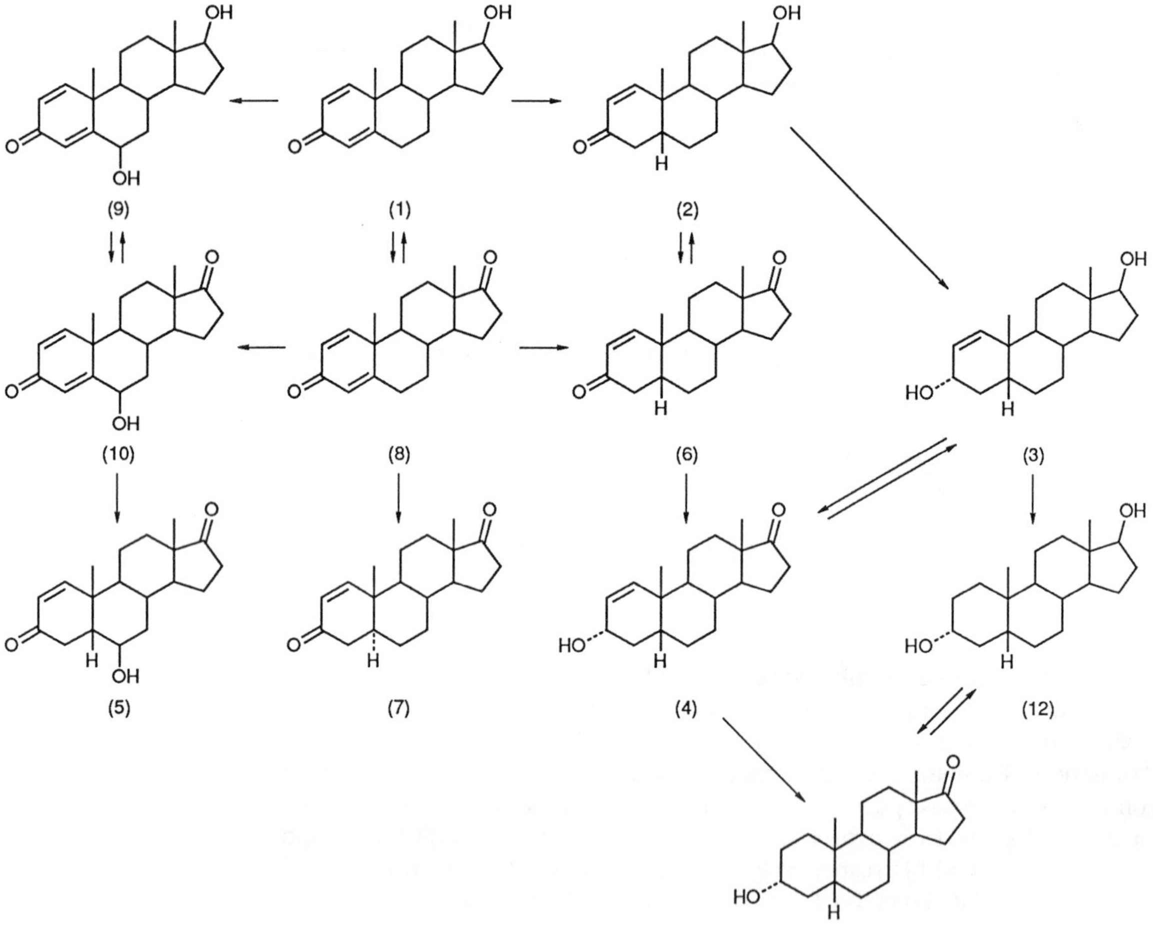

Urinary metabolites of (1) were isolated from urine following XAD-2 adsorption and enzymic hydrolysis with β-glucuronidase. The isolated metabolites were derivatized with N-methyl-N-trimethylsilyltrifluoro-acetamide/trimethyliodosilane and analysed by GC–MS. Boldenone (1) and four metabolites (2)–(5) were identified after hydrolysis of the urine. Five

390

other metabolites (6)–(10) were identified without enzymic hydrolysis. More than 95% of the metabolites were excreted as stable conjugates. The identification of the metabolites was based on their GC retention index, HPLC retention, EI mass spectrum, and chemical reactivity. Ten of the metabolites were interconvertible pairs formed by oxidation/reduction of the C-17 alcohol/ketone function. Other biotransformations consisted of reduction of the ring A diene ketone.

Reference

W. Schanzer and M. Donike, Metabolism of boldenone in man: Chromatographic/ mass spectrometric identification of urinary excreted metabolites and determination of excretion rates, *Biol. Mass Spectrom.*, 1991, **21**, 3.

Spironolactone

Use/occurrence:	Antihypertensive/diuretic drug
Key functional groups:	Acetylthio, androsten-3-one, steroid
Test system:	Rat liver microsomes 1 mM substrate)

Structure and biotransformation pathway

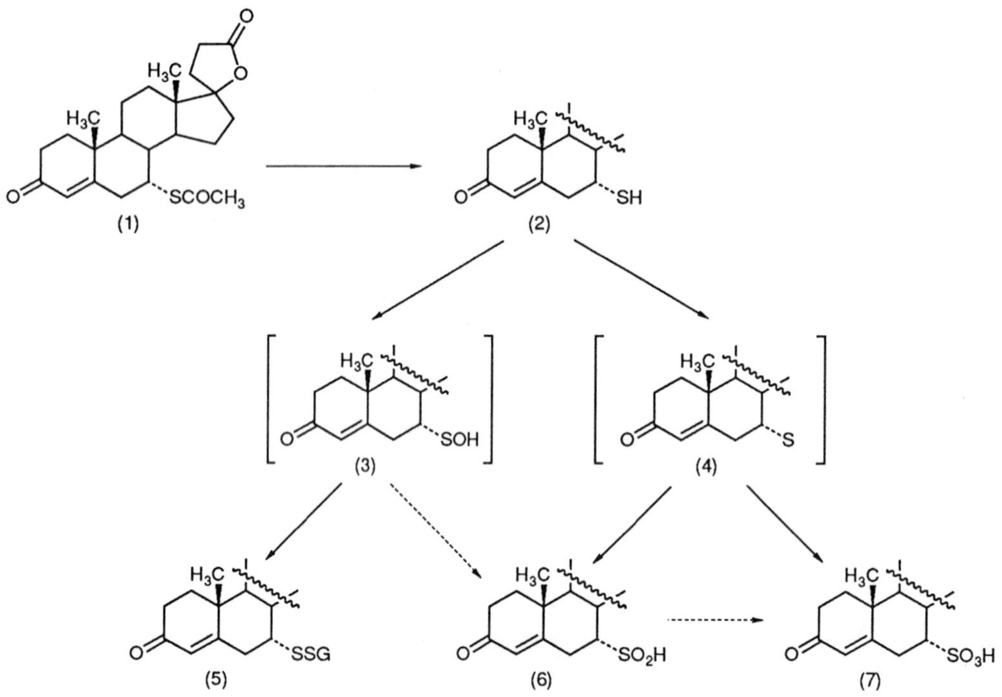

Microsomes of rats pretreated with dexamethasone were found to metabolize (1) to an electrophilic species that is trapped with glutathione. The glutathione adduct (5) so formed has been unambiguously identified by LC–MS and MS–MS techniques. Formation of the adduct was found to be dependent on glutathione, NADPH, and (1). The reaction was significant at pH 7.4, but was enhanced at pH 9, the pH optimum for the flavin mono-oxygenase- but not P-450-mediated reactions. Indeed the replacement of rat liver with purified hog liver flavin mono-oxygenase resulted in efficient conversion of (1) into (5). The two polar metabolites (6) and (7) have been previously reported; thus the intermediacy of (2), (3), and (4) was suggested.

Reference

C. J. Decker, J. R. Cashman, K. Sugiyama, D. Maltby, and M. A. Correia, Formation of glutathionyl-spironolactone disulfide by rat liver cytochromes P450 or hog liver flavin-containing monooxygenases: A functional probe of two-electron oxidation of the thiosteroid, *Chem. Res. Toxicol.*, 1991, **4**, 669.

Cinobufagin

Use/occurrence:	Cardiotonic agent, local anaesthetic, isolated from chinese toad venom
Key functional groups:	iso-Alkyl ester, androstane
Test system:	Rat liver microsomes

Structure and biotransformation pathway

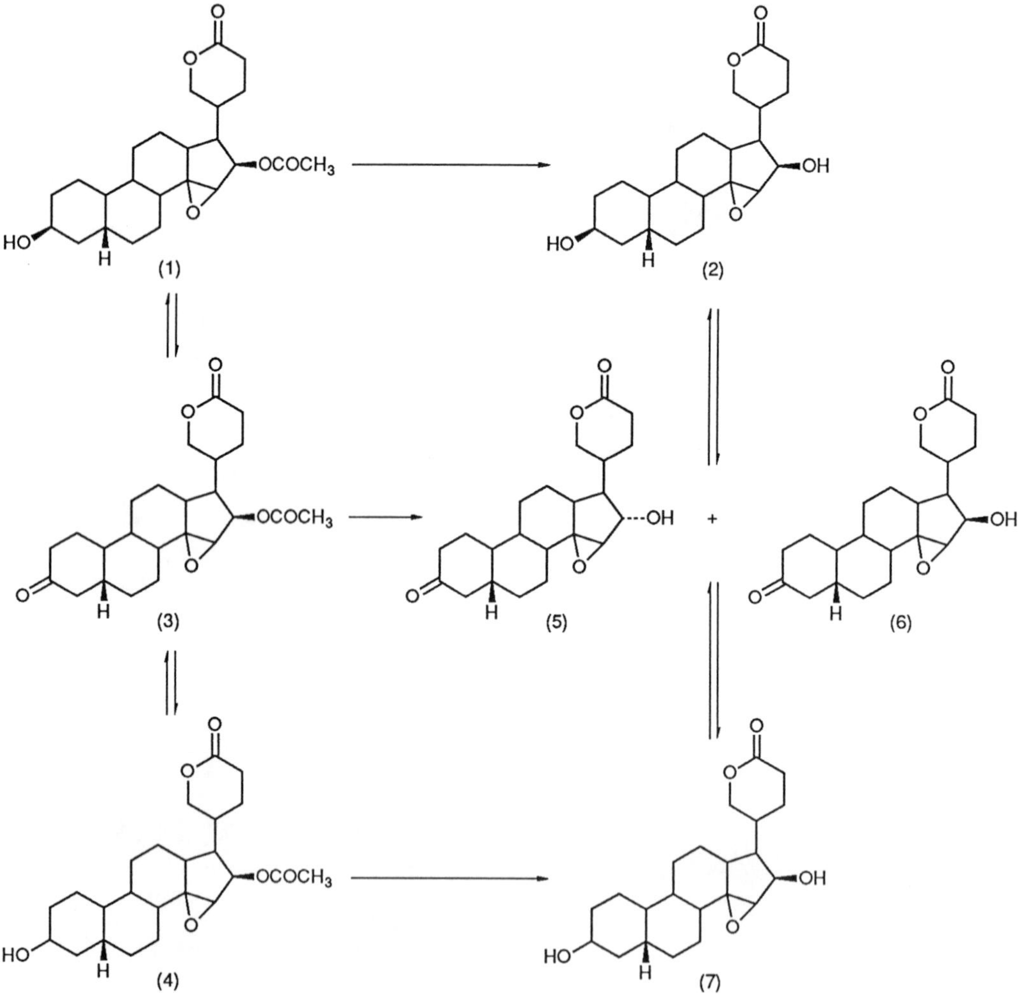

In this study four new metabolites of cinobufagin (1) were identified by LC–MS in addition to two previously known metabolites (2) and (7). Biotransformation pathways involved in the formation of these metabolites were deacetylation at the 16-position and epimerization at the 3-position *via* the 3-keto intermediate. NADPH was required for the formation of metabolites

394

(3)–(7) but not for metabolite (2). After 15 minutes incubation in the presence of an NADPH-generating system, (2) was the major metabolite (46% of initial cinobufagin) while (4) and (7) were also quantitatively important (*ca.* 10%). The other metabolites (3), (5), and (6) were formed in trace amounts (*ca.* 0.2%) only.

Reference

L. Zhang, T. Yoshida, K. Aoki, and Y. Kuroiwa, Metabolism of cinobufagin in rat liver microsomes, *Drug Metab. Dispos.*, 1991, **19**, 917.

Ginsenoside Rb$_2$

Use/occurrence: Natural product (Ginseng root)

Key functional groups: Glycoside

Test system: Rat (oral, 100 mg kg^{-1}), rat caecal contents (0.4 mg ml^{-1})

Structure and biotransformation pathway

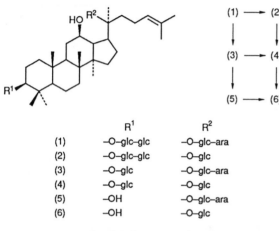

	R^1	R^2
(1)	–O–glc–glc	–O–glc–ara
(2)	–O–glc–glc	–O–glc
(3)	–O–glc	–O–glc–ara
(4)	–O–glc	–O–glc
(5)	–OH	–O–glc–ara
(6)	–OH	–O–glc

glc = β-D-glucopyranosyl
ara = α-L-arabinopyranosyl

Ginsenoside Rb$_2$ (1) is a compound found in the root of *Panax ginseng* which is used as a folk medicine. Ginseng saponins such as (1) are major components of the crude drug Ginseng Radix which is usually administered orally. The decomposition products of (1) were investigated in rat large intestine taken 6 hours following a single oral administration to rats and after incubation of (1) with rat caecal contents for 6 hours. Metabolites were separated by TLC and HPLC and identified by ^{13}C NMR.

Five metabolites of (1) were identified in both the *in vivo* and *in vitro* samples, all of which were formed by cleavage of sugar moieties from the parent compound. The major metabolites were (5), (20-*O*-[α-L-*arap*-(1 → 6)-β-D-glc]-20 S-protopanaxadiol), and (6), 20 S-protopanaxadiol. Intermediate hydrolysis products were also detected which suggested that decomposition began with cleavage of the terminal glucose at the C-3 hydroxy group [(1) → (3) → (5)] or the terminal arabinose of an oligosaccharide at the C-20 hydroxy group [(1) → (2) → (4) → (6)]. The order of the yield of the products was (5) > (6) > (4), which suggested the presence of a β-glucosidase, acting predominantly at the C-3 hydroxy group, in rat large intestine.

Reference

M. Karikura, T. Miyase, H. Tanizawa, Y. Takino, T. Taniyama, and T. Hayashi, Studies on absorption, distribution, excretion and metabolism of Ginseng saponins. V. The decomposition products of ginsenoside Rb_2 in the large intestine of rats, *Chem. Pharm. Bull.*, 1990, **38**, 2859.

Miscellaneous

Spirogermanium

Use/occurrence:	Immunomodulator and antimalarial drug
Key functional groups:	Cycloalkylamine, dimethylaminoalkyl, germano-organic
Test system:	Mouse (oral, 25 mg kg^{-1}), human and mouse liver microsomes

Structure and biotransformation pathway

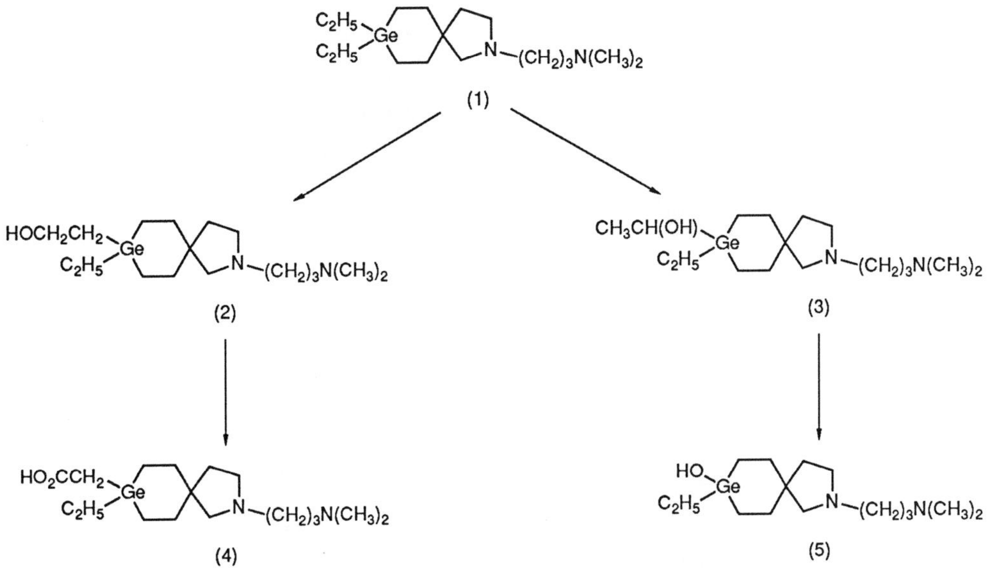

Two metabolites of spirogermanium (1) were detected by GC–MS following incubation with mouse liver microsomes, and as the spectroscopic data indicated that both were hydroxylated on an ethyl group attached to the germanium atom it was concluded that one was (2) and the other was (3). The same metabolites were similarly identified in solvent extracts of basified urine from mice dosed orally with (1). When these extracts were analysed by thermospray LC–MS using the loop injector (*i.e.* with no chromatographic separation), four germanium ion clusters were found, two of which corresponded to unchanged (1) and monohydroxylated metabolites. It was postulated that the other two corresponded to (4) and (5), but identification of these metabolites was only tentative. Extracts of human urine from a subject treated with (1) were analysed under the same LC–MS conditions, when it was found that the pattern of germanium clusters was identical to those in

401

mouse urine, suggesting that biotransformation of (1) in man was qualitatively similar to that in the mouse.

Reference

D. Garteiz, Z. H. Siddik, and R. A. Newman, *In vitro* and *in vivo* murine metabolism of spirogermanium, *Drug Metab. Dispos.*, 1991, **19**, 44.

N,N',N"-Triethylenethiophosphoramide (thio-TEPA)

Use/occurrence:	Cancer chemotherapeutic agent
Key functional groups:	Aziridine, thiophosphoramide
Test system:	Rat hepatic microsomes

Structure and biotransformation pathway

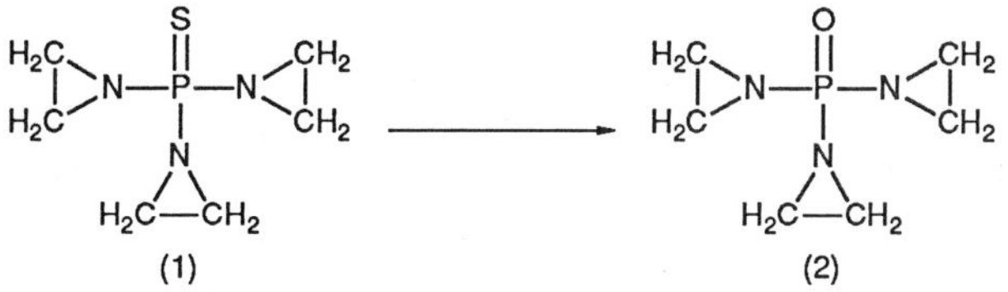

(1) → (2)

Following incubation of thio-TEPA (1) with microsomes from uninduced male and female rats, the compound underwent oxidative desulfuration to form *N,N',N"*-triethylenephosphoramide (TEPA) (2) as the only metabolite detected in the incubations following extraction and analysis by GC. The rate of formation of (2) was greater in the presence of microsomes from male rats than from females. An examination of the rate of formation of (2) in the presence of purified rat cytochrome P-450 enzymes showed that IIB1, the phenobarbital-inducible enzyme, was 4–5-fold more active than any other examined. However, forms IIC11 and IIC6 also produced (2) at significant rates, these being responsible for the oxidative desulfuration of (1) by uninduced rat microsomes.

Reference

S.-F. Ng and D. J. Waxman, *N,N',N"*-Triethylenethiophosphoramide (thio-TEPA) oxygenation by constitutive hepatic P_{450} enzymes and modulation of drug metabolism and clearance *in vivo* by P_{450}-inducing agents, *Cancer Res.*, 1991, **51**, 2340.

Chlorpyrifos

Use/occurrence: Organophosphate insecticide

Key functional groups: Chloropyridine, phosphorothioate

Test system: Channel catfish (intravenous, $123\ \mu g\ kg^{-1}$; dietary, $ca.\ 500\ \mu g\ kg^{-1}$)

Structure and biotransformation pathway

After oral administration to channel catfish, chlorpyrifos was well absorbed with a systemic bioavailability of 41% of the dose. The metabolites in the blood were unchanged chlorpyrifos (1) (> 60%) and trichloropyridinol (3) (40%). The terminal elimination half-life as calculated from blood data was 4.6 days.

The major route of excretion was via the urine (18.5% of the dose in 32 hours) with some excretion also via the bile. The main metabolite found in urine and bile was trichloropyridinol glucuronide (4) (73–79% in urine, 96–98% in bile) together with the free aglycone (3). Bile also contained small amounts of (1) and methoxytrichloropyridine (5). Urine contained two minor, unassigned metabolites which chromatographed close to O-ethyl-trichloropyridyl phosphorothioate (2), a metabolite earlier identified in goldfish. All metabolites were identified by a co-chromatography with standards on HPLC.

Reference

M. G. Barron, S. M. Plakas, and P. C. Wilga, Chlorpyrifos pharmacokinetics and metabolism following intravascular and dietary administration to channel catfish, *Toxicol. Appl. Pharmacol.*, 1991, **108**, 474.

Parathion, Methyl parathion

Use/occurrence:	Insecticides
Key functional groups:	Phosphorothioate
Test system:	Perfused rat liver

Structure and biotransformation pathway

$(RO)_2\overset{O}{\underset{}{P}}$—O—⟨benzene⟩—$NO_2$

(3) R = C_2H_5
(4) R = CH_3

$(RO)_2\overset{S}{\underset{}{P}}$—O—⟨benzene⟩—$NO_2$

(1) R = C_2H_5
(2) R = CH_3

HO—⟨benzene⟩—NO_2

(5)

HO_3SO—⟨benzene⟩—NO_2

(6)

GlucO—⟨benzene⟩—NO_2

(7)

Single pass perfusions with parathion (1) and methyl parathion (2) resulted in the appearance of paraoxon (3) and methyl paraoxon (4) respectively in the effluent. *p*-Nitrophenol (5) and its sulfate (6) and glucuronide (7) conjugates were also detected. HPLC comparison with reference standards was used to assign metabolites. The possibility of loss of an alkyl group was not investigated. Sex differences in metabolite profiles from both (1) and (2) were observed, but these could not account for known sex differences in susceptibility to the two compounds. Neither *S*-methylglutathione or *S*-*p*-nitrophenylglutathione was detected in effluent or bile of livers from either sex perfused with (2), suggesting that glutathione-dependent detoxification of this insecticide does not occur to any significant extent in intact liver.

Reference

H. X. Zhang and L. G. Sultatos, Biotransformation of the organophosphorus insecticides parathion and methyl parathion in male and female rat livers perfused *in situ*, *Drug Metab. Dispos.*, 1991, **19**, 473.

Compound Index

This Index is cumulative and covers Volumes 1–5. Volume numbers are in bold type.

L-648,051 **2**, 184
Labetalol **2**, 221; **5**, 188
Lamotrigine **5**, 293
Leiocarposide **3**, 158
Letosteine **2**, 325
Leukotriene B$_4$ **3**, 96
Levamisole **5**, 298
Lidocaine **3**, 181
Lisuride **3**, 247
Lonazolac **3**, 247
Lorazepam 3-acetate **5**, 326
Lovastatin **2**, 286; **4**, 205, 207, 209
Loxistatin **3**, 400; **5**, 376, 377
Lu253 **2**, 254
Lysodren **1**, 122
2-(L-Lysyloxy)-1,3-bis(2-ethoxy-
    carbonylchroman-5-yloxy)propane
    2, 139

Malotilate **3**, 276; **4**, 268
Mandelic acid **4**, 127
Mannose **1**, 141
Marcellomycin **1**, 254
MC-969 **5**, 208
MDL 27210 **5**, 374
MDL 74270 **5**, 316
Mebendazole **5**, 304
Medetomidine **5**, 252
Medroxyprogesterone acetate **5**, 386
Melphalan **1**, 202
Menogaril **4**, 217
l-Menthol **2**, 252
Mephentermine **3**, 198
Meptazinol **3**, 303
Merbarone **4**, 302; **5**, 292
2-Mercaptobenzothiazole **3**, 316
2-Mercaptobenzothiazole disulfide **3**,
    316
6-Mercaptopurine **4**, 355
Mespirenone **1**, 471
Metapramine **3**, 221
Metbufen **2**, 182
Methamphetamine **1**, 233
Methaphenilene **2**, 228
Methapyrilene **1**, 214, 216; **2**, 226; **4**,
    283
Methazolamide **4**, 271
Methenolone **5**, 387
Menthofuran **5**, 56
Methotrexate **1**, 364
4-Methoxyallylbenzene **2**, 175
16 α-Methoxycarbonylprednisolone **4**,
    422
Methoxychlor **3**, 186; **4**, 177

5-Methoxydimethyltryptamine **1**, 334
Methoxyflurane **2**, 97
1-Methoxyindole-3-carboxylic acid **5**,
    301
trans-2-(3'-Methoxy-5'-methylsulfonyl-
    4'-propoxyphenyl)-5-(3″,4″,5″-trim-
    ethoxyphenyl)tetrahydrofuran **5**,
    114
7-Methoxy-2-nitronaphtho [2,1-*b*]-
    furan **3**, 351
(*R/S*)-2-Methoxy-3-(octadecyl-
    carbamoyloxy)propyl 2-(3-thiazolio)-
    ethyl phosphate **2**, 140
Methoxyphenamine **3**, 199; **4**, 182
p-Methoxyphenyl azide **5**, 352
8-Methoxypsoralen **1**, 372
1-(2-Methoxypyridin-5-yl)-3,3-
    dimethyltriazene **5**, 358
2-(4-Methoxy-2-pyridylmethyl-
    sulphinyl)-5-(1,1,2,2-tetra-fluoro
    ethoxy)-1*H*-benzimidazole **3**, 311
9-Methyladenine **2**, 394; **3**, 340
N-Methyl-*N*-alkyl-*p*-chloroamides **5**,
    149
N-Methyl-4-aminoazobenzene **1**, 401
Methyl-n-amylnitrosamine **2**, 437; **3**,
    383, 385
Methylazoxymethanol **5**, 363
7-Methylbenz[*c*]acridine **5**, 41
N-Methylbenzamide **1**, 413
7-Methylbenz[*a*]anthracene **2**, 54
12-Methylbenz[*a*]anthracene **1**, 54
o-Methylbenzhydrylamine **2**, 430
3-Methylbenzotriazinone **4**, 359
Methylbenzylamine **1**, 412
Methyl t-butyl ether **4**, 87
Methyl carbamate **2**, 420
N-Methylcarbazole **5**, 329
Methyl chloride **3**, 72
3-Methylcholanthrene **3**, 48; **4**, 44, 45
3-Methylcholanthrene-2-one **4**, 50
3-Methylcholanthrylene **4**, 42
5-Methylchrysene **1**, 63, 65; **5**, 36
6-Methylchrysene **1**, 63; **5**, 36
Methylcyclohexane **2**, 30
Methyl 20-dihydroprednisolonate **4**,
    422
N-(1-Methyl-3,3-diphenylpropyl)
    formamide **4**, 175
4,4'-Methylenebis(2-chloroaniline) **2**,
    407; **3**, 169, 170
3,4-Methylenedioxymethamphetamine
    2, 214; **3**, 201; **4**, 181; **5**, 179, 181
Methylephedrine **4**, 183, 184

415

Key Functional Group Index

This Index is cumulative and covers Volumes 1–5. Volume numbers are in bold type.

Alkyl aryl ether **1**, 129, 242, 244, 246, 248, 293, 323, 397; **2**, 169, 232, 234, 375; **3**, 68, 151, 211, 213, 214, 215, 219, 408; **4**, 66, 68, 123, 147, 155, 170, 178, 195, 220, 364; **5**, 59, 122, 146, 163, 190, 194, 197, 199

Alkyl aryl sulfoxide **1**, 351; **3**, 311, 312

Alkyl aryl thioether **1**, 438; **2**, 381; **4**, 121

Alkyl carbamate **2**, 129; **3**, 353; **4**, 170, 305

Alkyl carbonate **4**, 179; **5**, 182, 183

Alkyl carboxamide **1**, 119, 295, 297, 318, 442; **2**, 101, 382, 384, 461; **3**, 74, 211, 353; **4**, 165, 275; **5**, 95, 151

iso-Alkyl carboxamide **4**, 95

Alkyl carboxylate **2**, 216, 372; **4**, 219, 314, 315, 429, 447; **5**, 386

tert-Alkyl carboxylate **4**, 210

Alkyl carboxylic acid **1**, 40, 196, 200, 371; **2**, 121, 135, 184, 193; **3**, 88, 90, 109, 110, 350, 397; **4**, 91, 104, 212, 373; **5**, 92, 210, 212, 320

iso-Alkyl carboxylate **3**, 428

iso-Alkyl carboxylic acid **1**, 143, 144; **2**, 122, 182; **3**, 91, 92, 93; **5**, 76, 77, 374

N-Alkyl cycloalkylamine **1**, 331

Alkylcyclohexane **1**, 38; **2**, 249, 252; **3**, 231, 232

Alkylcyclopentane **2**, 273

Alkyl ester **1**, 152, 197, 258. 260, 277, 290, 303, 305, 332, 338, 442; **2**, 89, 96, 137, 139, 186, 193, 211, 212, 289, 291, 293, 294, 295, 325, 353, 372, 455; **3**, 83, 102, 156, 159, 251, 305, 399, 400, 430, 432; **4**, 92, 93, 94, 132, 135, 247, 314, 315, 316, 429; **5**, 140, 160, 271, 273, 296, 297, 331, 374, 376, 377, 386

iso-Alkyl ester **1**, 492; **2**, 286, 349, 351; **3**, 276, 281; **4**, 201, 205, 207, 268; **5**, 210, 276, 294

Alkyl ether **1**, 227, 229, 231, 306; **2**, 209, 211, 212; **4**, 87, 251, 305; **5**, 259

Alkyl hydrazine **1**, 389

N-Alkylimidazole **4**, 259

Alkyl imide **2**, 371

N-Alkylimide **1**, 310, 376; **3**, 290; **5**, 287

Alkyl ketone **1**, 32, 150, 252, 261, 346, 422, 423, 455; **2**, 256, 258, 259,

261, 263, 299; **3**, 61, 345; **4**, 143, 392; **5**, 147, 216

N-Alkylmorpholine **4**, 339

Alkyl nitrate **3**, 87; **4**, 89, 90

Alkyl nitrile **2**, 310; **3**, 202, 388; **4**, 185, 187; **5**, 80

Alkyl N-oxide **2**, 426; **3**, 373, 375; **4**, 362; **5**, 251

Alkyl peroxide **3**, 249; **5**, 223

Alkylphenyl **1**, 39, 40, 172, 235, 237, 283, 310, 311; **5**, 129

iso-Alkylphenyl **1**, 47

Alkyl phosphate **2**, 140, 483; **3**, 112; **5**, 96

N-Alkylpiperazine **1**, 308; **3**, 151, 289, 300; **4**, 285, 287; **5**, 189, 190, 199, 310, 342

N-Alkylpiperidine **2**, 196, 408; **3**, 256; **4**, 275; **5**, 340

N-Alkylpurine **3**, 340

Alkylpyridine **2**, 380; **3**, 285; **4**, 319

Alkylpyrimidine **2**, 367; **3**, 319; **4**, 335

Alkyl quaternary ammonium **1**, 266; **5**, 316

Alkyl sulfamate **2**, 132

Alkyl sulfate **1**, 142

Alkyl sulfonate **1**, 146, 147; **2**, 125; **3**, 108

Alkyl sulfonic acid **2**, 126, 137

Alkyl sulfoxide **1**, 149

S-Alkyl thiocarbamate **2**, 130

Alkyl thiocarboxylate **2**, 467

Alkyl thioester **2**, 467; **3**, 324

Alkyl thioether **1**, 128; **2**, 129, 321, 325, 481; **3**, 336, 417; **4**, 264, 323

Alkyl thiol **2**, 126, 453, 454; **3**, 109, 110; **5**, 375

Alkylurea **3**, 247

Alkyne **3**, 430; **4**, 425

Allyl **3**, 290

Allyl amine **1**, 113, 278

Allylic alcohol **1**, 265; **2**, 299; **3**, 242; **4**, 125, 202, 220; **5**, 210, 235

Allylic methyl **4**, 205, 209, 210; **5**, 56, 57, 387

Amidine **1**, 410; **2**, 402; **5**, 122

Amidoxime **1**, 406

Amino acid **1**, 128, 202, 363, 364, 426, 437, 438, 439, 440, 445; **2**, 453, 454, 455, 458, 459, 461; **3**, 129, 342, 397, 399, 408; **4**, 180, 316, 325, 373, 407, 408, 409; **5**, 369, 372

Amino glycoside **3**, 234; **4**, 203, 249

Reaction Type Index

This Index is cumulative and covers Volumes 1–5. Volume numbers are in bold type.

benzodiazepine formation **1**, 201

benzoisoselenazone formation **4**, 146

benzoxazoline formation **5**, 140

chloroethylurea to imidazolidone **4**, 400

diamine to cyclic imine **3**, 99

dihydropyrrole formation **1**, 342

dithiocarbamate to ethylenethiourea **1**, 496

furan formation **1**, 109; **3**, 58, 59; **5**, 56

glutathione conjugate to tetra-hydrothiophene **2**, 124, 125

hydroxyalkene to tetrahydrofuran **1**, 446; **4**, 410

hydroxyepoxide **5**, 227

imidazole from nitroaryl amine **2**, 179

imidazopyridine formation **3**, 326

lactam formation **1**, 99; **2**, 351, 403; **4**, 316, 373

oxazolidinone **5**, 90

oxazolidone **5**, 90

lactone formation **3**, 59, 159

2-phytylhydroquinone **1**, 251

piperazinedione formation **2**, 458; **3**, 321

pyridinylalkyl aldehyde **1**, 219

pyrrolidine formation **2**, 310

pyrrolidone formation **1**, 342

spirolactam formation **2**, 310

tetrahydrothiophene formation **3**, 108

Cysteine/*N*-acetylcysteine conjugates

acrylate **2**, 89

acrylonitrile **1**, 116

akene **5**, 77, 78

alkene aldehyde **2**, 102

alkyl aryl sulfoxide **3**, 312

alkyl epoxide **2**, 91, 96

S-alkyl thiocarbamate **2**, 130; **5**, 90

alkyl thiol **2**, 454; **3**, 110

aminothiazole **2**, 314

arylalkene **3**, 65

aryl amine **1**, 391

aryl sulfonamide **4**, 271

benzoquinone **2**, 162

benzyl acetate **1**, 195

benzyl alcohol **2**, 319; **3**, 139, 141

benzyl chloride **3**, 140

bromoacetyl **3**, 74

bromophenyl **5**, 101

carboxylic acid thioester **3**, 324

chloroacetamido **2**, 198; **4**, 168

chloroalkene **2**, 104, 107, 111; **4**, 81, 82, 83; **5**, 69

chloroalkyl **5**, 63, 65

chloroalkyne **5**, 70

chlorophenyl **2**, 146, 153, 155; **3**, 115; **4**, 114, 168

chloropyridazinone **3**, 296

coumarin **5**, 143

epoxide **3**, 63

fluoroalkene **4**, 19

formamide **4**, 175

N-hydroxyacetyl **2**, 200

hydroxymethyluracil **2**, 365

isocyanate **4**, 97; **5**, 85

isothiocyanate **2**, 128; **4**, 98

2-mercaptobenzothiazole **2**, 388

N-methylformamide **1**, 408; **3**, 100

methylindole **5**, 300

naphthalene **2**, 37; **4**, 29

pyrrolizidine **4**, 315

trichloromethyl **5**, 259

N-Dealkylation

alkyl amide **3**, 104, 160, 161; **5**, 150

N-alkyl amide **1**, 148, 310; **2**, 232; **4**, 145

alkyl amine **2**, 403, 404; **3**, 302

alkyl tert-amine **2**, 219; **3**, 223, 375; **4**, 185, 272, 364

iso-alkylamine **3**, 215, 217, 302

iso-alkylamino **1**, 246; **2**, 375; **4**, 172, 195, 373

iso-alkylaminoalkyl **3**, 213

tert-alkyl amine **1**, 237

alkyl aryl amide **1**, 295, 297

alkyl aryl amine **2**, 179, 230, 369, 403; **3**, 184; **4**, 309, 446; **5**, 169, 253

N-alkyl barbiturate **4**, 305

N-alkyl imidazole **4**, 259

N-alkyl imide **3**, 290

N-alkyl morpholine **4**, 339

alkyl nitrosamine **4**, 393, 397; **5**, 362, 364

N-alkylpiperazine **1**, 308; **3**, 151, 289, 300; **4**, 285, 287; **5**, 189, 310

N-alkylpiperidine **1**, 295, 297; **2**, 196; **4**, 275; **5**, 340

N-alkylpurine **3**, 340

N-alkylpyrrolidine **3**, 193

alkyl sulfonamide **3**, 107

alkylurea **3**, 247

aminomethylthiophene **1**, 216
N-benzyl **1**, 212, 307
cyclopropylmethyl **1**, 266; **4**, 343
dialkyl carbamate **5**, 90
dialkyl amide **2**, 190, 234, 382; **3**, 211, 419; **5**, 149
dialkyl amine **2**, 215, 221, 323, 397, 399, 421; **3**, 219; **4**, 190; **5**, 178
dialkylaminoalkyl **1**, 129, 240, 342; **2**, 191, 232, 234, 405; **3**, 181, 211, 236, 353; **4**, 325; **5**, 193
dialkyl aryl amine **1**, 208; **2**, 226, 228
N-difluoromethyl **5**, 257
dimethylaminoalkyl **1**, 214, 216, 219; **4**, 283; **5**, 59, 62, 328
isopropylamino **1**, 205, 248
N-methyl alkyl amine **2**, 216; **3**, 202, 206
nitrosamine **2**, 438
phosphoroamidothioate **1**, 492
quaternary ammonium **1**, 266; **5**, 280
thienylmethylamino **1**, 214, 216; **4**, 283
O-Dealkylation
alkyl aryl ether **1**, 293; **3**, 68, 151, 211, 215; **4**, 147, 170, 178, 220; **5**, 122, 146, 163, 190, 193, 199
alkyl ether **1**, 227, 229; **2**, 119, 211, 212, 461
alkyloxyphenyl **1**, 197; **2**, 167, 196
aryloxyacetic acid **4**, 284
dialkyl ether **3**, 83, 249, 304, 347; **4**, 92, 305
S-Dealkylation
alkyl thioether **3**, 417; **4**, 325
aryl cysteine **3**, 342
dialkyl thioether **3**, 397
Deamination
alkyl amine **1**, 113, 232, 278, 293, 343; **2**, 214, 353, 403, 404; **3**, 101, 187, 195; **4**, 180, 195; **5**, 83, 178
amidine **1**, 410
tert-amine **1**, 229
amino acid **1**, 437, 439, 440; **3**, 129; **4**, 325, 409
aryl amine **1**, 187, 321; **2**, 206
benzyl amine **2**, 430; **5**, 176
cysteine **5**, 369
cytosine **4**, 299
dialkyl amine **3**, 217; **4**, 172
dialkylaminoalkyl **4**, 327, 344

dimethylaminoalkyl **3**, 131; **5**, 62, 80, 173
N-methyl alkyl amine **3**, 199, 201
phosphoroamidothioate **1**, 492
pyrimidine **1**, 316
Debenzylation
N-alkyl benzylamine **3**, 227
benzyladenine **2**, 394
benzyl amine **2**, 224, 230; **4**, 190
benzyl ether **2**, 250; **3**, 68, 232; **4**, 68; **5**, 331
N-benzylpiperidine **2**, 349; **4**, 148
benzyl thioether **2**, 112
dialkyl aryl amine **2**, 224
Decarboxylation
aryl carboxylic acid **2**, 349
imine carboxylate **1**, 325
Dechlorination
chloroalkene **4**, 83
chloroethane **1**, 121
chlorophenyl **1**, 159; **4**, 141
chloropyridazinone **3**, 296
trichloromethyl **1**, 121, 124
Defluorination
fluoroacetate **1**, 135
fluoroalkyl **4**, 74
fluorocytosine **1**, 313
Deformylation
N-formyl **1**, 410; **2**, 314; **5**, 255
N-formyl alkyl amine **3**, 187
Dehydration
sec-aklyl alcohol **5**, 95
allylic alcohol **2**, 299
hydroxycyclopentanone **2**, 135
hydroxypiperidine **5**, 267
pyrazinone to pyrazine **3**, 298
Dehydrobromination
bromoalkane **2**, 98
Dehydrochlorination
chloroalkane **2**, 98, 100; **4**, 429
trichloromethyl **1**, 124
Dehydroxylation
tert-alcohol **2**, 466
allylic alcohol **2**, 299
N-Demethylation
alkyl tert-amine **3**, 204; **4**, 185
sec-alkyl amine **5**, 179
tert-alkyl amine **1**, 227, 281
aryl amine **1**, 336
dimethylaminoalkyl **1**, 201, 208, 214, 219, 227, 229, 231, 235, 376; **2**, 167, 169, 186, 209, 224, 226, 228, 236, 238, 241; **3**, 131, 156, 223, 224; **4**, 184, 191, 249,

434

435

436

440

442

445